Lecture Notes in Physics

Springer-Verlag Berlin Heidelberg GmbH

The Editorial Policy for Proceedings

The series Lecture Notes in Physics reports new developments in physical research and teaching – quickly, informally, and at a high level. The proceedings to be considered for publication in this series should be limited to only a few areas of research, and these should be closely related to each other. The contributions should be of a high standard and should avoid lengthy redraftings of papers already published or about to be published elsewhere. As a whole, the proceedings should aim for a balanced presentation of the theme of the conference including a description of the techniques used and enough motivation for a broad readership. It should not be assumed that the published proceedings must reflect the conference in its entirety. (A listing or abstracts of papers presented at the meeting but not included in the proceedings could be added as an appendix.)

When applying for publication in the series Lecture Notes in Physics the volume's editor(s) should submit sufficient material to enable the series editors and their referees to make a fairly accurate evaluation (e.g. a complete list of speakers and titles of papers to be presented and abstracts). If, based on this information, the proceedings are (tentatively) accepted, the volume's editor(s), whose name(s) will appear on the title pages, should select the papers suitable for publication and have them refereed (as for a journal) when appropriate. As a rule discussions will not be accepted. The series editors and Springer-Verlag will normally not interfere with the detailed editing except in fairly obvious cases or on technical matters.

Final acceptance is expressed by the series editor in charge, in consultation with Springer-Verlag only after receiving the complete manuscript. It might help to send a copy of the authors' manuscripts in advance to the editor in charge to discuss possible revisions with him. As a general rule, the series editor will confirm his tentative acceptance if the final manuscript corresponds to the original concept discussed, if the quality of the contribution meets the requirements of the series, and if the final size of the manuscript does not greatly exceed the number of pages originally agreed upon. The manuscript should be forwarded to Springer-Verlag shortly after the meeting. In cases of extreme delay (more than six months after the conference) the series editors will check once more the timeliness of the papers. Therefore, the volume's editor(s) should establish strict deadlines, or collect the articles during the conference and have them revised on the spot. If a delay is unavoidable, one should encourage the authors to update their contributions if appropriate. The editors of proceedings are strongly advised to inform contributors about these points at an early stage.

The final manuscript should contain a table of contents and an informative introduction accessible also to readers not particularly familiar with the topic of the conference. The contributions should be in English. The volume's editor(s) should check the contributions for the correct use of language. At Springer-Verlag only the prefaces will be checked by a copy-editor for language and style. Grave linguistic or technical shortcomings may lead to the rejection of contributions by the series editors. A conference report should not exceed a total of 500 pages. Keeping the size within this bound should be achieved by a stricter selection of articles and not by imposing an upper limit to the length of the individual papers. Editors receive jointly 30 complimentary copies of their book. They are entitled to purchase further copies of their book at a reduced rate. As a rule no reprints of individual contributions can be supplied. No royalty is paid on Lecture Notes in Physics volumes. Commitment to publish is made by letter of interest rather than by signing a formal contract. Springer-Verlag secures the copyright for each volume.

The Production Process

The books are hardbound, and the publisher will select quality paper appropriate to the needs of the author(s). Publication time is about ten weeks. More than twenty years of experience guarantee authors the best possible service. To reach the goal of rapid publication at a low price the technique of photographic reproduction from a camera-ready manuscript was chosen. This process shifts the main responsibility for the technical quality considerably from the publisher to the authors. We therefore urge all authors and editors of proceedings to observe very carefully the essentials for the preparation of camera-ready manuscripts, which we will supply on request. This applies especially to the quality of figures and halftones submitted for publication. In addition, it might be useful to look at some of the volumes already published. As a special service, we offer free of charge LATEX and TEX macro packages to format the text according to Springer-Verlag's quality requirements. We strongly recommend that you make use of this offer, since the result will be a book of considerably improved technical quality. To avoid mistakes and time-consuming correspondence during the production period the conference editors should request special instructions from the publisher well before the beginning of the conference. Manuscripts not meeting the technical standard of the series will have to be returned for improvement.

For further information please contact Springer-Verlag, Physics Editorial Department II, Tiergartenstrasse 17, D-69121 Heidelberg, Germany

Wolfgang Kundt (Ed.)

Jets from Stars and Galactic Nuclei

Proceedings of a Workshop
Held at Bad Honnef, Germany, 3–7 July 1995

 Springer

Editor

Wolfgang Kundt
Institut für Astrophysik
Universität Bonn
Auf dem Hügel 71
D-53121 Bonn, Germany

Cataloging-in-Publication Data applied for.

Die Deutsche Bibliothek - CIP-Einheitsaufnahme

Jets from stars and galactic nuclei : proceedings of a workshop,
held at Bad Honnef, Germany, 3 - 7 July 1995 / Wolfgang
Kundt (ed.).

(Lecture notes in physics ; Vol. 471)
ISBN 978-3-662-14074-1   ISBN 978-3-540-49953-4 (eBook)
DOI 10.1007/978-3-540-49953-4
NE: Kundt, Wolfgang [Hrsg.]; GT

ISBN 978-3-662-14074-1

Typesetting: Camera-ready by the authors
Cover design: Springer-Verlag Design & Production
SPIN: 10520078   55/3142-543210 - Printed on acid-free paper

Preface

This volume of the 'Lecture Notes in Physics' is the outcome of a one-week workshop at BH (Bad Honnef, our German physics center near Bonn), from 3rd to 7th July 1995, devoted to the various jet sources of both galactic and extragalactic origin. It can be considered a follow-up of a NATO ASI at Erice in 1986.

The goal of the workshop was a better understanding of the family of astrophysical jet sources, viz. the (1) extragalactic radio sources powered by an AGN (active galactic nucleus), (2) bipolar flows from YSO (young stellar objects, or pre-T-Tauri stars), (3) jets from (young) binary neutron stars (and/or BH candidates), and (4) jets inside PNe (planetary nebulae), illuminated by a young (or forming) white dwarf. All four classes of jet sources show striking similarities, such as (i) elongated morphologies – thought to be generated by supersonic flows – consisting of a hot core, knots (Herbig–Haros, FLIERS), heads (bowshocks), and lobes (cocoons), typical jet opening angles of 10^{-2}, no jet branching ever during formation, (only at later stages, like in Cen A, NGC 315, 3C 236, and perhaps near our Galactic center [some 5 deg off]), and cocoon elongations of order $5:1$, (ii) jet/lobe power ratios of $10^{-2\pm2}$, (iii) sidedness, (iv) rapid core variability, (v) very broad spectra, and (vi) (often) superluminal motions of knots in their jets. 'Often' means that so far, galactic superluminal motion has been noticed in only the two BH candidates GRS 1915+105 and GRO J1655-40; but the YSO source HH 30 – if observed frequently enough (more often than daily) – may turn out to join the club.

Explanations of the jet phenomenon in the literature and in this volume are various – cf. the 'ramble, preamble, amble, bramble, and postamble' by Jim Pringle in 'Astrophysical Jets', edited by Burgarella et al in 1993. Key assumptions concern the nature of the (a) central engine, (b) jet substance, (c) jet focusing, and (d) boosting of the electron energies (Lorentz factors). Most models take the feasibility of in-situ acceleration (of electrons) for granted, i.e. are not afraid of violating the Second Law when assuming that mildly relativistic protons ($\gamma \lesssim 10$) can transfer their energy to extremely relativistic electrons ($\gamma \gtrsim 10^6$) whenever needed, with a high efficiency ($\gtrsim 0.3$). Other people consider this a Münchhausen trick (cf. p.6), and are thus forced to choose pair plasma as the jet substance, i.e. put the burden (d) of electron boosting on the central engine. They are faced with SS 433 as a flagrant counter example unless they vote for its 'preferred model' (below). And they have yet to convince the community that (ordinary) stellar winds cannot be focused into long, narrow jets ($\approx 10^{-2}$), as is often maintained.

Is there a universal central engine, item (a) above? If so, it must be able to function with as low a compactness as $GM/rc^2 \gtrsim 10^{-6}$, being demonstrated by YSO; in particular, a black hole does not qualify. As the probable cause of the

BLR (broad-line region), NLR (narrow-line region), ELR (extended emission-line region), and (perhaps) EES (extended emission shells), as well as of the jets (of the radio-loud subclass of QSOs, some 10%) and VHE γ-rays, the central engine must generate and eject both thermal and relativistic particle populations. At a modest level, even the Sun can do this. But YSOs – probably born with breakup rotational velocity at the center of their accretion disk – are largely more active than the present Sun. Does this point at a fast-rotating starlike object as the universal central engine, an updated version of a spinar, or magnetoid, or SMC? If so, an AGN may be nothing else but the nuclear-burning central region of a galactic disk, generating its relativistic leptons in high-voltage coronal discharges ($\gtrsim 10^{10}$ eV). This idea – a 'BD' (burning disk) as the central engine of an AGN – is given consideration in two chapters of this volume. Here the word 'burning' stands for what had more realistically been called 'continually detonating'.

Once the jet substance is assumed to be relativistic (and leptonic), i.e. different from the bulk circumnuclear matter, there remains alternative (c) of its confinement: magnetic or inertial? At BH, circular magnetic tensions were widely favoured, as in my 1979 model. Yet what controls their sign? Why doesn't the accretion disk advect both signs alternatingly, in rapid succession? Uniform toroidal fields would ask for large-scale return currents; can they explain the repeated re-focusing of the jets, observed in particular for the stellar ones? Inertial confinement looks like a much more robust mechanism – as proposed by Roger Blandford and Martin Rees in 1974 – and as is easily achieved for a cold beam, at the (sideways) speed of sound of the ambient medium.

After what I have said, it will be clear to the reader that this book does not offer a unanimous description of the jet phenomenon; it offers various interpretations by various authors. This friendly coexistence of different views culminates in the 'critical review of particle acceleration' by Klaus Meisenheimer, in the 'simple sums on burning disks' by Peter Scheuer, and in two different reports on SS 433. It is also highlighted in the 'epilogue' which reflects my own understanding of the various difficulties. Judging by the progress since 1986, our heterogeneous understanding of the jet phenomenon is likely to persist for a number of years to come.

At this point, I feel urged to thank the Volkswagen foundation, Joachim Debrus, and the 'staff' at Bad Honnef for having actively cooperated in making the workshop a memorable event in my life – and in the lives of most of the participants I am sure. I also thank Hans Ulrich Wolter, our motivated guest from Zweibrücken, for his toast during the barbecue. Finally, the preparation of this book – in LaTeX 'lamuphys documentstyle' – would not have been completed in time without the ever available and cheerful help by Carsten van de Bruck.

Bonn, February 1996   Wolfgang Kundt

Contents

1. Thomas Krichbaum	10. Uma S. Pandey	19. Silvano Massaglia
2. Bogdan Dimitrov	11. Klaus Meisenheimer	20. Nazar Ikhsanov
3. José Ma. Martí	12. Beverley Wills	21. René Vermeulen
4. Patrick Leahy	13. Heino Falcke	22. Martin Gaskell
5. Peter Scheuer	14. Jonathan Ferreira	23. Felix Mirabel
6. John Contopoulos	15. Robert Fosbury	24. Daniel Fischer
7. Stefan Wagner	16. Ski Antonucci	25. Steven Tingay
8. Garrelt Mellema	17. Hans Baumann	26. Carsten van de Bruck
9. Max Camenzind	18. Vasilii Gvaramadze	27. Sergei Bogovalov

Jets from Stars and Burning Disks

Wolfgang Kundt

Institut für Astrophysik der Universität Bonn

Abstract. The questions of the *jet-substance, jet-formation, and jet-focusing* - in all the (four) classes of the *binary-flow family* - are revisited. Unique answers are given in terms of extremely relativistic *pair plasma*, created in magnetospheric and/or *coronal discharges* of the *hot central rotator*, which is centrifugally pressurized and squeezed out according to the *twin-exhaust mechanism* (= inertial confinement by the ambient CSM) into vacuum channels for a *supersonic, cold* flow. The central rotator can be a *YSO, young binary neutron star, young binary white dwarf, or BD* (= explosively nuclear-burning central disk).

1 The Bipolar-Flow Family: Members and Non-Members

Jets are in fashion: suspicious sources with elongated morphologies are called 'jets'. Instead, 'jets' in hydrodynamics - and in this book - mean supersonic flows. Figure 1 shows four representative examples from the bipolar-flow family: (i) Cyg A, an extragalactic radio source powered by an AGN (= active galactic nucleus), (ii) HH 34, a bipolar flow inside the L 1641 molecular cloud in Orion, powered by a pre-T-Tauri star, (iii) SS 433, a (not always uncontroversial) binary neutron star inside the SNR W50 (sketched as a multi-frequency overlay), and (iv) the 'southern Crab' He 2-104 (at optical emission lines), a PN (= planetary nebula) powered by a forming white dwarf. All four astrophysical sources are thought to contain an antipodal pair of narrow, powerful, supersonic flows along their (approximate) symmetry axis which is responsible for the pair of embedding elongated 'balloons', or 'cocoons', or 'lobes'. The cocoons are probably blown by the jets, similar to a glass blower's craft. Some statistics and recent observational work on these sources is listed in the figure caption.

Having introduced, in figure 1, what are considered 'members' of the bipolar-flow family, or jet family, I now come to what I consider 'non-members'. First, there is the recent detection of an elongated X-ray emission feature near the center of the Vela SN shell, sketched in Figure 2, together with a similar (though not reliably established) feature in MSH 15-52. To me, these hot regions look like 'plumes', rising under buoyancy from within the pulsar nebula at the center. They do not show knots and/or heads, or indicate rapid variability, and can be understood without the action of a supersonic flow.

Next there are the 'chimneys' of the Crab SNR, and of Cas A, sketched in Figure 3, which have also been called 'spur', 'jet', or 'stem'. In my preferred interpretation (Kundt, 1990a), they represent outflows of (pressurized, relativistic) pair plasma from the SN explosion and/or from the central pulsar, most

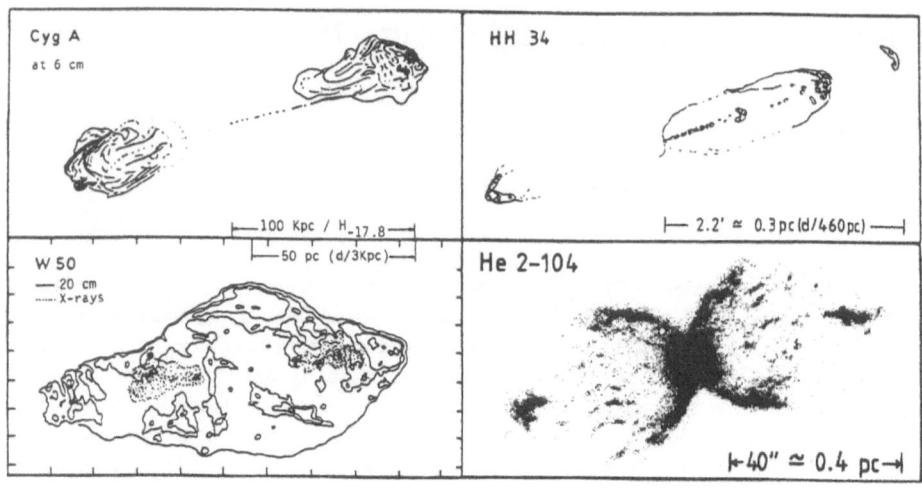

Fig. 1. Sketch of representatives of the Bipolar-Flow family, (one for each class). For recent maps of (some 10^2) extragalactic jet sources see Reid et al (1995), for multifrequency maps of *Cyg A*: Carilli et al (1991), and Leahy et al (1989) for low-frequency maps which show 'emission bridges', i.e. provide more complete coverage; further Barthel et al (1985) and Venturi et al (1993) for high resolutions. A (controversial) embedding of *HH 34* into a (5.8-times larger) parsec-scale 'superjet' is proposed by Bally & Devine (1994); nuclear radio detections of (some 20) bipolar flows are reported by Rodríguez & Reipurth (1994), and a molecular jet with 'bullets' in Bally et al (1993); see also Jochen Eislöffel's contributions. *SS 433*, a more-than controversial source and representative of over a dozen of neutron-star (and BH) binaries, is discussed separately in this volume by René Vermeulen and by myself. Dozens of PNe - the youngest members of the bipolar-flow family - are mapped by Gieseking (1985), Balick (1987), and by Schwarz et al (1992); see also Figure 5, and Garrelt Mellema's contribution. Shown above is *He 2-104*, the 'southern Crab'. The radio outflow from the symbiotic Mira variable R Aqr has been recently studied by Hollis et al (1992) and by Solf (1996).

likely at subsonic speeds, because we see neither knots nor heads, the signposts of particle deceleration.

More difficult to understand is a string of 11 semistellar condensations in the inner parts of the Crab, found by MacAlpine et al (1994), which move (more or less) radially at several times the speed of the surrounding (shell of) thermal filaments. These emission-line knots project, near the central pulsar, onto an elongated region of enhanced X-ray emission (of the 'pulsar nebula'). Of course, 'jet' is a shorthand notation for what has been detected. But most likely, the observations can be understood without any supersonic motion; and a more

3

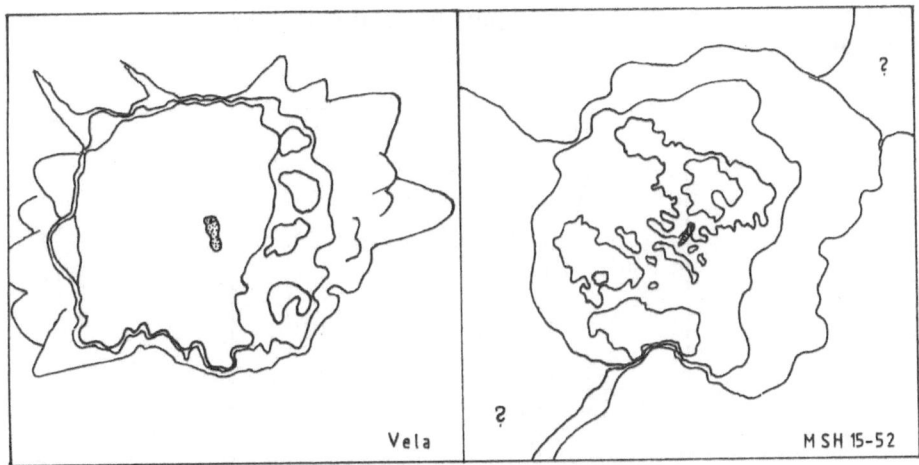

Fig. 2. The X-ray plumes of the Vela SN shell (Markwardt & Oegelman, 1995), and of MSH 15-52 (Bonn colloquium by Werner Becker).

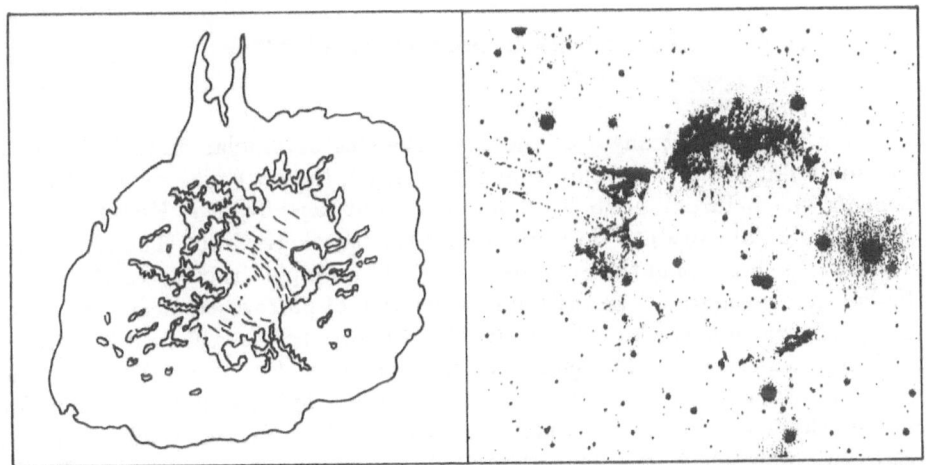

Fig. 3. The chimneys of the Crab, and of Cas A, and the string of fast knots in the Crab. Cf. MacAlpine et al (1994), Hester et al (1995), also Michel et al (1991), and Kundt (1990a).

neutral description - like 'surge', or 'flare', or 'string of knots' - would help the reader not to get confused by jargon.

Another phenomenon has entered textbooks under the name of 'bipolar flow' even though its kinematics are those of a Hubble flow ($v(r) \sim r$), or supernova flow, of age $10^{2.2 \pm 0.2} yr$: the outflow in Orion, centered on the Becklin-Neugebauer and Kleinmann-Low IR objects; see Figure 4, and Kundt & Yar (1995). Its morphology (from radio to IR) resembles a superposition of (many more than 20) radial Mach cones. Quite likely, it is the youngest known SNR in the Galaxy.

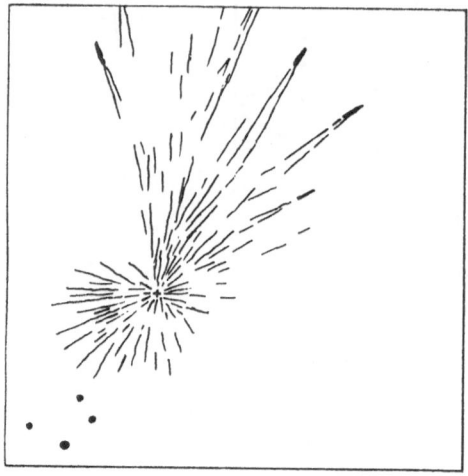

Fig. 4. SN flow in Orion, of age $10^{2.2\pm0.2} yr$.

What fraction - if any - of the PNe qualifies as bipolar flows ? Figure 5 presents a fair sample of them, from plain 'rings', (hollow) spheres, and lemons through the 'helix', 'dumbbell', 'hourglass', and faces (called 'Eskimo', 'owl', 'frosty Leo', 'cat's eye') to 'footprints' and the 'southern Crab'. The morphologies of more than 7% of the Schwarz et al (1992) sample - those near the end of the 'gallery' in figure 5 - are certainly indicative of narrow jets near their symmetry axis which have blown hourglass-shaped cocoons. As explained by Garrelt Mellema, planetary nebulae tend to be modeled by a slow wind (of the progenitor star) followed by a fast wind, both influenced by an accretion disk which curtails the winds in the orbit plane. Missing in those models is an additional relativistic flow which gets focused towards the angular-momentum axis by the inertia of the CSM (Kundt, 1987, 1993, 1996b; Icke et al, 1992), and creates Herbig-Haro-like knots called 'FLIERS'.

Of different morphological type are the (straight) radial 'streaks' seen in the Cat's-Eye nebula (fig. 5), which are absent in other bipolar flows but are reminiscent of the carefully studied Helix nebula where they have been explained - some 600 of them - as comet-like wind-blown trails at the inner edge of the slow windzone. The fast wind penetrates into the heavier, slow zone, in the form of radial spokes, and thereby erodes island-like globules - like cometary heads - at its inner edge: Huggins et al (1992), Meaburn et al (1992), Walsh et al (1993).

An independent hint at relativistic pair plasma as a building block of PNe is the blowout shown in Figure 6, observed some 2 days after outburst in Nova V603 Aql (1918), near maximum visible light. Two antipodal holes were blown into

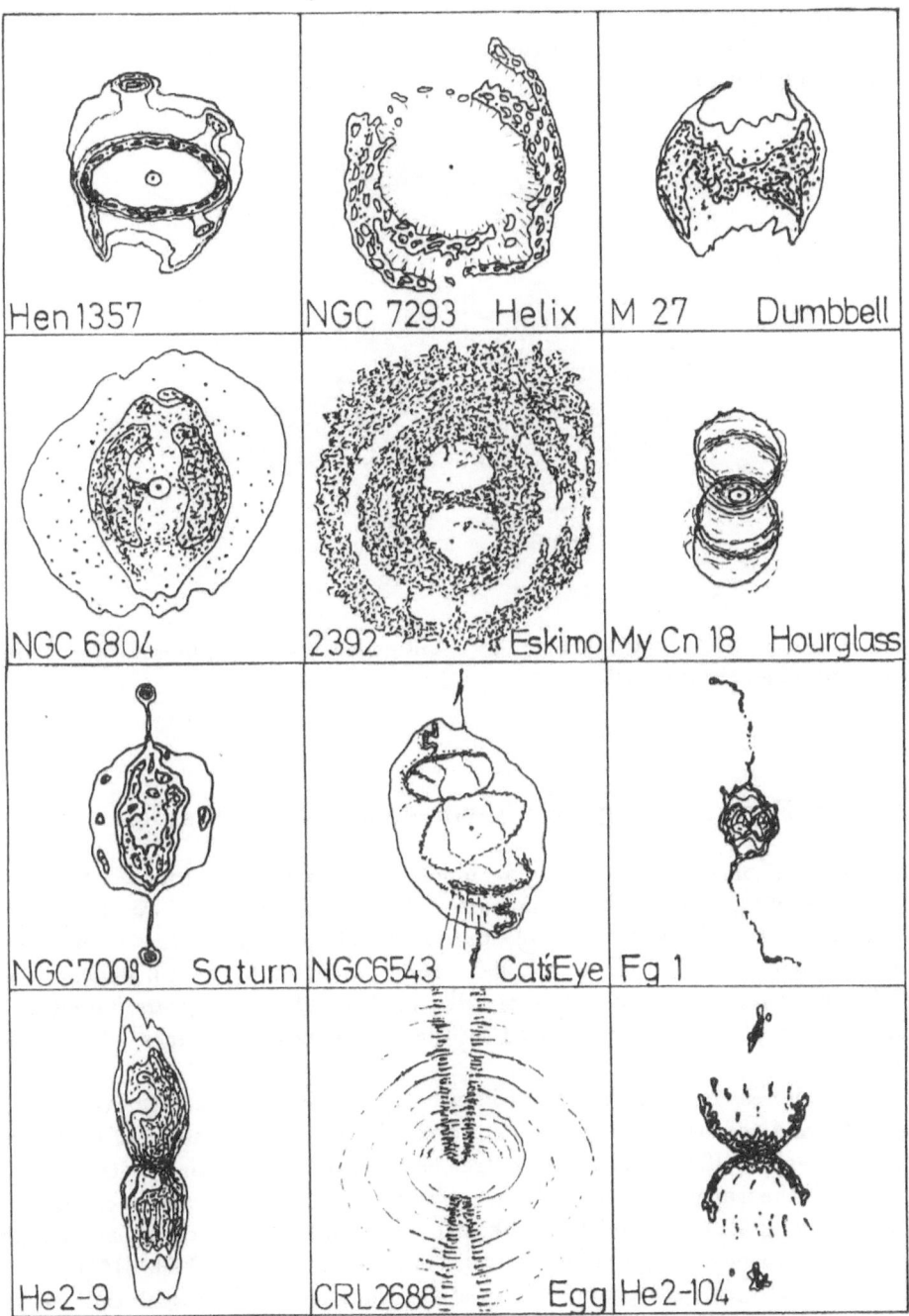

Fig. 5. Rogues' gallery of Planetary Nebulae, oriented such that the expected angular-momentum axis is in each case vertical; from Gieseking (1985), Bally (1987), Schwarz et al (1992), HST publications, and Mellema (this volume).

the thin, fast-expanding shell ($v/c \gtrsim 10^{-2}$), and the shell was post-accelerated (Weaver, 1974). The piston of this blowout, pushing at a speed of $\gtrsim 0.3c$, will surely have been relativistic.

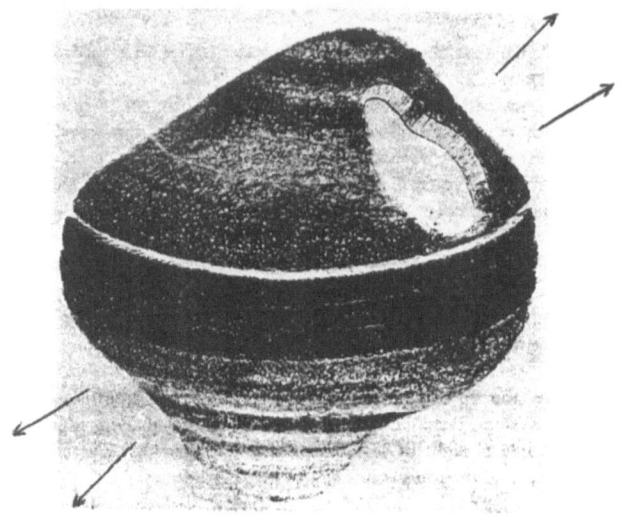

Fig. 6. Nova V603 Aql (1918) : Two antipodal holes are blown out of the fast-expanding, ejected shell.

2   Jet-Formation by the Twin-Exhaust Mechanism

A convincing mechanism of blowing bipolar flows has been proposed by Peter Scheuer (1974), whereby his 'tomatoes' will most likely (in my mind) have to be replaced by relativistic pair plasma, with bulk Lorentz factors γ in excess of 10^2: Kundt & Gopal-Krishna (1980, 1981), Kundt (1987, 1989), Begelman et al (1994). There has been a deep scepticism, throughout the years, about how enough pair plasma can be created by the various central engines, black holes or other.

For this reason, Fermi's idea of in-situ acceleration has been revived in the late 70s, and applied to a hoped-for conversion of the kinetic energy of (beamed) transrelativistic ISM ($\gamma \lesssim 10$) to extremely relativistic electrons, with Lorentz factors all the way up to some 10^7. To me, this suggestion - when large conversion efficiencies ($\gtrsim 0.1$) are required - looks like a violation of the Second Law: Kundt (1984). Independent arguments against it have been given by Falle (1990).

A way to grasp the problem is to study the (rigorous) integral ΔW of the energy gain of a charge e when traversing a finite interaction region, (preferentially in the comoving frame, so that all interactions stay in a bounded region):

$$\Delta W = e \int (\mathbf{E} + \beta \times \mathbf{B}) \cdot \mathbf{dx} \ll eB\Delta x \,, \tag{1}$$

in which **E** and **B** denote the total electric and magnetic fields in same region. In particular, these fields describe the interaction of the charge with (a large number of) locally generated waves (inside the assumed strong shock(s), and also in the unperturbed upstream medium). Whilst the charge traverses these waves, its energy both increases and decreases, each time with a net gain (or loss) of order the one received within half a cycle of that wave train, instead of that of N wave cycles boosting in phase. Typical (generic) gains are of order those during less than one cycle, smaller by N than what in-situ calculations predict. (Only exponentially few charges gain large energies, as is expected for a thermal equilibrium distribution). Simple estimates differ from those for stochastic gains by just the factor N. In-situ accelerations are expected to modify the electron-distribution function, but not to raise its level.

If bulk in-situ energy upgradings are against the Second Law, the observed relativistic electron population must be provided by the central engine - probably a magnetized rotator. For instance, the strong forming (relativistic) wave of a rotating magnet, near its speed-of-light cylinder, should have a vastly higher boosting efficiency than any of the possible downstream shocks: Gunn-Ostriker boosting, cf. Kundt (1986), Blome & Kundt (1988). Moreover, even the (slowly spinning) Sun is known to create relativistic negatrons and positrons in localized discharges at the base of its corona, via magnetic reconnections; (for the relevant production rates see Mastichiadis et al (1986)). For these reasons, I still prefer my 1980 proposal (with Gopal-Krishna) that the central engine provides the observed, extremely relativistic electron population.

An immediate problem is the one of γ-ray compactness: aren't the high-γ electrons destroyed in situ, by inverse-Compton collisions with the thermal photon bath (of photon energy $\lesssim 10^2 eV$)? This problem is severe for a black-hole engine, much less so for any other (less compact) one. It is less severe than it may appear because high-energy photons reproduce negatron-positron pairs on collision with the thermal photon bath whenever their energy product exceeds the pair-creation threshold (of $(MeV)^2$), so that the initial state is restored except for a halving of the energies, and a doubling of the number: one electron is reprocessed into one pair of (roughly) the same total energy. This pair avalanche reaches saturation when the energy drops to the creation threshold, and eventually escapes (for 'radio-loud' sources) in the form of VHE γ-rays plus a huge pair population, ready to blow the radio lobes (via two antipodal jets).

A second problem is how to form jets from a relativistic cavity, i.e. the focusing task. In 1974, Blandford & Rees gave a simple answer in terms of inertial confinement by the ambient medium: the twin-exhaust mechanism, see Figure 7, according to which the escaping relativistic fluid - in interaction with its CSM - creates two deLaval nozzles beyond which it turns supersonic. In 1979, I was not convinced of its reality, mainly for reasons of stability: the (confining) nozzle walls would easily break apart and fall down, and also would get hot and radiate; moreover, BLR pressures were assumed too small in those days, implying too large a nozzle radius. I have changed my mind (before 1986): the mechanism is indeed very robust. It works for any box with 2 opposite holes (as a container).

The nozzle walls do not perform much work if they confine pair plasma (rather than ISM) - because of its low inertia (for given pressure) - and besides, strong, variable MeV-radiation has meanwhile been observed from many - if not all - 'γ-ray blazars': Blom et al (1995). And fountain-like replenishment of wall material, near the center of a galactic bulge, should easily take care of the losses.

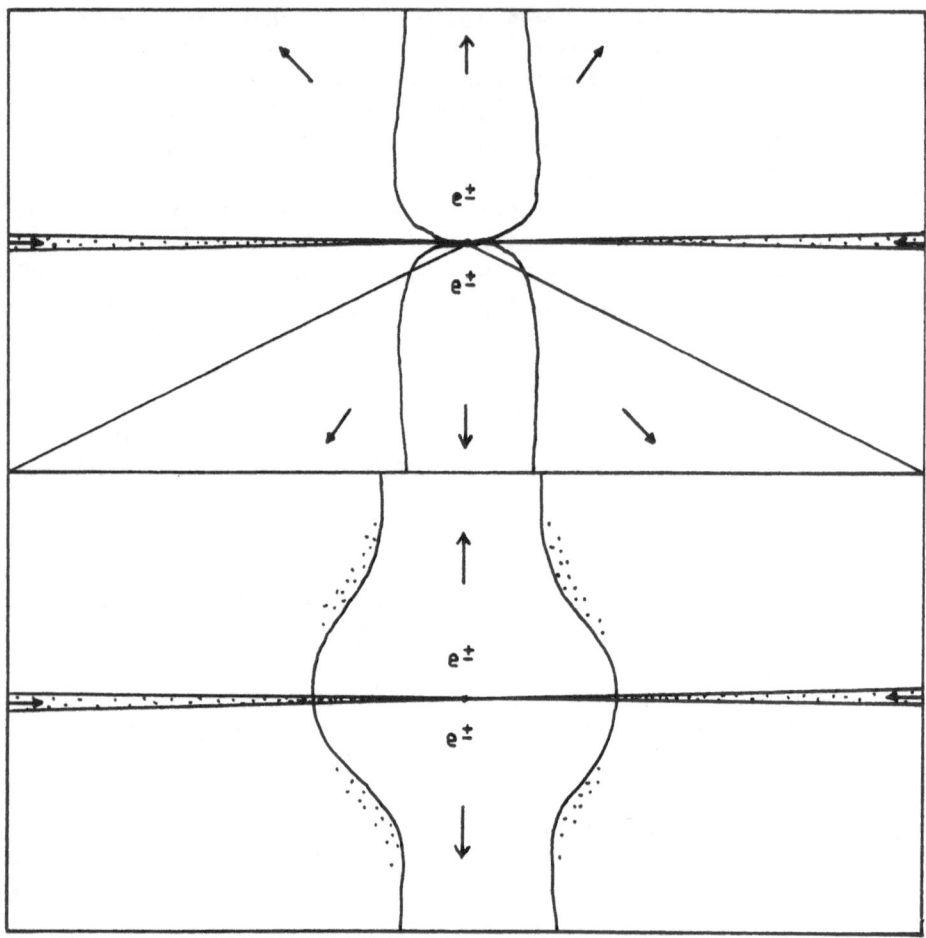

Fig. 7. Sketch of the Twin-Exhaust scenario, at 2 magnifications.

Inertial confinement has the (robust) property that it works all the way along the jet, out to the distant ramming 'head': a cold, relativistic beam is readily confined by the ISM, or IGM; cf. the numerical simulations by Icke et al (1992). No magnetic bandage (involving return currents on cosmic scales) is required. In our Galactic center, Figure 1d of my contribution on 'the Galactic Center' - showing Sgr A East plus environment - has exactly the twin-exhaust morphology.

According to Blandford & Rees (1974), its nozzle radius R

$$R = 0.5 \ (L_{jets}/p \ c)^{1/2} = 10^{17} cm \ (L_{46}/p_0)^{1/2} \qquad (2)$$

fixes the size of the BLR (= broad-line region), where L_{jets} is the power of the central engine fed into the jets, and p is the ambient pressure, ($p_0 :=$ $p/10^0 dyn cm^{-2}$).

BLRs of powerful AGN have indeed sizes up to $\lesssim 10^{17} cm$, as will be summarized below. Our own Galactic center, of jet luminosity $L \lesssim 10^{39} erg/s$ and pressure $p \approx 10^{-6.5} dyn/cm^2$ (Kundt, 1990b), should have $R \approx 10^{16.8} cm$, whereas the 'chimneys' of Sgr A East look $\lesssim 10^2$ times wider; quite likely, Sgr A East acts as a storage bubble, i.e. is almost air-tight because Sgr A* has declined in power, and R measures the size of its porosities. In the case of stellar outflows - for whose nuclear (optical) jets the HST has found widths of $\lesssim 10^{15} cm$ in HH 30 and HH 34 - equ. (2) can be applied with (mechanical) jet luminosities of $L \lesssim 10^{34} erg/s$ and (accretion) pressures $p \lesssim 10^{-8} dyn/cm^2$.

Filling the BLR (of an AGN) with a relativistic gas (rather than a $10^8 K$ gas) also solves the cloud-confinement problem encountered by Krolik et al (1981) - via static confinement - which has been a nightmare ever since it was published. And it removes the difficulty with the superluminal statistics encountered by René Vermeulen, which asks for "un-appealing" large bulk Lorentz factors ($\gamma \gtrsim 10^2$) in the VLBI jets.

What is the expected beam structure beyond the deLaval nozzle? Atomic and molecular beams in the laboratory are cooled by shooting at them with a laser beam: their collision losses reduce the disordered particle velocities (Maddox, 1995). A similar action is thought to be exerted by the light from the BLR. The ultimate, stable beam structure is expected to be an equi-pressure coexistence of pairs and fields whereby the magnetic fields are toroidal plus a possible beam-parallel component, and the electric (Hall) fields are radial, cf. Baumann (1993). This exact solution is (first-order) stable because toroidal magnetic fields can always be regenerated by (slightly) charge-asymmetric friction on the beam boundaries whereupon the corresponding electric fields are generated by the Hall effect, and conversely, excessive fields would post-accelerate the charges due to the Gunn-Ostriker effect. The motion of the (extremely relativistic) charges is a radiation-free $\mathbf{E} \times \mathbf{B}$-drift, apart from inverse-Compton losses on the background photon field, and on intruded neutral particles when they get re-ionized inside the beam. Note that there would be the possibility of a helical motion for a considerable uniform beam-parallel magnetic component. Such a uniform component is, however, most likely absent, because a stretching of magnetic flux loops in beam direction would always result in a net-zero total magnetic flux.

Sceptics have occasionally wondered how to stop an exactly charge-symmetric supersonic flow: neither purely electric nor purely magnetic fields can do that. But a solution is already indicated in Alfvén (1981). For a (fully symmetric) pair-plasma beam (with net charge and current zero), the toroidal magnetic fields which hit an obstacle get enhanced, like a pressed spiral spring. This leads to an increased radial (Hall) polarization, and to mirror charges in the (conducting) obstacle. The resulting enhanced fields deflect the arriving pairs sideways, as a combined $\mathbf{E} \times \mathbf{B}$- plus ∇B-drift. Maxwell's equations imply that the braking fields decrease exponentially upstream, on an e-folding scale of $< \gamma > m_e c^2/eB$.

In his critical assessment of 'particle acceleration in extended radio sources' (below), Klaus Meisenheimer lists three problems: (i) obstacles can drive strong plasma waves into the beam which destroy the regular field pattern; (ii) how to dispose of 'used' electrons, which have strongly radiated?; and (iii) does disappearance of the ordered morphological pattern of the jet in M 87 beyond knot C, or rather beyond knot G, mean destruction? My answers read: (i) No perturbation can move upstream a supersonic beam. The situation is different for oblique shocks, which can enter a beam from the side; but as argued above, beams are cooled (in the lab) by irradiating them with a laser. Streaming pair plasma finds back to order by establishing pressure balance with its frozen-in fields. (ii) All 'used' electrons are recycled, i.e. are re-accelerated by the guiding fields, because the (radiation-free, stable) $\mathbf{E} \times \mathbf{B}$-drift tolerates only a unique velocity (at a fixed point). Beyond a perturbation, all electrons share the remaining energy, (which causes a lower [average] bulk Lorentz factor). (iii) A jet is radiation-free if and only if it does not encounter obstacles. When the jet in M 87 gets invisible beyond knot G, I understand that it has found a straight (clean) channel segment. Of course, 'invisible' must refer to isotropic (thermal) radiation, not to beamed (non-thermal) radiation. But as both hammer and anvil get hot, a distorted beam is expected to emit a significant fraction of thermal radiation.

A few further estimates of BLR properties may help strengthen the reader's confidence in the proposed (twin-exhaust) jet-formation mechanism. The BLR is that environment of the central engine which is traversed by both its (thermal) radiation - the 'Big Blue Bump' in the spectrum - , its (explosively ejected) wind - the BLR cloudlets - , and by the (going-to-be) beam substance - relativistic pair plasma - which fills the volume. Beyond its 'walls' starts the NLR. As such, its size should be determinable by (3) the time-variability of its emission lines, (4) the strengths (equivalent widths) of its lines, and by (5) a sufficient weight of its confining walls - all of which should agree with the estimate (6) of the nozzle radius in eq (1). Moreover, (7) the BLR will more or less agree with the last-scattering region of the photon flood (Big Blue Bump plus VHE γ-rays); and it should be distinctly larger than its input pump, i.e. larger than (8) the speed-of-light radius of the central rotator. These six constraints express themselves in the form

$$r = c\Delta t/2 = 10^{16} cm \ (\Delta t)_{5.8} \ , \tag{3}$$

$$r \geq (L_{lines}/S_{lines})^{1/2} = 10^{16} cm \ (L_{44}/S_{12})^{1/2} \ , \tag{4}$$

$$r = GM\rho/p = 10^{16} cm \ M_{(8)}\rho_{-18}/p_0 \ , \tag{5}$$

$$r = 0.5 \ (L_{jets}/p \ c)^{1/2} = 10^{16} cm \ (L_{44}/p_{0.7})^{1/2} \ , \tag{6}$$

$$r \gtrsim 3c/\Omega \approx c(R/G\sigma)^{1/2} = 10^{15} cm \ (R_{13}/\sigma_{11.2})^{1/2} \ , \tag{7}$$

$$r \approx \left(\sigma_T r^2 \int dr \ n_\gamma \sin \Theta \right)^{1/2} \approx 10^{16} cm \ T_6^{3/2} \ , \tag{8}$$

11

where L_{lines} and L_{jets} denote the powers in the emission lines and jets; S_{lines} the Pointing flux in the emission region (estimated from line-intensity ratios); p the BLR pressure; ρ the BLR-wall mass density; Ω, σ, and R the inner disk's angular velocity, mass density, and radius; σ_T the Thomson cross section; $n_\gamma (\sim r^{-2})$ the BBB photon density, assumed blackbody of temperature T at radius $r = 10^{12} cm$; and $\sin \Theta \ (\sim r^{-1})$ a factor taking care of an increasingly reduced photon-photon scattering angle Θ with increasing distance from the central engine. As always, lower indices in numerical estimates denote logarithms of the assumed numbers (in cgs units), and $M_{(8)} := M/10^8 M_\odot$. Finally, above estimates have been made for a more typical AGN (than in eq (1)), with total luminosity $L = 10^{46} erg/s$, and $L_{jets}/L = 10^{-2\pm2}$.

Estimates (3) - (8) show that typical data for AGN are not inconsistent with a BLR size of order $10^{16} cm$; whereby earlier doubts in the BLR being an outflow region (rather than a rotating, infalling, or virialized one) seem to be declining: Kundt (1988), Arav et al (1995), Blandford (1995). Note that attempts to fit double-horned broad emission lines by expected line profiles from an optically thick (!) Keplerian disk, as by Chen et al (1989), are faced with the problem of the large equivalent widths of the lines.

Most of above estimates have been made for the class of extragalactic jet sources (rather than for the galactic ones); which of them have jets with large bulk Lorentz factors ($\gamma \gtrsim 10^4$)? As in my early work with Gopal-Krishna (1980), I am still convinced that the bulk Lorentz factors are as large as those inferred from the synchrotron spectra, because I do not see the possibility of in-situ electron upgrading. Large gammas are meanwhile observed through their $\lesssim$ TeV inverse-Compton radiation (von Montigny et al, 1995), through the occasional high (radio-) brightness temperatures (Lesch, below), through the statistics of superluminals, and sidedness (Kundt, 1989; O'Dea et al, 1988), through the extreme edge-brightening (of the jet in M 87: Owen et al, 1989), and through their inverse-Compton beacons reported by Bob Fosbury, see also Hippelein & Meisenheimer (1992) for the ELR (= UV cone); all this on top of the earlier reasonings (1987, 1989). And they are suggested by numerical simulations (Icke et al, 1992).

Most of the properties of the extragalactic jet sources are likewise encountered in the galactic ones; and I agree with Jim Pringle (1993) - and not with Heino Falcke (below) - that a uniform description, if desired, must part with a black-hole engine. Less obvious is the evidence for large (bulk) Lorentz factors in the stellar sources, cf. Blome & Kundt (1988), Reipurth & Heathcote (1993). Yet there are cases of (non-absorptive) sidedness also in the S 68 triple radio source in Serpens (Rodríguez et al, 1989), and in HH 111 (Rodríguez & Reipurth, 1994), which signal relativistic beaming. All the directly observed velocities in stellar outflows are extremely non-relativistic, $v \lesssim 1400 Km/s$ (Martí et al, 1995); but they are channel-wall velocities, imparted through ram-pressure thrusting by the quasi-weightless pair plasma. Indirect evidence (of extremely light flows, and inverse-Compton radiation) are the large velocity asymmetries (factor of 2 ± 0.5) observed in 50% of all (stellar) sources: Hirth et al (1994), the elongated

IR lobes seen by IRAS, and the preceding bowshocks of HH 34, 46/47, and 111 (Kundt, 1996b). And if the HST had looked at HH 30 not just twice but twice per day, it might have detected superluminal knot motion.

Another evidence for relativistic jet material in a Galactic source is the Orion 'streamer' in the L1641 cloud: Yusef-Zadeh et al (1990). I feel, however, reminded of CTB 80: is the streamer perhaps blown by a neutron star, i.e. do we deal with a pulsar nebula (Kundt & Chang, 1992), instead of a YSO? Another worry concerns the jet-blowing black-hole candidates: do they really involve black holes, or are they neutron stars inside of massive accretion disks (Kundt & Fischer, 1989; Krolik, 1984)? The list of neutron-star driven jet sources would thus grow beyond a dozen, involving not only SS 433, Cyg X-3, (the great annihilator) 1E 1740.7-2942 (Mirabel et al, 1992), and (the soft γ-ray repeater) SGR 1806-20, but also the two superluminals GRS 1915+105 and GRO 1655-40 (Tingay, below). In all cases, relativistic pair plasma is indicated.

Finally, the jet family includes the class of PNe with jets, of which the 'Southern Crab' (He 2-104), M 2-9, and the 'Egg Nebula' (CRL 2688) are the most obvious representatives, see figure 5 above. CRL 2688 may be the rare case of a twin-jet from a PN seen strongly broadened by projection (under $\lesssim 10°$ w.r.t. its axis). Direct evidence for (relativistic) pair plasma in PNe is (so far) indeed poor. But I see no other mechanism for driving $\lesssim$ pc-scale narrow outflows (cf. Mellema, below). The best direct evidence for a relativistic piston remains so far the blowout in Nova V 603 Aql (1918), figure 6 above.

3   Burning Disks at the Centers of AGN

The literature teaches that there are supermassive black holes at the centers of massive galaxies, dating back to Lynden-Bell (1969). Yet the evidence is declining: Kormendy & Richstone (1995), see also Rees (1992). Among alternative suggestions for the central engine in AGN are the ones by Pacini & Rees (1973), Kundt (1979, 1987, 1996a), Sorrell (1981), Camenzind & Courvoisier (1983), Camenzind (1986), Krügel & Tutukov (1986), and by Terlevich et al (1992, 1995). Difficulties of the BH model are (i) the missing remnants, (ii) the strong 'winds' ($\dot{M}_{in} = \dot{M}_{out}$: Collin-Souffrin, 1988), (iii) their γ-ray compactness, implying too strong a degrading of the relativistic pair plasma before its escape into the jets, and (iv) high metal enhancement of the outflow (Fe 10^2 times solar: Turnshek, 1988; also B.Wills and M.Brotherton, below).

The last item - metal enhancement - can be more easily extracted from the BAL spectra than from those of the BELs. Besides by Turnshek, it has been noticed by Hamann & Ferland (1992), and has urged Artymovicz et al (1993) and Arai & Hashimoto (1995) to think of ways how black holes can spew iron. My 1979 SMD model assumed that the 'cosmic' helium is, more likely, QSO-fabricated - an implication if one believes in nuclear energy as responsible for ejecting the ashes of the quasar fires. Recent worries about big-bang nucleosynthesis appear to point in the same direction: Hata et al (1995). A way out would be a cold big bang (rather than a hot one), which is thermodynamically at least as plausible as its hot rival, and lends itself more readily to the formation of structure: Layzer (1990).

Independently of whether or not black holes meet with difficulties, there is the worthwhile task of predicting how a self-gravitating (galactic) disk behaves near its center, see Figure 8. As is well-known from the extended literature on disks, matter spirals in due to viscosity, and all state parameters grow as inverse powers of r on approach of the center; (cf. Kundt, 1990c, 1996a). The condensing gas meets the star-formation threshold already at Milky-Way densities, but has a long way to go to reach the electron-degeneracy threshold - which it may or may never reach. Already when the star-formation hurdle is somewhat exceeded, a galaxy may turn starburst. Much later, on approach of the nucleus, Keplerian velocities can grow large enough to let coronal magnetic reconnections become important: the galaxy forms a LINER (= low-ionization nuclear emission-line region), (Ikhsanov & Pustil'nik, 1994; van Oss, 1994). On further condensation, midplane temperatures will grow towards stellar values, and the disk will start burning, like stars. Beyond a main-sequence ringlike region - the outer 'burning disk' (=: BD) - temperatures will quasi-regularly reach those of helium flashes and/or explosive hydrogen burning, and the disk is expected to turn 'active', i.e. to be regularly disrupted by nuclear detonations: it will form an AGN. Note that this explosive regime - the inner 'burning disk' - is reached long before the black-hole-formation density threshold.

How can a BD be modeled? Roxburgh (1993) has evaluated the instability criteria for plane-parallel stars: flat stars are more unstable than spherical stars, because they have a much larger relative burning volume, and less instantaneous (adiabatic and radiative) cooling. They overheat in hot spots, and detonate. Explosive hydrogen burning (towards iron) takes a few hundred seconds (Champagne & Wiescher, 1992), much less than buoyant escape (of hot cores) from near the disk's midplane ($> 10^3 sec$); nuclear burning can therefore go (almost) to completion, leading to large amounts of iron.

The scenario just described has certainly similarities with the that of the 'WARMER' model (Terlevich et al, 1992, 1995). It implies that AGN serve as (the often postulated) Pop III stars. But on top of nuclear burning, a BD experiences rapid rotation, at several % of the speed of light, and is thus expected to generate an abundant relativistic corona, via localized discharges plus photon scattering (above the pair-formation threshold). It thereby creates a non-thermal electron population which can ram the jets, and turn the QSO into a radio source. Radio-quiet QSOs - i.e. some 90% of all QSOs - will thereby result when the inner BD is sufficiently gas-enshrouded so that the newly born pairs are burnt in situ. (For quasi-periodic X-ray variability, down to 10 min, see: Halpern, 1993). Figure 9 sketches a snapshot of what a BD may look like.

In his parallel assessment of burning disks (below), Peter Scheuer encounters three serious problems: (a) fatally large binding energies (leading to subsequent black-hole formation), (b) a difficulty in moderating the accretion rate to reasonable values, and (c) instabilities of a thin disk which largely exceeds its Eddington

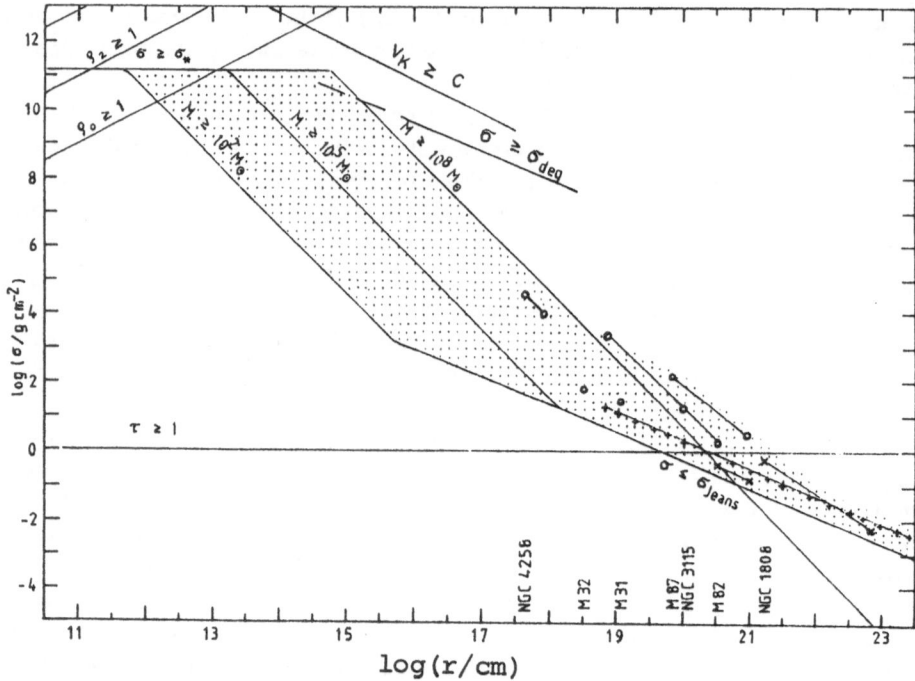

Fig. 8. Surface-mass density $\sigma := M(r)/\pi r^2$ versus radial distance r for galactic disks, from (Kundt, 1996a). The horizontal lines $\tau \geq 1$ and $\sigma \geq \sigma*$ mark limits for being opaque, and for behaving starlike. The falling lines $\sigma \leq \sigma_{Jeans}$, $\sigma \geq \sigma_{deg}$, and $v_K \geq c$ denote the onsets of star formation (Jeans instability), (midplane) degeneracy, and black-hole formation. Further (straight) lines show fixed enclosed mass M(r), and fixed volume-mass density ρ. Crosses, plusses, and circles have been taken from the literature.

luminosity. I should like to answer to them here. Problem (a) of too large a binding energy is the problem of the missing bottom line in his table 1, fault of Allen's table stopping with O5 stars:

$$\left(\frac{M}{M_\odot}, \frac{R}{m}, \frac{v}{km/s}, \frac{E_b}{E_n}, \frac{\dot{M}}{M_\odot/yr}\right) = (\lesssim 10^5, \approx 10^{11}, \lesssim 10^4, \lesssim 0.05, 10^2). \qquad (9)$$

I.e. a BD gets much hotter than O5 stars, and radiates much above its Eddington rate, hence has a much lower binding energy than a close packing of stars. (The relative burning volume is much larger). Problem (b) arises when one ignores the radial gradient of the angular velocity in the viscosity law, which can vanish. Popham & Narayan (1991) show that an accretion disk adjusts its inward mass transport to its inner boundary conditon. And problem (c) is as difficult to grasp as the Cheshire cat: what bad can a burning disk do? When it settles, it heats up, and newly ignites (in particular long before collapsing); when patches detonate,

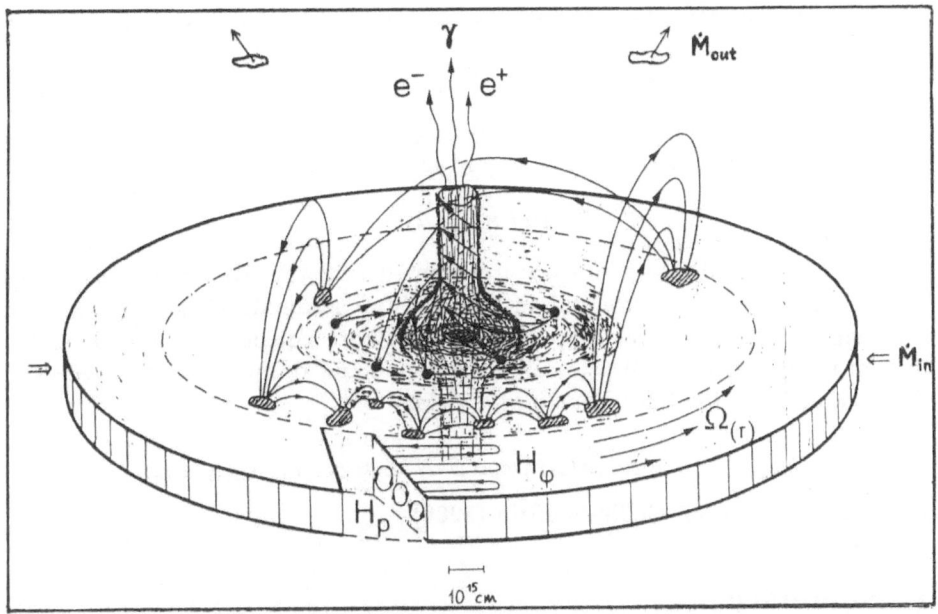

Fig. 9. Snapshot of a 'Burning Disk', secured from - and modified after - Ikhsanov &
Pustil'nik (1994).

they contribute to the BLR, extinguish the fire locally (like explosives thrown
into an oil fire), and alleviate the binding-energy problem. A BD is certainly not
a stationary something, but may have well-defined steady-state properties.

Such steady-state properties are estimated by Kundt (1996a); for completeness' sake, I will repeat some of them here: The radius r of the (inner) BD must
be the effective blackbody radius for given (BBB) luminosity L and surface temperature T_s:

$$r = (L/2\pi\sigma_{SB}T_s^4)^{1/2} = 10^{12.7} cm \; L_{46}^{1/2}T_6^{-2}, \qquad (10)$$

i.e. is of solar-system size for soft X-ray temperatures. The average height $H :=$
$2h$ follows from a balance of its columnal weight $2\rho gh$ by the thermal pressure
gradient (corresponding to a central temperature T_c):

$$H = (kT_c/\pi Gnm^2)^{1/2} = 10^{11.2} cm \; (T_8/n_{24})^{1/2}. \qquad (11)$$

For an average mass density $\rho \; (\approx 1g/cm^3)$ of the BD (which has been found
consistent with convective heat transport between T_c and T_s), the mass $M(r)$
inside of radius r, and its peripheral velocity v, follow as

$$M(r) = \pi r^2 \sigma = 10^{4.7} M_\odot \; r_{13}^2\sigma_{11.5} , \qquad (12)$$

and

$$v \lesssim (10rG\sigma)^{1/2} = 10^4 Kms^{-1}(r_{13}\sigma_{11})^{1/2}. \qquad (13)$$

The (small) mass $M(r)$ implies a (short, though unobservable) refilling timescale t_{fill}:

$$t_{fill} = \pi r^2 \sigma c^2 \epsilon / L = 10^{3.6} yr \ r_{13}^2 \sigma_{11.5} \epsilon_{-2.3} / L_{46} \ , \qquad (14)$$

where $\epsilon := L/\dot{M}c^2$ is the efficiency of hydrogen burning - assumed to happen for half of the available fuel - , and a largely super-Eddington luminosity

$$L/L_{Edd} \gtrsim 10^3. \qquad (15)$$

Crucial for the BD model - when compared with the BH - is a steady-state balance between mass inflow and outflow. For (approximately) spherical outflow (through the BLR) at velocity $c\beta$, the latter rate equals $4\pi r^2 m_p n c\beta$, and a column number density $N := n\Delta r$ – corresponding to marginal opaqueness in the continuum at the BLR boundary – yields

$$\dot{M}_{out}/\dot{M}_{in} = 4\pi\epsilon(r/\Delta r)m_p c^3 \beta r N / L \approx \beta_{-1} r_{16} N_{24.5}/L_{46} \ , \qquad (16)$$

a reassuring result; cf. Collin-Souffrin (1988).

4 Summary

The (uniform) description of the bipolar-flow family exposed in 1987 is (still) preferred, involving a magnetized rotator as the central engine which generates the extremely relativistic pair plasma required for jet formation, and evidenced by various observational details throughout the family.

Acknowledgements
I wish to thank Ski Antonucci for an insider's remark, Carsten van de Bruck for help and insight whenever I needed it, Daniel Fischer for much information, Michaela Kraus for involved discussions on PNe, Karl-Heinz Mack for references, and Klaus Meisenheimer for cheerful provocation at BH.

References

Alfvén H. (1981): *Cosmic Plasma*, Astrophys. and Sp. Sci. Libr. **82**, Reidel, p. 46.
Arai K., Hashimoto M. (1995): A&A **302**, 99.
Arav N., Korista K.T., Barlow T.A., Begelman M.C. (1995): Nat **376**, 576.
Artymowicz P., Lin D.N.C., Wampler E.J. (1993): ApJ **409**, 592.
Balick B. (1987): AJ **94**, 671.
Bally J., Devine D. (1994): ApJ **428**, L65.
Bally J., Lada E.A., Lane A.P. (1993): ApJ **418**, 322.
Barthel P.D., Schilizzi R.T., Miley G.K., Jägers W.J., Strom R.G. (1985): A&A **148**, 243.
Baumann H. (1993): *Astrophysikalische Jet-Phänomene und ihre Deutung als Paarplasma-Strahlen*, Diplomarbeit, Bonn.
Begelman M., Rees M.J., Sikora M. (1994): ApJ **429**, L57.

Blandford R.D. (1995): Nat **377**, 477.

Blandford R.D., Rees M.J. (1974): MNRAS **169**, 399.

Blom J.J., Bennett K., Bloemen H., Collmar W., Diehl R., Hermsen W., Iyudin A.F., Schönfelder V., Stacy J.G., Steinle H., Williams O.R., Winkler C. (1995): A&A **298**, L33.

Blome H.J., Kundt W. (1988): Ap&SS **148**, 343.

Camenzind M. (1986): A&A **156**, 137.

Camenzind M., Courvoisier T.J.-L. (1983): ApJ **266**, L83.

Carilli C.L., Perley R.A., Dreher J.W., Leahy J.P.(1991) ApJ **383**, 554.

Champagne, A.E., Wiescher M. (1992): ARA&A **42**, 39.

Chen K., Halpern J.P., Filippenko A.V. (1989): ApJ **339**, 742.

Collin-Souffrin S. (1988): Ap&SS **149**, 175.

Falle S.A.E.G. (1990): in NATO ASI C **300**, ed. W. Kundt, Kluwer, p. 303.

Gieseking F. (1985): SuW **8**, 448.

Halpern J.P. (1993): Nat **361**, 203.

Hamann F., Ferland G. (1992): ApJ **391**, L53.

Hata N., Scherrer R.J., Steigman G., Thomas D., Walker T.P.(1995): Phys. Rev. Lett. **75**, 3977.

Hester J.J., Scowen P.A., et al. (21 authors) (1995): ApJ **448**, 240.

Hippelein H., Meisenheimer K. (1992): A&A **264**, 472.

Hirth G.A., Mundt R., Solf J. (1995): ApJ **427**, L99.

Hollis J.M., Dorband J.E., Yusef-Zadeh F. (1992): ApJ **386**, 293.

Huggins P.J., Bachiller R., Cox P., Forveille T. (1992): ApJ **401**, L43.

Icke V., Mellema G., Balick B., Eulderink F., Frank A. (1992): Nat **355**, 524.

Ikhsanov N.R., Pustil'nik L.A. (1994): ApJS **90**, 959.

Kormendy J., Richstone D. (1995): ARA&A **33**, 581.

Krolik J.H. (1984): ApJ **282**, 484.

Krolik J.H., McKee C.F., Tarter B. (1981): ApJ **249**, 422.

Krügel E., Tutukov A. (1986): A&A **158**, 367.

Kundt W. (1979): Ap&SS **62**, 335.

Kundt W. (1984): JA&A **5**, 277.

Kundt W. (1986): in *Cosmic Radiation*, ed. M.M. Shapiro, NATO ASI C **162**, Reidel, pp. 57, 67.

Kundt W. (1987): *Astrophysical Jets and their Engines*, NATO ASI C **208**, Reidel, p.1.

Kundt W. (1988) Ap&SS **149**, 175.

Kundt W. (1989): in *Hot Spots in Extragalactic Radio Sources*, eds. K. Meisenheimer, H.-J. Röser, Lecture Notes in Physics **327**, Springer, pp. 179, 275.

Kundt W. (1990a): *Neutron Stars and their Birth Events*, NATO ASI C **300**, Kluwer, p.1.

Kundt W. (1990b): Ap&SS **172**, 109.

Kundt W. (1990c): Ap&SS **172**, 285.

Kundt W. (1993): Foundations of Phys. **23**, 931.

Kundt W. (1996a): *Galactic Nuclei*, Ap&SS , $\gtrsim$ March.

Kundt W. (1996b): *Outflows from Young Stars*, in *Disks and Outflows around Young Stars*, eds. H.J.Staude et al, Springer, to appear.

Kundt W., Chang H.-K. (1992): Ap&SS **193**, 145.

Kundt W., Fischer D. (1989): JA&A **10**, 119.

Kundt W., Gopal-Krishna (1980): Nat **288**, 149.

Kundt W., Gopal-Krishna (1981): Ap&SS **75**, 257.

Kundt W., Yar A. (1995): *Fireworks in Orion*, Bonn preprint.

Layzer D. (1990): *Cosmogenesis: the Growth of Order in the Universe*, Oxford Univ. Press.

Leahy J.P., Muxlow T.W.B., Stephens P.W. (1989): MNRAS **239**, 401.

Lynden-Bell D. (1969): Nat **223**, 690.

MacAlpine G.M., Lawrence S.S., Brown B.A., Uomoto A., Woodgate B.E., Brown L.W., Oliversen R.J., Lowenthal J.D., Liu Ch. (1994): ApJ **432**, L 131.

Maddox J. (1995): Nat **375**, 531.

Markwardt C.B., Oegelman H. (1995): Nat **375**, 40.

Martí J., Rodríguez L.F., Reipurth B. (1995): ApJ **449** 184.

Mastichiadis A., Marscher A.P., Brecher K. (1986): ApJ **300**, 178.

Meaburn J., Walsh J.R., Clegg R.E.S., Walton N.A., Taylor D., Berry D.S. (1992): MNRAS **255**, 177.

Michel F.C., Scowen P.A. Dufour R.J., Hester J.J. (1991): ApJ **368**, 463.

Mirabel I.F., Rodríguez, L.F., Cordier B., Paul J., Lebrun F. (1992): Nat **358**, 215.

O'Dea C.P., Barvainis R., Challis P.M. (1988): AJ **96**, 435.

Owen F.N., Hardee P.E., Cornwell T.J. (1989): ApJ **340**, 698.

Pacini F., Rees M.J. (1973): Sci. Am.**288**, 98.

Popham R., Narayan R. [1991]: ApJ **370**, 604.

Pringle J.E. (1993): in *Astrophysical Jets*, eds. D. Burgarella, M. Livio, C.P. O'Dea, Sp.Tel. Sci. Inst., Symp. Ser. **6**, Cambridge, p.1.

Rees M.J. (1992): in *Physics of Active Galactic Nuclei*, eds. W.J. Duschl, S.J. Wagner, Springer, p. 662.

Reid A., Shone D.L., Akujor C.E., Browne I.W.A., Murphy D.W., Pedelty J., Rudnick L., Walsh D. (1995): A&AS **110**, 213.

Reipurth B., Heathcote S. (1993): in *Astrophysical Jets*, eds. D. Burgarella, M. Livio, C.P. O'Dea, Sp.Tel. Sci. Inst., Symp. Ser.**6**, Cambridge, p.35.

Rodríguez L.F., Curiel S., Moran J.M., Mirabel I.F., Roth M., Garay G. (1989): ApJ **346**, L85.

Rodríguez L.F., Reipurth B. (1994): A&A **281**, 882.

Roxburgh I.W. (1993): MNRAS **264**, 636.

Scheuer P.A.G. (1974): MNRAS **166**, 513.

Schwarz H.E., Corradi R.L.M., Melnick J. (1992): A&AS **96**, 23.

Solf J. (1996): A&A , submitted.

Sorrell W. (1981): **291**, 394.

Terlevich R., Tenorio-Tagle G., Franco J., Melnick J. (1992): MNRAS **255**, 713.

Terlevich R., Tenorio-Tagle G., Różycka M., Franco J., Melnick J. (1995): MNRAS **272**, 198.

Turnshek D.A. (1988): in *QSO Absorption Lines: Probing the Universe*, eds. J.C. Blades, D.A. Turnshek, C.A. Norman, Cambridge Univ. Press, p.17.

Van Oss R.F. (1994): in *Fragmented Energy Release in Sun and Stars*, ed. G.H.J. van den Oord, Kluwer, p. 309.

Venturi T., Giovannini G., Feretti L., Comoretto G., Wehrle A.E. (1993): ApJ **408**, 81.

Von Montigny C., Bertsch, D.L., et al.(21 authors) (1995): ApJ **440**, 525.

Walsh J.R., Meaburn J. (1993): The Messenger **73**, 35.

Weaver H., (1974): Highlights of Astron. **3**, 509.

Yusef-Zadeh F., Cornwell T.J., Reipurth B., Roth M.: (1990): ApJ **348**, L61.

Black Hole, Jet, and Disk: The Universal Engine

Heino Falcke

Department of Astronomy, University of Maryland, College Park, MD 20742-2421, USA, email: hfalcke@astro.umd.edu

Abstract: In this paper I review the results of our ongoing project to investigate the coupling between accretion disk and radio jet in galactic nuclei and stellar mass black holes. We find a good correlation between the *UV bump luminosity* and the radio luminosities of AGN, which improves upon the usual [OIII]/radio correlations. Taking mass and energy conservation in the jet/disk system into account we can successfully model the correlation for radio-loud and radio-weak quasars. We find that jets are comparable in power to the accretion disk luminosity, and the difference between radio-loud and radio-weak may correspond to two natural stages of the relativistic electron distribution – assuming that radio weak quasars have jets as well. The distribution of flat- and steep-spectrum sources is explained by bulk Lorentz factors $\gamma_j \sim 5 - 10$. The absence of radio-loud quasars below a critical optical luminosity coincides with the FR I/FR II break and could be explained by a powerdependent, "closing" torus. This points towards a different type of obscuring torus in radio-loud host galaxies which might be a consequence of past mergers (e.g. by the temporary formation of a binary black-hole). Interaction of the jet with the closing torus might in principle also help to make a jet radio-loud. Turning to stellar-mass black holes we find that galactic jet sources can be described with the same coupled jet/disk model as AGN which is suggestive of some kind of universal coupling between jet and accretion disk around compact objects.

1 Introduction

1.1 The AGN zoo

Non-stellar activity in galactic nuclei is generally thought to be produced by a powerful engine located at the dynamical center of the galaxy. Because of the high luminosity of active galactic nuclei (AGN) concentrated in a small volume, it has been argued that those engines are powered by accretion onto a massive black hole. To the pleasure of observers this activity appears in many different forms and flavors which has led to a proliferation of object classes based on specific properties in one or the other wavelength band. The most important ones are:

a) *Seyfert galaxies* of type 1 (narrow and broad emission lines) and type 2 (narrow lines only) with luminosities up to several 10^{44} erg/sec, corresponding to accretion rates of $\lesssim 10^{-2} M_\odot$/yr

b) *Quasars*, with broad and narrow emission lines and a peak in their spectral energy density distribution (SED) in the UV ("UV bump"), the luminosities are in the range $10^{44..48}$ erg/sec corresponding to accretion rates of $10^{-2..+2} M_\odot$/yr, some quasars have strong radio emission and are labeled radio-loud, otherwise radio-weak (or quiet)

c) *Radio galaxies*, with powerful jets, steep-spectrum radio lobes and compact, flat-spectrum cores; low-power sources have edge-darkened, smoke-trail like lobes (FR I) while high-power sources have well collimated jets and terminate in hotspots (FR II), the optical spectrum of the core usually shows only narrow emission lines

d) *Blazars*, sources completely dominated by variable, non-thermal (synchrotron) emission from a compact core (BL Lacs), if a quasar spectrum is still seen then one has a highly polarized quasar (HPQ) or optically violently variable (OVV), those sources have the highest probability of showing high-energy ($\gtrsim$GeV) emission.

In summary, the different signs of nuclear activity, which are rarely seen in one object together (except perhaps 3C 273), are: a luminous thermal bump in the SED, broad and narrow high-excitation emission lines, a compact radio core, a powerful radio jet and lobes, and variable high-energy (x-ray to gamma-) emission.

1.2 The unification

The diversity of objects has provoked the foundation of a "unification church"[1] (Antonucci 1993 and refs. therein, see also Antonucci – this volume) and a sporadic Counter-Reformation which, however, has not yet rallied its forces effectively. The ingredients to make the unification work are relativistic beaming in a jet and obscuration of the central engine by a molecular torus such that objects appear different if seen from different aspect angles: a Seyfert 1 galaxy becomes a Seyfert 2 by obscuration, a radio-loud quasar becomes a FR II radio galaxy by obscuration and an HPQ by beaming, likewise a FR I radio galaxy becomes a BL Lac by relativistic beaming.

A third ingredient to unify object classes, which I consider very important but is seldom mentioned explicitly, is the power (or in the black hole picture the accretion rate) of the nuclear engine. Thus, by increasing the power, normal galaxies might turn into Seyferts and then into quasars, provided that the majority of normal galaxies has a central black hole. As the FR I/FR II separation

---

[1] This religious allusion appears reasonable considering that both ideas – despite being basically correct – are seemingly not accessible through reasoning but require the experience of a personal conversion.

is by morphologhy *and* power, it is likely that by decreasing the accretion rate, a FR II radio galaxy turns into an FR I and then into a quiescent elliptical. Even though this already gives a pretty and simple picture, a few fundamental questions remain: a) if high-power (FR II) galaxies and quasars, and FR I and FR II are connected, how are low-power (FR I) and quasars/Seyferts connected, and b) what makes the difference between a radio-loud and a radio-weak AGN (the $1,000,000 question)? Remembering that radio-loud AGN occur only in elliptical galaxies, the latter question is even more tantalizing. Finally, it should be realized, that the influence of source evolution is difficult to assess; shape and size of the obscuring torus may well depend on the evolutionary stage of the AGN and the host galaxy.

1.3 The universal engine – a *simple* Ansatz

If the central engine is indeed associated with a black hole, it has the unpleasant property of being so small that it is almost inaccessible by observational means and hence open to wild theoretical speculations. Currently there is no way to prove or disprove that all central engines are completely different or absolutely identical and one is forced to choose a basic Ansatz for the nature of the engine which allows one to draw further conclusions and test them against observational data. Strangely, in Astronomy the burden of proof is usually on those who postulate a simpler solution (like the unified schemes) while Occam's Razor should force one to start with the simplest theory until experimentally disproved. Consequently our Ansatz for the nature of the central engines in AGN should be that those engines are all very similar and governed by a few parameters only. Such an engine would be a black hole which accretes matter within an accretion disk, producing a jet at the black hole/disk boundary layer flowing out along the rotation axis (alternative engines are discussed in the contributions by Scheuer, Kundt, and Sorrell – this volume). As the escape speed from a black hole is relativistic, those jets would have to be relativistic as well, as the jets are produced by the disk, one expects a strong coupling between jet and disk, and as most of the power in an accretion disk is relased close to the center, the jet can be very powerful, and finally, as "a black hole has no hair", there are not many parameters that can vary from one engine to another, and the main parameter of the engine is expected to be the accretion rate. This "simple Ansatz" also implies that jets are a natural companion to accretion disks, and both are necessary and symbiotic features in the accretion process onto the compact central object (Falcke & Biermann 1995; hereafter FB95).

2. Jet-disk coupling

A coupled jet-disk system has to obey the same conservation laws as all other physical systems, i.e. energy and mass conservation. We can express those constraints by specifying that the total jet power Q_{jet} is a fraction $2q_j < 1$ of the accretion power $Q_{disk} = \dot{M}_{disk}c^2$, the jet mass loss is a fraction $q_m < 1$ of the disk accretion rate $\dot{M}_{disk}$, and the disk luminosity is a fraction $q_l < 1$ of Q_{disk} ($q_l = 0.05 - 0.3$ depending on the spin of the black hole). The dimensionless jet power q_j and mass loss rate q_m are coupled by the relativistic Bernoulli equation (FB95). For a large range in parameter space the total jet energy is dominated by its kinetic energy such that one has $\gamma_j q_m \simeq q_j$, in case the jet reaches its maximum soundspeed $c/\sqrt{3}$ the internal energy becomes of equal importance and one has $2\gamma_j q_m \simeq q_j$ ('maximal jet'). The internal energy is assumed to be dominated by the magnetic field, turbulence and relativistic particles. We will constrain the discussion here to the most efficient type of jet where we have equipartition between the relativistic particles and the magnetic field and between internal and kinetic energy – we will later see that other, less efficient models (see FB95) would fail.

Knowing the jet energetics, we can describe the longitudinal structure of the jet by assuming a constant jet velocity (beyond a certain point) and free expansion according to the sound speed ($\simeq c/\sqrt{3}$). For such a jet, the equations become very simple, the magnetic field is given by

$$B_j = 0.3\,G\,Z_{pc}^{-1}\sqrt{q_{j/l}L_{46}}$$

and the particle number density is

$$n = 11\,\mathrm{cm}^{-3}L_{46}q_{j/l}Z_{pc}^{-2}$$

(in the jet restframe). Here Z_{pc} is the distance from the origin in pc, L_{46} is the disk luminosity in 10^{46} erg/sec, $2q_{j/l} = 2q_j/q_l = Q_{jet}/L_{disk}$ is the ratio between jet power (two cones) and disk luminosity which is of the order 0.3 (FMB95) and $\gamma_{j,5} = \gamma_j/5$ ($\beta_j \simeq 1$). If one calculates the synchrotron spectrum of such a jet, one obtains locally a self-absorbed spectrum that peaks at

$$\nu_{ssa} = 20\,\mathrm{GHz}\,\mathcal{D}\,\frac{\left(q_{j/l}L_{46}\right)^{2/3}}{Z_{pc}}\left(\frac{\gamma_{e,100}}{\gamma_{j,5}\sin i}\right)^{1/3},$$

integrating over the whole jet yields a flat spectrum with a monochromatic luminosity of

$$L_\nu = 1.3 \cdot 10^{33}\,\frac{\mathrm{erg}}{\mathrm{s\,Hz}}\,\left(q_{j/l}L_{46}\right)^{17/12}\mathcal{D}^{13/6}\sin i^{1/6}\gamma_{e,100}^{5/6}\gamma_{j,5}^{11/6},$$

where $\gamma_{e,100}$ is the minimum *electron* Lorentz factor divided by 100, and $\mathcal{D}$ the *bulk* jet Doppler factor. At a redshift of 0.5 this luminosity corresponds to an unboosted flux of ~ 100 mJy. The brightness temperature of the jet is

$$T_b = 1.2 \cdot 10^{11} \, \text{K} \, \mathcal{D}^{4/5} \left(\frac{\gamma_{e,100}{}^2 q_{j/l} L_{46}}{\gamma_{j,5}^2 \beta_j} \right)^{1/12} \sin i^{5/6}$$

which is almost independent of all parameters except the Doppler factor. An important factor that governs the synchrotron emissivity is of course the electron distribution, for which we have assumed a powerlaw distribution with index $p = 2$ and a ratio 100 between maximum and minimum electron Lorentzfactor. As we are discussing here the most efficient jet model we also assume that all electrons are accelerated (i.e. $x_e = 1$ in FB95), hence the only remaining parameter is the minimum Lorentzfactor of the electron distribution $\gamma_{e,100}$ determining the total electron energy content. In order to reach the magnetic field equipartition value, which is close to the kinetic jet power governed by the protons, we have to require $\gamma_{e,100} \sim 1$. It cannot be higher because otherwise the power in electrons would exceed the total jet power, and it cannot be much lower because we would not reach equipartition. Such a high, low-energy cut-off in the electron energy distribution was suggested already by Wardle (1977) and Celotti & Fabian (1993) for other reasons.

If radio-interferometric techniques were not yet developed today, and we would have been asked to predict what kind of jet sources we would expect to see, we would have needed only very few simple considerations:

a) *'total equipartition' everywhere*, i.e. equipartition between the luminosity radiated by the disk and expelled by a jet wind, equipartition between internal energy and kinetic energy, and equipartition between relativistic particles and magnetic field

b) *relativistic speed*, because, if the jet is produced close to the black hole, relativistic escape speeds are required,

c) *disk luminosity* (UV-bump), which is a measurable quantity

Thus, using $L_{disk} \sim 10^{46}$ erg/sec and $\gamma_j \sim 5$, we could have predicted pc-scale radio cores at cm-wavelengths, with brightness temperatures of 10^{11} K and fluxes of 100 mJy and more. But of course, nobody would have believed us as those assumptions are obviously too simplified...

3. UV/radio correlation

3.1 Estimating the disk luminosity

Now, we will have to validate some of our assumptions and test the jet-disk coupling derived above. For this we have to estimate the disk luminosity of quasars as precisely as possible and compare it to their radio cores. The best studied quasars sample so far is the PG quasar sample (Schmidt & Green 1983). For most sources in this sample Sun & Malkan (1989), using optical and IUE data, fitted the UV bump with accretion-disk models and a few more were available in the archive (Falcke, Malkan, Biermann 1995, hereafter FMB95). There are also excellent photometric (Neugebauer et al. 1987) and spectroscopic data (Boroson & Green 1992) available, but unlike the broadband UV-bump fits, emission lines and continuum colors do not give a direct estimate for the bolometric UV luminosity (L_{disk}) and we need to calibrate those values to the UV bump luminosity using the sources which have a complete set of data available, yielding

$$\lg(L_{disk}/\text{erg s}^{-1}) = 2.85 + \lg(L_{[OIII]}/\text{erg s}^{-1}),$$
$$\lg(L_{disk}/\text{erg s}^{-1}) = 2.1 + \lg(L_{H\beta}/\text{erg s}^{-1}),$$
$$\lg(L_{disk}/\text{erg s}^{-1}) = -0.4M_b + 35.90.$$

With those correlations one should be able to estimate the "disk luminosity" for almost any quasar. If several indicators are available, we can combine them (assigning appropriate weights) to get the final estimate, this then gives a fairly reliable estimate of L_{disk} and reduces the scatter in the correlations considerably as it also reduces the effects of the orientation dependence of some lines (e.g. [OIII]). In the next step, we can compare those disk luminosities with VLA radio cores (Kellermann et al. 1989, Miller et al. 1993) and total radio emission. In addition to the optically selected sample in FMB95 I have now also included quasars from the southern 2 Jansky sample (Morganti et al. 1994, Tadhunter et al. 1994) which are predominantly flat-spectrum quasars, and steep-spectrum, lobe dominated quasars from Bridle et al. (1994), Akujor et al. (1995), and Reid et al. (1995) which had emission lines readily available (Steiner 1981, Jackson & Browne 1991, Wills et al. 1993). Thus, the number of radio-loud quasars is increased considerably – the results are shown in Figures 1&2.

3.2 Different types of radio sources

For those kinds of optical/radio correlation it is very important what kind of radio source one is talking about. Here I distinguish between four cases represented by different symbols: a) radio-weak quasars with weak, diffuse or unresolved radio emission, b) core dominated, flat-spectrum sources, c) FRÍI type steep spectrum sources, and d) compact steep spectrum (CSS) or irregular radio sources. It is quite obvious that the radio/optical correlation may be very different for all these sources – at least in total flux. CSS and irregular sources usually have jets which are strongly interacting with a dense environment inside the galaxy, and their radio output is expected to be strongly modified by this interaction. Flat spectrum sources are usually dominated by their relativistically boosted jet with the inclination being a very sensitive parameter and cannot be compared with the steep-spectrum emission of lobe-dominated sources, and finally steep-spectrum radio-loud and regular radio-weak sources have a completely different radio morphology and hence must also be treated separately.

This is highlighted in Fig. 1, where one can see that radio-loud and radio-weak sources are clearly separated. The undisturbed, steep-spectrum FR II sources do show a relatively tight correlation, the CSS and irregular quasars scatter around and do not show any correlation, and the core-dominated sources are mainly located at the upper end of the radio distribution consistent with being relativistically boosted – with the exception of a few radio-intermediate quasars (RIQ) located in the gap between radio-loud and radio-weak quasars.

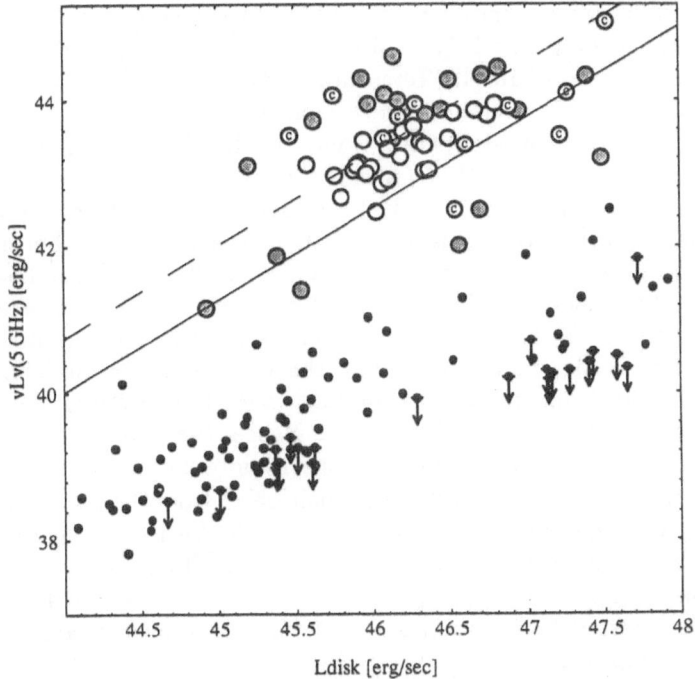

Fig. 1. Total radio luminosity vs. disk (UV-bump) luminosity for quasars (including a complete optical and a radio-selected sample). The shaded circles are core-dominated, flat-spectrum sources, open circles are steep-spectrum (FR II type) sources, circles labeled 'c' are CSS sources and filled points are radio-weak (diffuse or unresolved) sources. Only undisturbed radio-loud FR II type and radio-weak sources show a tight correlation. Flat-spectrum sources are boosted to the upper end of the distribution except 6 radio-intermediate quasars which might be boosted radio-weak quasars (see text). The total emission of CSS does not show a tight correlation with disk luminosity. The solid line is the (oversimplified) model for the lobes from FB95 (see also FMB95).

3.3 Boosted radio-weak quasars?

The RIQ in the gap are all optically selected PG quasars, and their radio fluxes are typically a few ten up to a few hundred mJy. From their R-ratio between radio and optical flux, they are neither clearly radio-loud nor clearly radio-weak, but all are unresolved with the VLA. Using the Effelsberg 100m telescope, we measured the fluxes of those sources at 11 and 2.8 cm to look for variability and spectral slopes (Falcke, Sherwood, Patnaik 1996, in prep.). Even though the data is not yet fully evaluated it is quite obvious that 6 of those sources have flat-spectrum cores and at least 5 are variable. In order to distinguish them from the 2 probable CSS sources which can also be found in the radio-intermediate 'gap', we will label them flat-spectrum radio-intermediate quasars (FIQ).

Some of the FIQ were known as variable sources before, e.g. III Zw 2 which varies between 40 mJy and 1 Jy at 5 GHz and has a brightness temperature probably well in excess of 10^{11}K (Teräsanta & Valtaoja 1994)– ususally a sign of relativistic boosting. The presence of a variable flat-spectrum radio core (without or with weak extended emission) alone is usually already regarded as a good sign for relativistic boosting. In FMB95 (with the knowledge of only the 3 FIQs in the $z < 0.5$ sample), we have argued that, if the variability and the core prominence is due to relativistic boosting, it is unlikely that those quasars are boosted radio-loud cores: at a given optical luminosity their allegedly boosted radio-cores are much weaker than the usual distribution of flat-spectrum radio quasars and as bright or even weaker than the cores of (unboosted) lobe-dominated quasars. Their total flux is also much lower than the total flux of radio-loud quasars, demonstrating that – like radio weak quasars – they lack anything similar to the radio lobes expected for radio-loud quasar. Hence, the only parent population they could have been boosted from are the radio-weak quasars. As shown in FMB89 the relative number of FIQs and their offset from the parent population is consistent with moderate Lorentzfactors of 3-5. An interesting test for the radio-weak blazar hypothesis for the FIQ will be VLBI observations and an investigation of their host galaxies – one would expect to find at least a few of them in spiral host galaxies, as opposed to radio-loud host galaxies which are exclusively in ellipticals.

3.4 Modelling the UV/radio distribution

The fact that we find such good correlations between the UV-bump and the radio luminosities also tells us a lot. Especially for the radio-weak quasars it shows clearly that the nuclear and the *extended* radio emission is AGN related. It would be very difficult to find an argument that the extended emission is produced by starbursts and explain the tight UV/radio correlation unless one is willing to postulate that the UV itself is produced by a starburst (as Terlevich et al. 1992). The total radio emission of the undisturbed, lobe-dominated radio-loud quasars, which is clearly jet-related, scales with the UV-bump as well, which indeed suggests a direct link between the radio-jet producing mechanism and the UV source.

We can now use the jet-disk model derived in Sec. 2 and compare it to the UV/radio correlation for the cores. Because we have carefully calibrated the line emission and the continuum fluxes and scaled them to the UV-bump luminosity, we are able to apply an actual physical model with absolute numbers to the distribution and hence can apply the mass and energy conservation laws to it.

To simplify the discussion, we will use only the most efficient model, where the internal energy is comparable to the kinetic energy and dominated by the magnetic field and relativistic particles. If we would start with a normal plasma jet, where the number of particles is limited by the mass conservation, and assume that all electrons are accelerated from the thermal pool ($\gamma_e \sim 1$) into a powerlaw distribution, we find that we can well explain the radio luminosity of

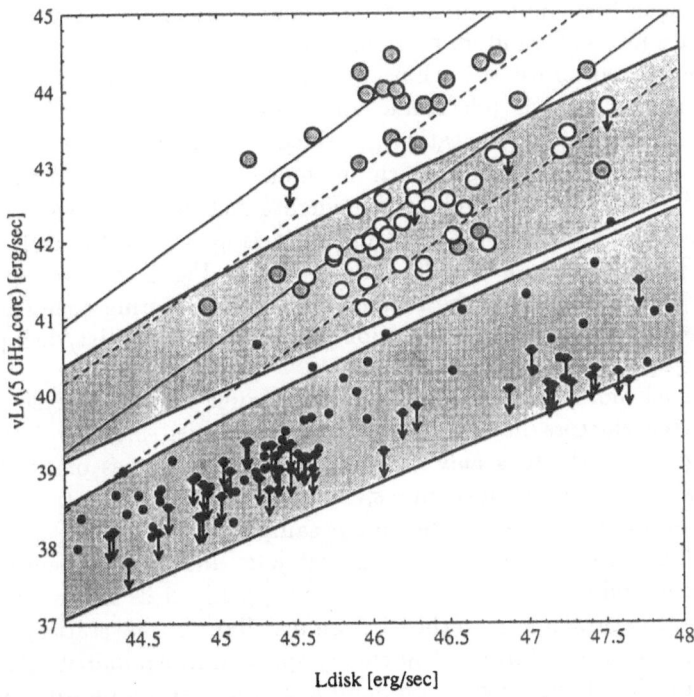

Fig. 2. The same as Fig. 1 (CSS are not explicitly marked) but now the radio core flux is plotted. The shaded bands represent the radio-loud and radio-weak jet model where the width is determined by relativistic boosting. The dashed line represents sources at the boosting cone (inclination $1/\gamma$) and the solid line represents $0°$ inclination – corresponding to the maximum possible flux. The position of flat-spectrum and steep-spectrum sources and the radio-loud/radio-weak separation can be naturally accounted for with the coupled jet-disk model plus boosting.

the radio-weak quasars (Fig. 2, lower band), however, fail to explain the radio-loud quasars by a large margin if we demand $2q_{j/l} < 1$. The reason for this is that in such a model the total energy of the electrons is still just a small fraction of the total energy dominated by the kinetic energy of the ions ("protonic model"), and most of the electrons are found at low energies where they do not contribute to the radio flux. To bring the electrons in equipartition, one either has to create additional pairs (100 times more then electrons) or inject them at a high energy where γ_e is of the order 100 ("electronic model"). Such models are the only ones, which are capable of explaining the UV/radio correlation for radio-loud quasar cores and they do require the 'total equipartition' mentioned in Section 2.

Interestingly, related energetical arguments earlier have led the editor of this book to speculate that high electron Lorentz factors must be present in radio jets (Kundt & Gopal-Krishna 1980). He interpreted this as evidence for ultrarelativistic bulk Lorentz factors, but as demonstrated in Fig. 2 the spread of radio core luminosities in jets (due to the anisotropy of relativistic boosting) is too narrow for such high bulk Lorentz factors and therefore those Lorentz factors must indicate internal, random motions of the electrons.

In order to reproduce the whole distribution of the radio cores, with the two equipartition models (electronic and protonic), we have to specify only two parameters: the jet-disk ratio $q_{j/l}$ which we assume to be constant, and the proper velocity of jet for which we make the powerlaw Ansatz $\gamma_j \beta_j = 6((2/6)^{1/0.15} + L_{46})^{0.15}$. This allows a moderate increase of the jet velocity with power, where the typical quasar Lorentzfactor is $\gamma_j \sim 6$ at $L_{46} = 1$, but never becomes subrelativistic, i.e. smaller than $\gamma_j \sim 2$. In Fig. 2 the two equipartition models are shown, depicting the regions within the boosting cone (dashed and solid line) and the unboosted population (shaded band); the jet/disk parameter used here is $2q_{j/l} = 0.3$ (two-sided jet).

Many conclusions can be drawn from this simple kind of analysis:

a) The Lorentzfactors have moderate values between 5-10, and there is no evidence for *stationary bulk* Lorentzfactors far in excess of 10 in quasars. Those sources – if seen face on – should have a radio/UV ratio much higher than seen in any of the sources in our sample.

b) Radio-loud sources are utmost efficient jets, and the differences between radio-loud and radio-weak sources are remarkably close to the difference between two natural stages of the electron distribution: one starting at thermal energies, the other shifted up in energy-space until equipartition is reached.

c) The radio jets have powers comparable to the disk luminosity, hence they must be produced in the very inner parts of the disk close to the black hole where the bulk of the gravitational energy is released.

d) The magnetic flux in the jet is much higher than the maximum possible radial magnetic flux in an accretion disk and hence the magnetic field for the jet must be produced locally at the footpoint and because of the high efficiency must be related somehow to the dissipation process in the disk (FB95).

4. Unified unification

4.1 The void of FR I quasars

There is another very interesting observation to be made in Figure 1. While the radio-quiet quasars spread out over the whole luminosity interval from $10^{44} - 10^{48}$ erg/sec, radio-loud quasars are predominantly found in the range $L > 10^{46}$ erg/sec, and none is below $L \sim 2 \cdot 10^{45}$ erg/sec. One may argue that at the lower luminosities, the larger elliptical host galaxies become visible and therefore the sources are not classified as quasars, however, this falls short of explaining the void over 2 orders of magnitude. The alternative explanation is that in fact below a critical power, radio-loud quasars lose their typical quasar characteristics, i.e. the broad (and narrow) emission lines and the UV bump. A hint why this may be so comes from the radio morphological data we have for the PG quasars: all radio-loud sources are either of FR II type or compact, none has a typical FR I structure, and indeed it is part of the radio-astronomers folklore that FR I radio sources never show up as quasars. *Consequently, we can identify the void of radio-loud quasars below a certain optical luminosity with the FR II (high-power) to FR I (low-power) transition.* This link between transition of radio morphology and disappearence of optical emission is difficult to understand, especially as the emission line properties of radio-weak quasars – which are almost identical to those of radio-loud quasars – do not show any change at this critical power. Thus just a change (or disappearance) of the accretion disk with decreasing power seems unlikely. A change of the engine would also violate the simplicity of our Ansatz and therefore is not the preferred option here.

4.2 The closing torus

The question now is whether we can explain the behaviour of the radio-loud sources qualitatively without having to postulate different central engines. A minor modification to the unified scheme may indeed do this job: if the opening angle of the obscuring torus is not constant but power-dependent, the torus could approach the jet opening angle at low powers. In this case the central engine would be obscured for almost all aspect angles, and the interaction between the jet and the torus could start the disruption of the low-power jet and initiate its morphological transition (Falcke, Gopal-Krishna, & Biermann 1995).

4.3 Observational consequences

Consequently we would not expect to see broad emission lines from a FR I type radio source as the broad-line region is expected to be inside the 1-100 pc scale torus and be completely obscured. If the opening angle is smaller than the boosting cone, even for boosted FR Is (i.e. BL Lacs) most of the emission line region would be obscured. In fact, one would have to wonder if broad emission line clouds could survive at all in the narrow funnel of the torus if it extends down to the smallest scales. The narrow lines would also be strongly suppressed as the escaping ionizing continuum that produces the NLR region itself is suppresssed. The same is true for broad lines in polarized (scattered) light: as there is not much optical light escaping from the nucleus there will also not be much light that could be scattered.

The best wavelength regime to test the "closing torus" scenario would therefore be the IR where most of the energy of the central source should be re-emitted, predominantly at 10-20μm. The FR I should therefore have an IR output comparable to FRII radio galaxies and quasars scaled to the same engine power. As already in FR II galaxies and quasars, more than half of the energy is absorbed, the *relative* increase in the IR luminosity for FR I, where almost everything is absorbed, is less than a factor two and difficult to detect. However, because of the different shapes of the tori, the IR spectrum itself (e.g. 10μm silicate feature) might be different (see Pier & Krolik 1992). Also NIR spectroscopy might reveal the presence of a quasar engine in FR I radio galaxies; a first pilot study is currently on the way.

Another observational effect concerns the ratio between quasars and radio galaxies in low-radio-frequency selected (orientation independent) samples. This ratio should not be constant but depend on the power of the central engine as it reflects the width of the torus opening. Such a powerdependence is indeed observed (and often used as an argument against the quasar/galaxy unification; Lawrence 1991; Singal 1993).

4.4 Making the jet radio-loud

The fact that the jet may interact with the torus opens the field for many interesting speculations and future studies. If we imagine a powerful, magnetized, relativistic jet scraping along the inner parts of a dense torus (or a cloud therein, or a cloud blown off its surface) we may anticipate the formation of a violent shear layer between jet and the external medium. The interaction might induce highly oblique shocks where particles are accelerated and thus would make the jet radio loud. If the shear-layer is very thin, magnetized and dense, collisions of the ions, carried by the jet at relativisitic speeds, with the external matter would lead to hadronic cascades and consequently to the production of gamma rays and to the injection of pairs at $\gtrsim 35$ MeV – this could also make the jet radio-loud.

In this context it is interesting to note that AGN jets cannot start as radio-loud jets. The synchrotron losses and the inverse Compton losses of the relativistic particles with the UV photons from the disk would be catastrophic and lead to a complete dissipation of the relativistic electrons with high Lorentz factors in its inner parts. A simple extrapolation of the radio emission to the black hole scale would also predict a radio luminosity in excess of the jet power. The typical scale where the losses become less severe and the electron injection can happen is $z \gtrsim 10^{16-17}$ cm (FB95). This is close to the scale often quoted for the production of gamma rays and the scale for the BLR. One starts to wonder whether gamma-ray emission, electron injection and jet torus or jet/BLR interaction have something to do with each other.

4.5 Torus and host galaxy

In the whole discussion we have ignored the nature of the torus and its origin and why it should be powerdependent. None of these questions can be readily answered. The torus may be just a very thick accretion disk with a steep funnel, or it may be composed of molecular clouds in turbulent motion. The powerdependence of its opening might be due to heating and depletion of its inner parts by the central engine or due to the jet itself that drills through a more or less sphercial dust distribution and carries matter outwards, thereby opening the "torus".

In any case a very high (gas, magnetic, or turbulent) pressure is needed to maintain the thickness of the torus. Besides the electron injection discussed above, interaction of the jet with the torus itself, or with winds or clouds produced by it, may therefore also have a confining effect at the inner torus scale (sub-pc to pc). A jet without such a closing torus and without the jet/torus interaction would neither be well collimated nor have the efficient injection of electrons/pairs to make it radio-loud. *Hence it is in principle possible to attribute the radio-loud/radio-quiet dichotomy to different environments at the pc scale rather than to differences in the engine itself.*

The advantage of this concept is that it is easier to relate the pc scale environment rather than the engine properties to the host galaxy. It was proposed that the merging of galaxies may lead to a spin-up of the central black hole by black hole merger during the creation of an elliptical galaxy (Wilson & Colbert 1995). However, it is by no means clear why spiral galaxies should not have a rotating black hole ab initio. Nevertheless, the idea of mergers being a necessary prerequisite for the production of a radio-loud jet is quite tempting and supported by observations (e.g. Heckman et al. 1986). And indeed it seems unavoidable that such a merger sooner or later leads to the formation of a single black hole. There is, however, a critical separation between the two merging holes – again at the pc scale – where neither gravitational friction nor gravitational radiation is very efficient (Begelman, Blandford, Rees 1980), and the binary may stay there for quite a while. So, merging may lead to black hole coalescence, but it definitively will also change the pc-scale structures of the stars and the dust in the central bulge in a way which will be very different from those in spiral galaxies!

5. Starved and stellar-mass black holes

5.1 Sgr A* and its siblings

An interesting consequence of our approach is that it is to first order independent of the scale. If jets and disks are symbiotic features, this may apply to almost any kind of accretion disk with a compact central mass. The equations in Sec. 2 do not depend on the mass of the central object but mainly on the mass accretion rate (i.e. the disk luminosity) and consequently we can use the same scheme for stellar-mass black holes and for the starved black holes in inactive galactic nuclei such as in our own Galaxy. In fact, the jet/disk symbiosis was initially developed to explain the Galactic Center source Sgr A* (Falcke et al. 1993a&b; Falcke 1996a&b) and the jet/disk model still remains a viable explanation for this compact source. Also the radio core of other weakly active galaxies like M81 and M31 seem to follow the same rules (see Falcke 1994, Figure 8.1).

5.2 Galactic jet sources

The sources which received most attention recently are the galactic jet sources associated with either neutron stars or black hole candidates. Two of those sources show apparent superluminal motion indicating relativistic speeds, and they have properties similar to extragalactic jets (Mirabel & Rodriguez 1994; Hjellming & Rupen 1995). Besides those sources, we also know a few other sources in the Galaxy (1E1740-2942, SS433, GRS 1758-258) which show clear jets but without superluminal motion yet detected, and a few x-ray binaries do show flat spectrum radio cores which might be related to a jet. To compare the radio cores of all these sources with their disk luminosity, one has to use x-ray data because the disk spectrum of low-mass black holes is shifted to higher energies. In Fig. 3 the galactic jet sources are shown in a diagram similar to Fig. 2 but extending now down to very low luminosities (Falcke & Biermann 1996). As one can see, no change of the basic parameters is necessary to roughly predict the range of radio fluxes expected for these sources at a luminosity which is 6-10 orders of magnitude lower than in the supermassive AGN black holes. There is even a hint for a dichotomy between radio-loud and radio-weak sources among the galactic jets, but the statistics are not yet good enough. The fact that those equations and the formulation of the symbiosis principle that implied the presence and the luminosity of the galactic superluminal jet sources (FB95; Falcke 1994) were suggested *before* the discovery of those sources, demonstrates the predictive power of this principle — and there is yet a lot of parameter space in Fig. 3 to be filled.

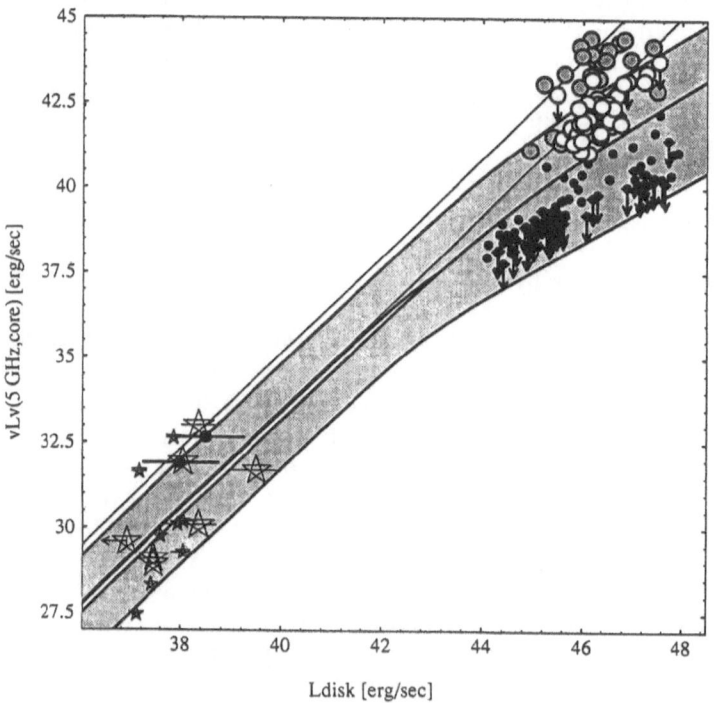

Fig. 3. The same model and data as in Fig. 2 for quasars but now extended to lower powers where galactic jet sources are found. Here the disk luminosity corresponds is interpreted as the x-ray luminosity. Big stars represent confirmed jet sources, while small stars represent x-ray binaries. The big black dots are Sgr A* and M31*.

6. Summary

This work could be summarized by stating that at present it is *not possible* to show observationally that the central engines in AGN are essentially different (i.e. on a scale of 10-100 R_g). Postulating that jets and disks around compact objects are symbiotic and universal features is sufficient to account for most of the observed effects and allows several interesting conclusions:

- The jet/disk coupling explains the UV/radio correlation for quasars and the x-ray/radio flux of stellar-mass black holes.
- Radio-loud jets are utmost efficient and have total powers comparable to the disk luminosities.
- The distribution of flat- and steep-spectrum quasars within the UV/radio correlation reflects relativistic boosting with $\gamma_j \sim 5$, and a population of flatspectrum radio-intermediate quasars (FIQ) suggests the presence of relativistic jets in radio-weak quasars.
- The pc-scale environment ("the torus") may change the jet properties drastically. For example, a closing torus in radio-loud galaxies may explain the transition from FR II jets to FR I jets and the weakness of emission lines in FR I and BL Lacs. The jet/torus interaction may in principle also help to make a jet radio-loud.
- There is no fundamental difference in the parameters between jets from stellar-mass und supermassive black holes.

The fact that stellar-mass black holes and AGN can be described with the same simple jet/disk model (Fig. 3) suggests that there is a universal correlation between radio emision and disk luminosity that spans the whole luminosity range from a few hundred solar luminosities up to several $10^{14} L_\odot$, and hence there may be quite a few other sources (e.g. Seyfert galaxies and nearby galactic nuclei) that follow the same trend but have not yet been discussed in this respect.

References

Akuyor C.E., Lüdke E., Browne I. et al. 1994, A&AS 105, 247

Bridle A.H., Hough D.H., Lonsdale C.J., Burns J.O., Laing R.A. 1994, AJ 108, 766

Antonucci R. 1993, ARAA 31, 473

Antonucci R. 1996 – this volume

Begelman M.C., Blandford R.D., Rees M.J. 1980, Nat 287, 307

Boroson T.A., Green R.F. 1992, ApJS 80, 109

Celotti A., Fabian A. 1993, MNRAS 264, 228

Falcke H. 1994, PhD thesis, RFW Universität Bonn

Falcke H. 1996a, to appear in: "Unsolved Problems of the Milky Way", IAU Symp. 169, L. Blitz & P.J. Teuben (eds.), Kluwer, Dordrecht, p. 163

Falcke H. 1996b – this volume

Falcke, H., Biermann, P. L. 1995, A&A 293, 665 (FB95)

Falcke, H., Biermann, P. L. 1996, A&A in press

Falcke, H., Biermann, P. L., Duschl, W. J., Mezger, P. G. 1993a, A&A 270, 102

Falcke, H., Mannheim, K., Biermann, P. L. 1993b, A&A 278, L1

Falcke, H., Malkan, M., Biermann, P.L. 1995a, A&A 298, 375 (FMB95)

Falcke, H., Gopal-Krishna, Biermann, P.L. 1995b, A&A 298, 395

Heckman T.M., Smith E.P., Baum S.A. et al. 1986, ApJ 311, 526

Hjellming R. M., Rupen M.P. 1995, Nat 375, 464

Jackson N., Browne I.W.A. 1991, MNRAS 250, 414

Kellermann K.I., Sramek R., Schmidt M., Shaffer D.B., Green R. 1989, AJ 98, 1195

Kundt W., Gopal-Krishna 1980, Nat 288, 149

Lawrence A. 1991, MNRAS 252, 586

Miller P., Rawlings S., Saunders R. 1993a, MNRAS 263, 425 (MRS)

Mirabel I.F., Rodriguez 1994, Nat 371, 46

Neugebauer G., Green R.F., Matthews K. et al. 1987, ApJS 63, 615

Pier E., Krolik J.H. 1992, ApJ 401, 99

Reid A., Shone D.L., Akujor C.E. et al. 1995, A&AS 110, 213

Schmidt M., Green R. 1983, ApJ 269, 352

Steiner J.E. 1981, ApJ 250, 469

Sun W.H., Malkan M.A. 1989, ApJ 346, 68 (SM89)

Singal A.K. 1993, MNRAS 262, L27

Tadhunter C.N., Morganti R., di Serego Alighieri S., Fosbury R.A.E., Danziger I.J. 1993, MNRAS 263, 999

Teräsanta, H., Valtaoja, E. 1994, A&A 283, 51

Terlevich R., Tenoria-Tagle G., Franco J., Melnick J. 1992, MNRAS 255, 713

Wardle J.F.C. 1977, Nat 269, 563

Wills B.J., Netzer H., Brotherton M.S., et al. 1993, ApJ 410, 534

Wilson A.S. & Colbert E. 1995, ApJ 438, 62

Simple sums on burning discs

P.A.G. Scheuer

Mullard Radio Astronomy Observatory,
Cavendish Laboratory
Madingley Road
Cambridge CB3 0HE

Can nuclear power replace gravitational power in Active Galactic Nuclei? To test the proposition that it might, I present some order-of-magnitude calculations, based on the premise that we wish to be able to account for standard powerful quasars with luminosities around 10^{39} watts and not just for Seyfert galaxies *et sim.*

I conclude that there are energetic difficulties in (a) blowing away the spent nuclear fuel (which will otherwise form a black hole) and (b) keeping the accretion rate down to a reasonable figure, unless the Shakura-Sunyaev α is $\ll 10^{-4}$.

1   Introduction

Last summer I had the benefit of a personal introduction to the Burning Disc idea from Wolfgang Kundt, and the mental picture that I carried away was of a plane-stratified hydrogen-burning star (as opposed to the better-known spherically symmetrical kind), supported in the plane of the disc by rotation. As this workshop approached, it seemed appropriate to carry out the calculations that a simple-minded astronomer might think of, and Wolfgang kindly consented to my showing you what numbers came out of these sums.

For the purpose of these highly simplified calculations I used a highly simplified model, a close-packed monolayer, of radius R and total mass M, consisting of spherical stars (Figure 1) each of mass M_* and radius h, the half-thickness of the disc. (The hope is that the space left between the close-packed spheres will go some way towards making up for the fact that pressure and temperature will be increased above those in similar isolated stars.)

Following what, rightly or wrongly, I thought Wolfgang had in mind, I imposed the following requirements:

1. The disc shall radiate mostly nuclear (rather than gravitational) power.
2. No black hole shall be formed.

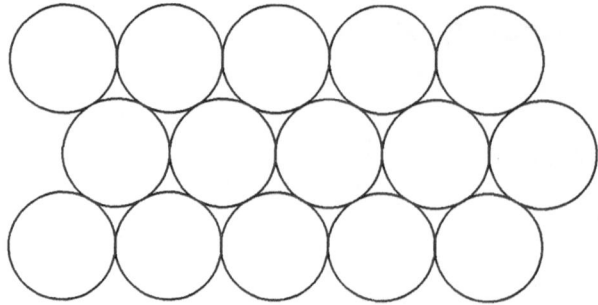

Figure 1: A very simple model for a burning disc.

In order to produce the power of a quasar such a disc must consume $10 - 100 M_\odot$ per year, because the efficiency of nuclear power is much less than the the efficiency of an accretion disc around a black hole. All this mass must then be expelled again, for the spent nuclear fuel will finally form a black hole if it is allowed to accumulate.

How much Burning Disc do we need to account for 10^{39} watts of luminosity? Turning to Allen's 'Astrophysical Quantities', one finds that we need 2.5×10^{12} suns, or 2×10^8 B0 stars or 8×10^6 O5 stars, and therefore the disc radii and total masses listed in Table 1.

2   Stability

The first piece of bad news for such a model is the instability of thin self-gravitating discs, well-known in the context of galactic structure, and one of the reasons for believing that flat galaxies have massive haloes.

3   Throwing out the spent nuclear fuel

At the most fundamental level of principle, there is always enough energy to throw out the spent fuel, since the requisite gravitational energy was released in the formation of the disc. However, if we believe that the gravitational energy of infall was radiated away, it makes sense to ask whether there is enough nuclear energy per kg of fuel to give that kg the escape velocity from the disc.

The escape velocity from the centre of a uniform disc is

$$v_{esc} = \sqrt{\pi G M / R};$$

Table 1: Properties of burning discs delivering 10^{39} W of nuclear power, as a function of the type of component star. For comparison with the tabulated escape velocities: a kinetic energy of 0.7% of Mc^2 corresponds to a speed of 3.6×10^4 km/s.

number of stars	disc mass, $M/M_\odot$	disc radius /m	escape speed from centre of disc km s^{-1}	binding energy of disc /$(0.007Mc^2)$	$\dot{M}/M_\odot$ yr^{-1}
2.5×10^{12} ⊙	2.5×10^{12}	1×10^{15}	9×10^5	190	$1.4 \times 10^8\alpha$
2×10^8 B0	3.5×10^9	8×10^{13}	1.3×10^5	4	$4.5 \times 10^7\alpha$
8×10^6 O5	3.2×10^8	4×10^{13}	6×10^4	0.8	$2 \times 10^7\alpha$

for the edge of the disc it is $\sqrt{2/\pi}$ less. From Table 1 it is clear that the required escape speed exceeds c for a disc of solar-type stars; more significantly, even for discs of very early type stars (which are partially supported by radiation pressure, and not too far from the Eddington limit) it is still above the maximum of 36000 km/s that nuclear burning can provide.

Let us consider the same problem in a slightly different way. The mechanical energy of a rotating disc of uniform surface density consists of

$$\text{gravitational energy} = -\frac{8}{3\pi}\frac{GM^2}{R}$$

$$\text{kinetic energy} = \frac{4}{3\pi}\frac{GM^2}{R} \qquad \text{--virial theorem!}$$

$$\text{hence net energy} = -\frac{4}{3\pi}\frac{GM^2}{R}$$

This is a slightly better way than using escape speed from the centre; it takes proper account of the kinetic energy of rotation as well as taking a proper average over the disc. (For *thin* discs the heat and turbulent energy supporting the disc vertically must be negligible by comparison. A well-ordered magnetic field perpendicular to the disc could extend to heights comparable with the disc radius before its intensity begins to fall rapidly, and could therefore store energy comparable with the rotational energy.) Table 1 shows again that the net binding energy exceeds that available from nuclear burning for solar-type and B0 stars; for O5 stars there is *just* enough nuclear energy to throw out the fuel, provided we don't want to use any substantial part of the nuclear energy to generate radiation.

The moral appears to be that we need burn nuclear energy *quickly*, so that we don't need to have a very large mass in the disc at any one time.

Could supernova explosions in the disc help? They certainly explode very fast, and, as they are non-stationary, they are not subject to the Eddington limit. Unfortunately we then remember that what counts is the energy out per unit time that the mass spends in the disc, i.e. we must include the waiting time that the supernova-to-be spends on the main sequence. Now (my stellar evolution friends tell me) supernovae start fast evolution when they have used 0.25 to 0.35 of their hydrogen, so

$$\frac{\text{SN energy rate}}{\text{stellar energy rate}} \leq \frac{1 - 0.25}{0.25} = 3$$

(Of course many types of supernovae release gravitational energy too, but we have to avoid those like the plague because they leave behind compact objects which would eventually combine into a big black hole ...!!) So in terms of nuclear power per kg we gain not more than a factor of 3. Even worse, supernovae relaese only a few percent of their energy as radiation; the rest goes into kinetic energy, so the nuclear power *radiated* by supernova explosions is at best an order of magnitude *less* than the nuclear power radiated while the star is on the main sequence.

We gain a little bit - not more than a factor 10 in luminosity, according to the standard 'unification' models only a factor 3 - if we believe that the radiation from quasars is beamed along the rotation axis. As the escape speed scales $\propto \text{luminosity}^{1/4}$ the gain is small.

We could also reduce the gravitational potential for a given total luminosity by spreading out the stars instead of insisting on a close-packed disc. We should then have something more like a starburst. The obvious disadvantage of that is that the usual models of jet formation (e.g. the lectures by Camenzind, Contopoulos and Ferreira at this workshop) do not apply.

(The remainder of this section is the result of discussion with Prof. Kundt since the workshop.) Our very simple model close-packed stars does not correspond at all closely to a plane-stratified 'star'; in the latter nuclear burning would occur throughout the central plane. Other things being equal, a much smaller mass then generates nuclear energy at the required rate (10^{39}W), but any quasi-steady radiator is still subject to the Eddington limit ($8 \times 10^7 M_\odot$ by the standard formula but $1.6 \times 10^8 M_\odot$ for a flat disc). If we apply only the Eddington limit and the condition that the nuclear energy exceeds the net gravitational binding energy we find

$$R > \frac{\frac{4}{3\pi}GM^2}{0.007Mc^2} > \frac{4G}{3\pi \times 0.007c^2}\frac{L}{(L/M)_{\text{Edd}}} > 1.4 \times 10^{13}\text{m},$$

which is similar to the radius for close-packed O stars (Table 1), and orders of magnitude greater than the size for close-packed hydrogen-burning cores of such stars.

At this point we should note a profound difference between a flat star and a spherical star. For spherical stars, progression to 'earlier' type means progression to greater mass. However, earlier type stars have *less* mass *per unit area*. Consider what that implies: if we perturb a bit of uniform disc so that, say, the central temperature is increased locally and nuclear burning goes faster, then that local patch finds itself with too much mass per unit area. We cannot then invoke the excellent servo-mechanism that stabilizes ordinary spherical stars.

Put another way: in a spherical star, increased central heating leads to radial expansion, hence to diminished hydrostatic pressure p and hence diminished temperature T ($T \propto p/\rho \propto \rho g R/\rho \propto GM/R$). Result: stability. In a flat star, increased heating in the central plane leads to 'vertical' expansion only, hence no decrease in hydrostatic pressure and an increase in temperature. Result: instability.

These considerations indicate that we ought not to think of a disc-like 'star' with a continuous nuclear-burning central plane. Perhaps we should rather think about a collection of little patches (drawn out by differential rotation?) with frequent nuclear explosions occurring wherever and whenever the nuclear instability strikes. The advantage of providing the luminosity by nuclear explosions is that, being non-steady, they are not subject to the Eddington limit. The picture begins to resemble the multiple supernova scenario once advocated by Geoffrey Burbidge.

4 Accretion rates

Let us estimate the accretion rate $\dot{M}$, in the spirit of a Shakura-Sunyaev-type calculation: that is, we assume at least a rough balance between the rate of advection of angular momentum inward and the viscous torque transporting angular momentum outward. Then

$$\dot{M}\,\Omega r^2 = 2\pi r h \times \text{ viscous stress } \times r$$

where $2h$ is the thickness of the disc at radius r (so that h is the radius of one of the constituent 'stars'), Ω its angular velocity, and we shall use Σ for the mass per unit area, G for the Newtonian gravitational constant, g for the acceleration due to gravity perpendicular to the disc and ρ for the mass density. Thus

$$\dot{M}\,\Omega r^2 = 2\pi r^2 h \; \alpha \; \frac{1}{2}\rho g h$$

and the average g in the disc is roughly $\pi G\Sigma$. Also, in a uniform self-gravitating disc,

$$\Omega^2 r \approx G(\pi r^2 \Sigma)/r^2; \quad \text{combining these we find}$$

$$\dot{M} = \pi^{3/2}\alpha r^{1/2}G^{1/2}\Sigma^{3/2}h$$

The numerical values of $\dot{M}$ for the crude uniform disc models (at a radius comparable with the radius R of the disc) are shown in Table 1, and it is evident that an accretion rate comparable with the required fuel input (a few $100M_\odot$ per year) requires extremely low values of the Shakura-Sunyaev coefficient α. The above formula also shows that steady uniform disc models are not self-consistent: $\dot{M}$ is much larger at large radii than at small radii, indicating that matter will soon accumulate at small radii. (Assuming the mass-radius relation $M_* \propto h$ we should find that $h \propto r$, $\Sigma \propto 1/r$ gives an accretion rate independent of r, but we should not take such a model seriously. For $M_* \propto h^{4/3}$, which is a better approximation to the data in Allen's Astrophysical Quantities, $\dot{M} \propto r^{1/2}$, the dependences on M_* and h cancelling out. Only the order-of-magnitude estimates in Table 1 can be regarded as fairly robust.)

5   Recent evidence for Black Holes in AGN

[Here I described the observation by Tanaka et al. (1995) of an X-ray line whose extraordinary broad line profile is just what the authors expect for an accretion disc extending to within a few Schwarzschild radii of a black hole.]

6   Conclusions

The Burning Disc model faces at least three serious problems when we attempt to apply it to powerful quasars. These problems are not necessarily fatal, but serious detailed work would be needed to convince a simple astronomer that they can be overcome. At all events, I would claim that, in the case of powerful quasars, the gravitational energy of a disc cannot be just a small perturbation on the nuclear energy.

References

1 C.W. Allen: *Astrophysical Quantities*, 2nd ed., §100. The Athlone Press (1962)
2 Y. Tanaka et al.: *Nature* **375** 659 (1995)

Coherent Emission and Intraday Variability of Active Galactic Nuclei

Harald Lesch

Institut für Astronomie und Astrophysik der Universität München,
Scheinerstraße 1, 81679 München, Germany

Abstract: An alternative view of intense variability in active galactic nuclei is considered. The extreme brightness temperatures $T_B \sim 10^{17-19}$ K far beyond the Compton-limit of 10^{12} K in the radio regime of intraday variable sources raise the question whether incoherent radiation mechanisms are really able to provide these intensities. Even relativistic boosting models require Doppler factors which are significantly larger than the observed values. Thus, we investigate the role of coherent plasma processes to account for the huge brightness temperatures.

1 Introduction

Observations in the radio (Quirrenbach et al. 1989a,b) and the optical (Wagner et al. 1990) reveal variability in flat spectrum sources on time scales shorter than 1 day. The amplitude changes are between 25% in the radio and 100% in the optical. Correlated variations of the total flux density S and the polarized flux density P in the radio range indicate that the variability is intrinsic. Additionally a rapid swing of the polarization angle has been observed, when the variable emission decreases (Quirrenbach et al. 1989b).

Furthermore the optical observations exhibit for the BL Lac 0954+65 an apparant luminosity fluctuation of 10^{46} erg s^{-2} during the steepest increase (assuming isotropic emission) (Wagner and Witzel 1995).

The apparent brightness temperatures in the radio range up to $T_B \simeq 10^{19}$ K exceed the Compton limit of 10^{12} K significantly. This led to the model that at least the radio variability has its origin in shock waves which propagate with relativistic speeds in an inhomogeneous jet (e.g. Blandford 1990). Relativistic Doppler boosting of the radiation then leads to the observed huge brightness temperatures, whereas the actual intrinsic brightness temperatures T_B^{int} are lower than the Compton limit, i.e

$$T_B \simeq \Gamma^3 T_B^{\text{int}}. \tag{1}$$

Doppler factors $\Gamma \sim 10$ consistent with observed superluminal expansion are indeed adequate to accommodate most of the compact source variability.

However, the reported intraday variability requires Doppler factors an order of magnitude larger. We note that Romero et al. (1995) reported a rapid radio variability of a blazar (PKS 0537-441) on time scales of 10^4 sec, which suggests a brightness temperature of 10^{21} K. Thus, we like to propose an intrinsic model that might explain the variability apart from the steady emission.

The described observations indicate that a very effective radiation mechanism is working in the flaring object, which can provide the small rise time for the variable emission, its enormous brightness temperatures and its high luminosities. The requirement of an effective radiation process is equivalent to the requirement of an efficient acceleration process. The enormous flaring activity within a day needs a lot of relativistic particles, which dissipate effectively their kinetic energy in radiation.

It is well known that coherent emission is much more efficient in transferring particle energy into radiation than incoherent processes. Our suggestion of a role of coherent processes stems from the experience with solar flares and with plasma processes in the magnetosphere and in laboratory plasma devices. Seldom in these situations are single-particle emission processes energetically important; usually collective emission mechanisms completely overwhelm the incoherent processes (Melrose et al. 1984; Hewitt et al. 1981; Kato et al. 1983).

In Sect. 2 we describe the generation of relativistic electron beams by magnetic reconnection. The excitation of Langmuir turbulence by such beams is treated in Sect. 3. With these ideas in background we consider in Sect. 4 the transfer of the Langmuir turbulence into electromagnetic waves via coherent emission. Section 5 contains the discussion and conclusions.

2 Magnetic activity in galactic nuclei

Galactic nuclei contain magnetic fields. Whatever the central engine is, we know from observations that in very compact radio sources relativistic electrons emit polarized synchrotron radiation. The indicated magnetic fields in the very centre of the host galaxies are either advected inwards via mass accretion and/or are produced via dynamo action by the centrally rotating gas motions. Especially the extremely energetic outflows, extending to scales beyond the outer edge of the host galaxy in the form of strongly collimated jets, present highly magnetized outflows of relativistic particles. There is substantial evidence that magnetic fields are involved in the jet driving mechanism and the collimation of the flow, as well (cf. Lesch et al. 1989; Camenzind 1990; Blandford 1990).

In general, magnetic fields are subject to the laws of electrodynamics. Whenever a conductor moves relative to a magnetic field, electric currents are induced which themselve induce toroidal magnetic fields. Magnetic fields which are spatially or timely variable correspond to electric fields and/or currents. Since the electromagnetic energy is dissipated via currents and electric fields, any perturbed magnetic field line is equivalent to some dissipation, depending on the conductivity of the gas. This relation leads to the concept of Ohm's law, which relates the current density j to the conductivity σ, the electric field E and the inductive term $\mathbf{v} \times \mathbf{B}$, where $\mathbf{v}$ denotes the gas velocity:

$$\frac{\mathbf{j}}{\sigma} = \left[\mathbf{E} + \frac{\mathbf{v} \times \mathbf{B}}{\mathbf{c}} \right] \tag{2}$$

As long as the conductivity is high, the presence of motion $\mathbf{v}$ must generate E fields that balance the $\mathbf{v} \times \mathbf{B}/c$ term rather than balance the j/σ term. These E fields are actually generated by changing B fields, and such changes work out so that the lines of force representing $\mathbf{B}$ move rigidly with the fluid $\mathbf{v}$. This concept of *frozen-in-field lines* depends on the value of σ. The conductivity is given by

$$\sigma = \frac{\omega_{pe}^2}{4\pi\nu_{coll}} \tag{3}$$

where ν_{coll} is the collision frequency and $\omega_{pe} = 5.6 \cdot 10^4 \, s^{-1} \sqrt{n_e}$ denotes the electron plasma frequency, depending on the number density n_e of the electrons.

For Coulomb-collisions we have

$$\sigma = 6.3 \cdot 10^{10} T^{3/2}. \tag{4}$$

For temperatures above 10^3 K, the conductivity of cosmical plasmas is high enough to ensure the frozen-in condition. The dissipation rate of electromagnetic energy is

$$Q = \frac{j^2}{\sigma}. \tag{5}$$

The current density is related to spatial variations of the magnetic field by Maxwell's equation

$$\nabla \times \mathbf{B} = \frac{4\pi}{c} \mathbf{j}. \tag{6}$$

The dissipation rate is very small as long as the conductivity is provided by Coulomb collisions only and the variations of the magnetic field happen on large spatial scales.

However, if the magnetic field lines are subject to strong plasma motions, the situation changes drastically. The frozen-in field lines may be twisted, stretched, compressed etc.., and thereby strong currents are induced. Especially in the case of shear motions (like in the case of differential rotation), or turbulent gas flows, the magnetic energy is heavily dissipated via **magnetic reconnection**.

Reconnection is an intrinsic property of a magnetized plasma with shearing and/or turbulent motions. The encounter of magnetic field components with different polarity corresponds to parallel electric currents, which attract each other. In that sense any kind of plasma motion can trigger reconnection.

If the plasma pressure between opposite fields $\pm B$ is insufficient to keep the fields apart (e.g. by pushing the different flux systems apart), the plasma squeezed from between them and the two fields approach each other. The field gradient steepens and eventually the current density $\frac{c}{4\pi} \nabla \times B$ becomes so large that there is strong dissipation. The problem of reconnection is to know how the dissipation of the currents with density $j = en_e v_d$ is provided (e is the charge and v_D denotes the drift velocity of the electrons).

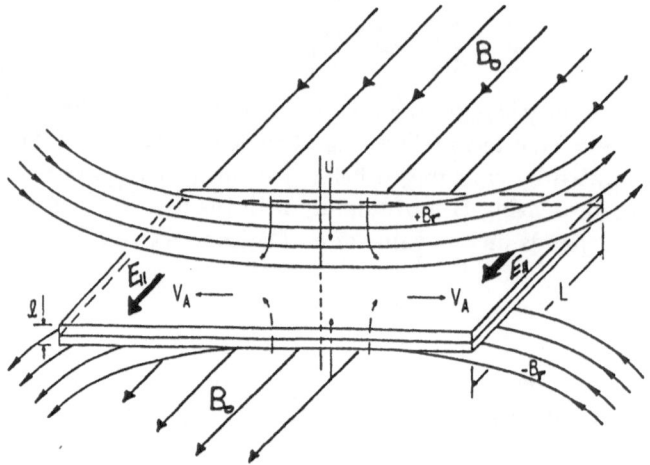

Fig. 1. A perspective view of two attracting flux tubes, which could have their origin in unbalanced forces in a magneto-active accretion disk. When the magnetic field gradient is strong enough, rapid dissipation sets in and an electric field E is induced which accelerates the particles. This electric field is perpendicular to the reconnecting components of the magnetic field, B_r but parallel to that magnetic field component B_0 which is not directly involved in the annihilation process. Thus it presents a field aligned potential drop $\propto EL$. L denotes the characteristic length of the reconnection zone.

If the current density is constant, enhanced dissipation is equivalent to the reduction of the electrical conductivity σ.

Thus strong dissipation not only means an enhancement of the current density but also a reduction of the conductivity. Indeed both effects support each other. This can be understood as follows: when the conductivity is high, the magnetic field lines are "frozen into" the plasma. Field lines embedded within a volume element of plasma are carried along by the moving plasma. Any two plasma elements that are threaded by the same lines of force will remain threaded in this way, i.e. two plasma elements that are threaded by different lines of force can never be threaded together. Thus a high conductivity prevents the field lines from merging. Reconnection occurs only in those plasma regions where the electrical conductivity is drastically reduced below its classical Coulomb value. The reduction of the electrical conductivity is provided by plasma instabilities, driven by the induced current.. If the plasma is locally unstable, i.e. if the current density exceeds a critical value $j > j_{crit}$ or $v_d > v_{crit}$ (e.g. Huba 1985) microscopic instabilities will excite waves which enhance the collision frequency by wave-particle interactions; they lead to *anomalous conductivity*. The effective σ becomes anomalously low, greatly enhancing the dissipation and reconnection of the lines of force. Then the concept of "frozen in" field lines is no longer valid and the plasma moves relative to the field lines. When this occurs, strong electric fields along the reconnection length L are induced. It is the dissipation of the electric field E which transforms the current energy into particle energy.

Schindler et al. (1988, 1991) have shown that in three dimensions the reconnection process is always related to a magnetic field aligned electric potential $U = -\int E_\parallel ds$ (see Fig 1). The induced electric field is always parallel to that field component which is not directly involved in the reconnection process. This result was used to consider the acceleration of particles in cosmical plasmas and it was concluded that parallel electric fields associated with non-ideal plasma flows can play an important role in cosmic acceleration. The actual value of $E_\parallel$ depends on the details of the microscopic instability, which is responsible for the deviations of the plasma from the ideal high conductivity state. It has been shown that in the case of the slow reconnection mode first proposed by Parker (e.g. Parker 1979) it is possible to get an estimate of the maximum Lorentz factor particles can attain via parallel electric potentials (Lesch 1991, Lesch and Pohl 1992, Lesch and Reich 1992).

Starting from the assumption of a stationary Ohmic dissipation in a three-dimensional reconnection sheet with an area $\sim L^2$ and a thickness l, the dissipation surface density in the sheet lj^2/σ is just to devour the influx of magnetic energy $uB_0^2/8\pi$. u is the approaching velocity of the field lines. Conservation of fluid requires that the net magnetic field inflow balances the outflow

$$uL = v_A l. \tag{7}$$

In terms of the magnetic Reynolds number $R_M = \frac{2Lv_A}{\eta}$, where $\eta = \frac{c^2}{4\pi\sigma}$, one gets (Parker 1979)

$$l = \frac{2L}{\sqrt{R_M}} \tag{8}$$

and

$$u = \frac{2v_A}{\sqrt{R_M}} . \tag{9}$$

The thickness of a reconnection sheet is also defined by Maxwell's equation $\nabla \times \mathbf{B} = \frac{4\pi}{c}\mathbf{j}$, which means $l \simeq \frac{cB}{4\pi j}$. The accelerating electric field is given by $E \simeq \frac{u}{c}B$. Now the whole problem is shifted to the microscopic level of description. As mentioned above the particle collisions are not efficient enough to produce a significant magnetic diffusivity (or to decrease the electric conductivity). The collision frequency is too small. Current driven plasma instabilities are involved, whose excited frequency depends on the drift velocity between electrons and ions. We choose in the following the wave mode which has the lowest instability threshold where the drift velocity v_d is equal to the ion thermal velocity $v_{thi} = \sqrt{\frac{k_B T_i}{m_i}}$ the lower hybrid (LH) wave $\omega_{LH} \simeq 4.2 \cdot 10^5 B$. LH waves depend on the magnetic field strength, which means they are also suitable for three-dimensional reconnection where the magnetic field is not zero in the reconnection zone. A fully developed LH-instability results in an effective collision frequency of the order of ω_{LH} (Sotnikov et al. 1978). We note that the central role of LH-waves in reconnection zones has been established for magnetospheric activity (e.g. Shapiro et al. 1994)

Inserting u into the electric field

$$E = \frac{\mathbf{v} \times \mathbf{B}}{c} \sim uB/c \qquad (10a)$$

and l into L, using $\nu_{coll} \simeq \omega_{LH}$ and with

$$\gamma m_e c^2 \simeq eEL \qquad (10b)$$

one obtains the maximum Lorentz factor electrons can achieve in a three-dimensional reconnection zone, where the conductivity is determined via lower-hybrid waves (Lesch 1991)

$$\gamma_{LH} \simeq 6 \cdot 10^3 B \simeq 6 \cdot 10^5 \left[\frac{B}{100\,G} \right]. \qquad (11)$$

Magnetic reconnection is a very efficient acceleration mechanism. All low-energetic particles which enter the reconnection region are accelerated. The energy distribution function of the electrons will be quasi-monoenergetic since the achievable energy depends on the length of the reconnection zone, i.e. it is a linear accelerator. Since the final energy distribution will show an accumulation of particles either at the maximum energy or at the energy where radiation losses and acceleration exactly cancel, we obtain an energy distribution which exhibits pronounced low and high energy cutoffs, i.e. a relativistic electron beam (REB) (Lesch and Schlickeiser 1987).

3 The excitation of Langmuir waves by a relativistic electron beam

It is well known that a distribution function which exhibits a positive momentum gradient $\frac{\partial f}{\partial p}$ is unstable (for example Davidson 1972 and references therein). In particular electron distribution functions with $\frac{\partial f}{\partial p} > 0$ excite the high frequency electrostatic Langmuir wave at frequencies almost equal to the electron plasma frequency

$$\omega_{pe} = \sqrt{\frac{4\pi e^2 n_e}{m_e}} \simeq 5.6 \cdot 10^4\, s^{-1} \left[\frac{n_e}{1\,\mathrm{cm}^{-3}} \right]^{1/2}. \qquad (12)$$

A plasma in which a relativistic electron beam propagates exhibits a positive momentum gradient and is unstable (Kaplan & Tsytovich 1973). The instability responsible for wave excitation is called the *beam-plasma instability* and operates on the time scale (Lesch & Schlickeiser 1987) (Γ_{lin} denotes the linear growth rate of the instability)

$$t_{lin} \simeq \Gamma_{lin}^{-1} \simeq \frac{n_e}{n_b} \frac{\gamma}{\omega_{pe}}, \qquad (13)$$

where n_b is the number density of relativistic electrons.

The excited Langmuir waves slow down the beam through resonant wave particle interactions towards the plateau distribution function (with $\frac{\partial f}{\partial p} \simeq 0$) in the resonant momentum intervall (for the details we refer to Davidson 1972). For our purposes it is only necessary to estimate the field energy of the waves being generated during the relaxation process of the beam towards the plateau distribution function. A rough estimate is provided by the difference between the energy density of the relativistic electrons before and after the relaxation process (Lesch & Schlickeiser, 1987) ($\bar{\gamma}$ is the mean Lorentz factor)

$$\Delta W \simeq n_b \bar{\gamma} m_e c^2 - 4\pi c \int_{p_{\min}}^{p_{\max}} p^3 f(t = t_{ql}) dp. \tag{14}$$

t_{ql} denotes the time scale for the relaxation of the beam into a plateau distribution given by

$$f(t = t_{ql}) \simeq \frac{3}{4\pi} \frac{n_b}{p_{\max}^3 - p_{\min}^3}. \tag{15}$$

Using $f(t = t_{ql})$ in ΔW gives (using $p = \gamma m_e c$)

$$\Delta W \simeq \frac{1}{4} n_b \gamma m_e c^2. \tag{16}$$

In the final state of the relaxation process a large fraction ($\simeq 25\%$) of the energy of the relativistic electrons is transferred into plasma waves. To find the density of the oscillation energy during the relaxation process, we should note that the oscillation energy density flux $v_g W_L$ (where v_g is the group velocity of the Langmuir waves while W_L is their energy density) becomes comparable with the energy density flux of the beam $\gamma n_b m_e c^3 / 4$. Since the group velocity for Langmuir waves excited by a relativistic electron beam is equal to $3v_{the}^2 / c$, from the given condition one finds that

$$W_L \simeq \frac{\gamma n_b m_e c^2}{12} \frac{m_e c^2}{k_B T_e}. \tag{17a}$$

W_L presents the energy reservoir for any kind of coherent radiation process which transfers beam energy into radiative energy.

When the condition (Lesch and Schlickeiser 1987)

$$\frac{W_L}{n k_B T_e} \geq \frac{k_B T_e}{m_e c^2} \tag{17b}$$

is fulfilled the fate of the plasma waves depends on the ratio of beam particle density to background particle density, the plasma temperature and the Lorentz factor. If

$$\frac{n_b}{n_e} \geq \frac{k_B T_e}{\gamma m_e c^2} \tag{17c}$$

the waves are nonlinearly scattered via wave-wave interactions and develop into localized electrostatic solitons if the initial energy density is strong enough (Lesch and Schlickeiser 1987). When energetic electron streams are scattered in regions of such strong Langmuir turbulence they radiate coherently (Krishan and Wiita 1990; Weatherhall and Benford 1991; Lesch 1991).

This coherent radiation, which is called "collisionless bremsstrahlung" and which we describe in the next section will dominate plasma radiation or synchrotron radiation if the beam is dense and highly relativistic (Weatherhall 1988). Experiments show that the critical beam density for the onset of intense coherent beam radiation is (Benford 1992)

$$n_b \simeq 10^{-2} n_e. \tag{18}$$

Before we proceed, we mention that propagating electron beams can trigger reconnection due to the beam excited plasma turbulence (Spicer et al. 1986). In magnetic field structures which are initially stable to reconnection, the beam excited plasma turbulence reduces the conductivity, which leads to magnetic reconnection as discussed above. As was discussed for solar flares, the inductive triggering of reconnection may generate a network of dissipation regions in which particles are effectively accelerated (Anastasiadis and Vlahos 1994)

4. Collisionless bremsstrahlung

As the electrons pass through the Langmuir turbulence they encounter the spatially varying electrostatic wave fields, causing them to oscillate and therefore radiate electromagnetic waves as in a free-electron laser (Baker et al. 1988). In what follows we describe qualitatively the interaction of a relativistic beam in a strongly turbulent plasma.

We consider the situation that a charge clump in a relativistic electron beam passes a charge clump in a plasma. This beam-plasma instability drives density perturbations in the beam and plasma of the same amplitude $\delta n_b \sim \delta n_e \sim \delta n$. The wavelength of the density modulations is also the same $\lambda_b \sim \lambda_e \sim \lambda$, where λ is the observed wavelength of the instability.

For a beam clump containing N electrons, Larmor's formula yields the power emitted by the clump as it is undergoing an acceleration,

$$P = \frac{2}{3} \frac{N^2 e^2}{c} \gamma^6 \left[(\dot{\beta})^2 - (\dot{\beta} \times \beta)^2 \right], \tag{19}$$

where $\beta = v/c$. Taking the beam to be relativistic we consider only accelerations perpendicular to the velocity, which are much more producers of radiation than parallel accelerations. In the direction perpendicular to the beam, the acceleration of each beam electron when that electron is within the electric field E of a plasma electron clump is given by

$$\frac{dv_\perp}{dt} = - \left(\frac{e}{\gamma m_e} \right) E. \tag{20}$$

Since all the electrons in a clump undergo the same acceleration, P becomes

$$P = \frac{2}{3}\frac{N^2 e^4}{m_e^2 c^3}\gamma^2 E^2, \tag{21}$$

where $\beta \simeq 1$ was used. To get an estimate for the strength of the electric field of an electron density clump in the plasma, we assume that the clump is a sphere with radius $r = \lambda/2$ containing a charge density $e\delta n$. Integrating Coulomb's law $\nabla \cdot \mathbf{E} = 4\pi en$, we obtain the maximum value $E = (2\pi/3)e\delta n\lambda$. The number N of the electrons in the clump is given by δn times the volume of a clump, yielding $N = \delta n(\pi/6)\lambda^3$. Inserting these E and N into P yields

$$P = \frac{2\pi^4}{3^5}\frac{e^6}{m_e^2 c^3}\gamma^2 \delta n^4 \lambda^8 \tag{22}$$

for the power emitted by one beam clump. Clearly, to maximize the power estimate, the maximum value of the electrostatic length scale λ has to be taken. This length is the plasma skin depth $\lambda = c/\omega_{pe}$, which is also the length scale of the beam-driven Langmuir wave. Using $\omega_{pe}^2 = 4\pi e^2 n/m_e$ one reaches

$$P = \frac{1}{2^7 3^5}\frac{m_e^2 c^5}{e^2}\gamma^2 \left(\frac{\delta n}{n}\right)^4 = 2.8 \cdot 10^{12}\,\mathrm{erg\,s^{-1}}\gamma^2 \left(\frac{\delta n}{n}\right)^4 \tag{23}$$

as the maximum power per clump. An estimate of the maximum power from a plasma is given by the maximum power per clump times the number of clumps in that plasma. The number of clumps is given by the total volume occupied by clumps divided by the total volume of one clump. The occupied volume is fV_{plasma}, where f is the fraction of the volume filled with clumps and where V_{plasma} is the volume of the plasma. The volume of the clump is $(4\pi/3)(\lambda^3/2)^3$, so the number of clumps is $6fV_{\mathrm{plasma}}/\pi\lambda^3$. With the maximum wavelength $\lambda = c/\omega_{pe}$ one finally reaches

$$P = \frac{\pi^{1/2}}{640}ec^2 m_e^{1/2} n^{3/2}\gamma^2 \left(\frac{\delta n}{n}\right)^4 fV_{\mathrm{plasma}}, \tag{24}$$

which gives

$$P = 3.6 \cdot 10^{-5}\mathrm{erg\,s^{-1}} \left(\frac{n}{cm^{-3}}\right)^{3/2}\gamma^2 \left(\frac{\delta n}{n}\right)^4 fV_{\mathrm{plasma}}. \tag{25}$$

Fig. 2 presents the emitted power of a coherently radiating source in dependence of the density n_e calculated with $\delta n \sim n_b$ and the necessary condition for the onset of strong Langmuir turbulence $n_b \geq 10^{-2}n_e$ (Benford 1992).We note that with coherent beam radiation one easily reaches enormous luminosities without relativistic boosting by a macroscopic relativistic flow.

The characteristic emitted frequency range is between the plasma frequency ω_{pe} and $\gamma^2\omega_{pe}$. The collisonless bremsstrahlung mechanism provides a broad spectrum, depending on the energy of the relativistic beam electrons (Fig. 3).

Analyzing the collective emission properties (Benford 1992), one recovers several spectral features, well known from synchrotron radiation:

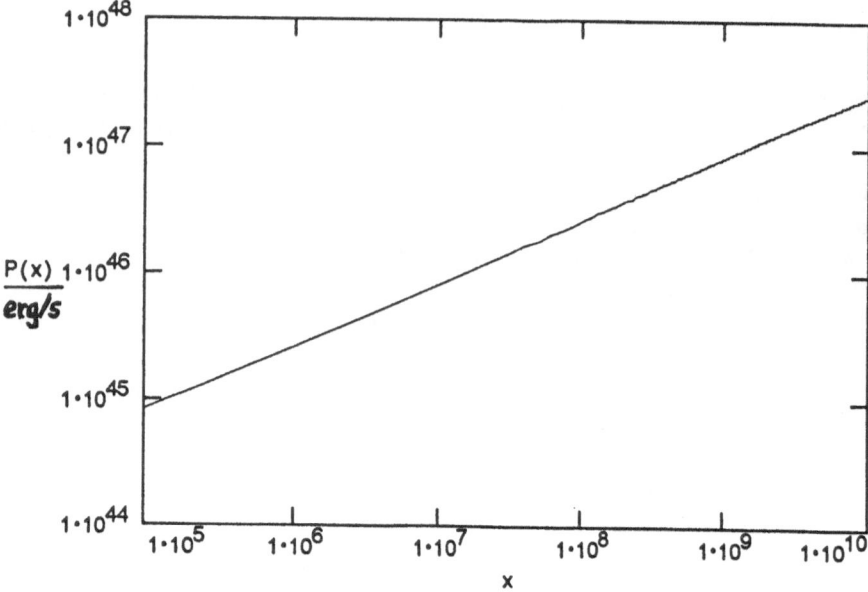

Fig. 2. The coherently emitted power P in erg/sec of a volume with radius of one light day. Depending on the number of background particles measured by $x = \frac{n_e}{cm^6}$ the luminosity can easily reach values which, in the case of incoherent synchrotron radiation need relativistic boosting. The volume filling factor f is weighted with the density x via $f = 1/x$, which takes into account that high density regions have a much smaller filling factor than low density regions, i.e. for a density of $10^6 \, cm^{-3}$ particles we have a filling factor of 10^{-6}. The adopted Lorentz factor is $\gamma = 10^4$.

1. The spectrum rises as $\nu^{2.5}$ at low frequencies and decreases as ν^{α} above the self-absorption frequency. The spectral index is related to the energy distribution of the radiating electrons $f(\gamma) \propto \gamma^{-x}$ via $\alpha = (1 - x)/2$. A quasi-monoenergetic distribution would result in a $\nu^{0.33}$ spectrum with an exponential cutoff (Lesch and Reich 1992).

2. Emission occurs in a $1/\gamma$ cone along the propagation, and spectral changes occur simultaneously throughout the spectrum.

3. Linear polarization features are the same as for synchrotron radiation but there is no circular polarization (Windsor and Kellogg 1974).

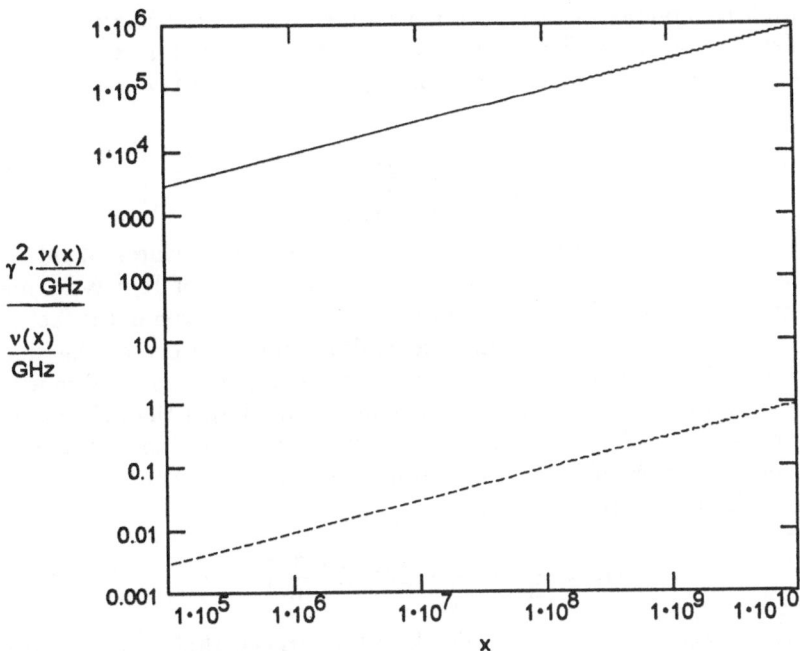

Fig. 3. The emitted frequency of the collisionless bremsstrahlung mechanism in GHz in dependence on the density $x = \frac{n_e}{cm^3}$. The variability spectrum is bounded by the plasma frequency and γ^2 times the plasma frequency. We used a Lorentz factor of 10^4. The dotted line is the plasma frequency ω_{pe} and the solid line is $\gamma^2 \omega_{pe}$.

5 The brightness temperature

As argued in the introduction the high brightness temperatures T_B deduced from radio observations may require some form of coherent emission. The quantitative implications of this requirement are subject to considerable uncertainty both from the observational side, related to the estimation of T_B, and also from the theoretical side due to the uncertainty concerning the emission mechanism.

Observationally, the brightness temperature is determined via

$$T_B = \frac{c^2 S_\nu}{2k_B \nu^2},\qquad(26)$$

where k_B is Boltzmann's constant. The brightness temperature is manifestly the temperature of a black body radiating spectral intensity I_ν in the Rayleigh-Jeans (classical) part of the Planck spectrum. $S_\nu = \int I_\nu d\Omega$ denotes the spectral flux density (with Ω as an element of solid angle at the position of the source).

The brightness temperature of a variable source is called *variability brightness temperature* T_{var}. An unresolved radio source changes its flux S in a time $\sim t_{\mathrm{var}}$. If its distance is D, then one can assume that its linear size is $\sim c t_{\mathrm{var}}$ and one obtains

$$T_{\mathrm{var}} = \frac{S_\nu D^2}{2k_{\mathrm{B}}\nu^2 t_{\mathrm{var}}^2}. \tag{27}$$

S_ν is now taken at the position of the observer. In the business of intraday variable radio sources this T_{var} exceeds the Compton limit of 10^{12} K by many orders of magnitude. The Compton limit finds its physical reason in the fact that the brightness temperature of **incoherent radiation** cannot exceed the kinetic temperature of the radiating relativistic electrons. If all the emission from within the source were to contribute a brightness temperature larger than the electron temperature $\sim \gamma m_e c^2/k_{\mathrm{B}}$, then the source will become self-absorbed and only its outer shell is observable. In other words the brightness temperature of an incoherent radiation source is limited to

$$T_{\mathrm{B}} \leq \frac{\gamma m_e c^2}{3k_{\mathrm{B}}} \sim 5 \cdot 10^{12} \, \mathrm{K} \left[\frac{\gamma}{10^3} \right]. \tag{28}$$

On energetic grounds Begelman et al. (1994) argued that a more realistic limit for the maximal intrinsic brightness temperature of an incoherent source is 10^{11} K. Readhead (1994) draws a similar conclusion by analyzing data from powerful extragalactic radio sources.

One explanation for brightness temperatures exceeding the Compton limit (Eq. 28) is relativistic boosting (e.g. Blandford 1990). As observed in many active galactic nuclei, the radio emission originates in structures which move with Doppler factors Γ up to 10. The frequency in the source frame is $\nu\prime = \Gamma^{-1}\nu$. The brightness temperature will be boosted by a similar factor. If the source size is R, so that its proper variability timescale is R/c, then an estimate of the observed variability time is $t_{\mathrm{var}} \sim R/c\Gamma$. Now the observed flux can be expressed as

$$S_\nu = \frac{2k_{\mathrm{B}}\nu^2(\Gamma T\prime)R^2}{c^2 D^2}. \tag{29}$$

This leads to the relation

$$T_{\mathrm{var}} \sim \Gamma^3 T\prime. \tag{30}$$

Doppler factors of $\Gamma \sim 10$ consistent with the observed superluminal motion are just good enough for temperatures up to a few 10^{15} K, but what is observed is up to 10^{19} K (Wagner and Witzel 1995). Begelman et al. (1994) examined that question and proposed that Doppler factors of the order of 30-100 have to be involved in order to reach at least 10^{16} K. They also claimed that these highly relativistic jets have never been observed and that their radiative output has to be very inefficient. It should be emphasized that the ultrahigh brightness temperatures cannot be explained by Doppler factors of the order of 100.

We note that Phinney (1987) and Abramowicz et al. (1990) have argued that the maximum Lorentz factor of electron-proton jets is limited by the drag force of the intense radiation field at the origin of the jet, which is supposed to start close to the central object and accretion disk. The maximum bulk Lorentz of such jets is

$$\Gamma_{\max} \sim \left[\frac{m_{\mathrm{p}}}{m_e}\right]^{1/3} \sim 12.$$ (31)

If their conclusion would be true (and so far there is no observation in contradiction to that prediction (e.g. Wagner and Witzel 1995)), the extremely high brightness temperatures of the intraday variable sources can only be produced by coherent radiation processes.

We note that limit (31) is not appropriate for electron-positron jets (Kundt and Gopal Krishna (1981; 1984).

Now we come to the attainable brightness temperature of a coherent emission mechanism:

First one supposes that coherence is due to emission by relativistic particles with number density n and $\gamma \gg 1$. Then the maximum Lorentz factor a **coherent** source can attain is $\sim \gamma m_e c^2/k_B$ times the effective number of particles that radiate in phase. As noted in the previous section, this number cannot exceed the total number of particles per coherence volume nV_{coh}. In the case of collisonless bremsstrahlung we have ($\lambda = c/\omega_{\mathrm{pe}}$ and n corresponds to the beam number density n_b)

$$N = nV_{\mathrm{coh}} \sim n\lambda^3 = 3.7 \cdot 10^{19}\frac{n_{\mathrm{b}}}{n_e}\frac{1}{\sqrt{n_e}}.$$ (32)

Taking into account the necessary condition for the onset of strong Langmuir turbulence $\frac{n_b}{n_e} \simeq 10^{-2}$, we reach $10^{10} < N < 10^{15}$ for a density range from $1 - 10^{10}$ thermal particles per cm^3.

Thus, from the coherent radiation process described above the brightness temperatures

$$T_{\mathrm{B}}^{coh} \simeq N\frac{\gamma m_e c^2}{k_{\mathrm{B}}}$$ (33)

follow, which are much higher than 10^{12} K. A fully coherent source (like a pulsar) would result in brightness temperatures of the order of 10^{22-27} K. However, in the AGN case the observed brightness temperature is the result of an incoherent superposition of coherent emitters and therefore the final brightness temperature is considerably lower than the maximal achievable value. In the case of intraday variable sources only factors of $N = 10^{6-7}$ are necessary to explain the observed values of T_{B}, whereas the actual local value may be much larger. But the efficiency is weakened by the incoherent superposition of the coherently radiating fragments. This inefficiency is also taken into account by the volume-filling factor f in Eq.(25).

The coherent radiation depends sensitively on the density values and the ratio between beam density and thermal background density. The coherent emission cannot be absorbed, since its frequency is well above the plasma frequency. The radiating particles are supposed to be beam particles, which posses a very anisotropic energy distribution function, i.e. the particles do not gyrate, since initially their pitch angle is zero. Thus, there is also no absorption by gyro-resonance e.g. (Melrose 1986). As in the case of pulsar emission, there is no absorption problem, since the radiation originates from the region with highest densities and plasma frequencies, respectively.

6 Discussion

We have designed a scenario which qualitatively treats the problem of intra-day variability in active galactic nuclei. The observationally deduced brightness temperatures 10^{18-19} K raise the question whether the received radiation is of incoherent origin. As is well known, synchrotron and inverse Compton radiation are limited by the kinetic temperature of the radiating particles, i.e. the maximal intrinsic brightness temperatures are of the order of 10^{11-12} K. Relativistic motions allow for higher brightness temperatures since the radiation intensity is boosted in the forward direction. This amplification results in an increase of the observed brightness temperature by a factor Γ^3. Since the observed values are $\Gamma \leq 20$ the allowed brightness temperatures are about $8 \cdot 10^{15}$ K. **We have a gap of two to three orders of magnitude between the allowed brightness temperatures of incoherently radiating sources and the observations.**

Coherent radiation easily accounts for the huge brightness temperatures. The effective particle acceleration by magnetic reconnection gives rise to dense relativistic electron beams, which excite Langmuir turbulence. The finite electron beams interact strongly with these waves, producing coherent radiation with large intensities. The underlying physical mechanism is well known from plasma experiments, it is called free-electron laser. the emitted frequency range covers several decades, depending on the Lorentz factors of the relativistic electrons. It starts with emission at the plasma frequency and stops at γ^2 times the plasma frequency. Thus, coherent emission can explain the simultaneous variability in the radio and optical.

A central argument against coherent radiation processes concerns the influence of thermal particles onto the brightness temperature of the emission: the high energy density of the Langmuir waves corresponds to a high number density of Langmuir photons, i.e. their brightness temperature is much higher than the gas temperature. For such high photon number densities one has to consider the effect of *induced Compton scattering* (Levich & Sunyaev 1970; 1971). This process leads to photon scattering towards lower frequencies and sets a constraint on T_B, given by (Levich & Sunyaev 1971)

$$T_B \simeq \frac{m_e c^2}{k_B(\tau_T[1+\tau_T])} = 10^{12}\,\mathrm{K} \left[\frac{n_e}{10^6\,\mathrm{cm}^{-3}}\right]^{-1} \left[\frac{R}{10^{15}\,\mathrm{cm}}\right]^{-1}, \qquad (34)$$

where $\tau_T \simeq \sigma_T n_e R$ denotes the Thomson depth, with the Thomson cross section σ_T and the source radius R. However, induced Compton scattering is only efficient if the photon field is isotropic and covers a wide energy range. If the photon field is anisotropic the efficiency of induced Compton scattering is almost zero (Zeldovich et al. 1972). Since beams of relativistic electrons are both, almost monoenergetic and strongly anisotropic (especially within the source itself), the influence of induced Compton scattering on the brightness temperature of the beam emission is negligible.

We think that in any kind of AGN coherent processes could account for at least a part of the variability since reconnection and the accompanying acceleration of electrons to relativistic energies should appear in any plasma where the magnetic fields are turbulently mixed. Especially the inductive coupling of turbulent magnetic fields via wave excitation by relativistic electrons can trigger further reconnection by reducing the electrical conductivity. A "firework" of reconnection sites can be built up which easily accounts for the variability of active galactic nuclei, even up to X-rays (Maraschi 1989) This can be seen by the following example: we increase the magnetic field which is annihilated in the reconnection zones from 100 Gauss to 200 Gauss. This leads to a Lorentz factor of the relativistic electrons (see Eq. (11)) of 10^6, which gives a maximum frequency of $2 \cdot 10^{21}$ Hz, i.e. well in the γ-ray regime. Since the higher field strengths are expected to appear at smaller radii, the variability in the X-rays should occur on smaller time scales than the variability at lower frequencies, which indeed has been observed (Hayakawa 1991; Remillard et al. 1991).

It is a pleasure to thank G. Benford and W. Kundt for important suggestions and a careful reading of the manuscript.

References

Abramowicz, M.A., Ellis, G.F.R., Lanza, A.: ApJ **361**, 470 (1990)

Anastasiadis, A., Vlahos, L.: ApJ **428**, 819 (1994)

Baker, D.N., Borovsky, J.E., Benford, G., Eilek, J.E.: ApJ **326**, 110 (1988)

Begelman, M.C., Rees, M.J., Sikora, M.: ApJ **429**, L57 (1994)

Benford, G.: ApJ **391**, L59 (1992)

Blandford, R.D. in *Active Galactic Nuclei*, Saas-Fee Advanced Course 20, Eds. T.J.L. Courvosier and M. Mayor, Springer, Berlin, p.161 (1990)

Camenzind, M.: in *Reviews in Modern Astronomy* **3**,p.54 (1990)

Davidson, R.C.: *Methods in Nonlinear Plasma Theory*, Benjamin, New York (1972)

Hayakawa, S.: Nature **351**, 214 (1991)

Hewitt, R.G., Melrose, D.B., Ronnmark, K.G.: PASA **4**, 221 (1981)

Huba, J.D.: in *Unstable Current Systems and Plasma Instabilities in Astrophysics*, IAU 107, eds. M.R. Kundu, G.D. Holman, Reidel, Dordrecht, p. 315 (1985)

Kaplan, S.A, Tsytovich, V.N.: *Plasma Astrophysics*, Pergamon Press, Oxford (1973)

Kato, K.G., Benford, G., Tzach, D.: Phys. Fluid **26**, 3636 (1983)

Krishan, V., Wiita, P.J.: MNRAS **246**, 597 (1990)

Kundt, W., Gopal-Krishna: ASS **75**, 257, (1981)

Kundt, W., Gopal-Krishna: A& A **136**, 167, (1984)

Lesch, H.: A&A **245**, 18 (1991)

Lesch, H., Schlickeiser, R.: A& A **179**, 93 (1987)

Lesch, H., Appl, 'S., Camenzind, M.: A& A **225**, 341 (1989)

Lesch , H., Pohl, M.: A& A **254**, 19 (1992)

Lesch, H., Reich, W.: A& A **264**, 493 (1992)

Levich, E.V., Sunyaev, R.A.: Astrophys. Lett. **7**, 69 (1970)

Levich, E.V., Sunyaev, R.A.: Sov. Astr. **15**, 363 (1971)

Maraschi, L., Maccaraco, T., Ulrich, M.H.: *BL Lac Objects*, Lecture Notes in Physics, Springer, Berlin (1989)

Melrose, D.B., Hewitt, R.G., Dulk, G.A.:JGR **89**, 897 (1984)

Melrose, D.B: *Instabilites in space and laboratory plasmas*, Cambridge University Press (1986)

Parker, E.: *Cosmical Magnetic Fields*, Clarendon Press, Oxford (1979)

Phinney, E.S.: in *Superluminal Radio Sources*, eds.J.A. Zensus & T.J. Pearson, Cambridge University Press, p. 301 (1987)

Quirrenbach, A., Witzel, A., Krichbaum, T., Hummel, C.A., Alberdi, A., Schalinski, C.: Nature **337**, 442 (1989a)

Quirrenbach, A., Witzel, A., Quian, S.J., Krichbaum, T., Hummel, C.A., Alberdi, A.: A& A **226**, L1 (1989b)

Readhead, A.C.S.: ApJ **426**, 51 (1994)

Remillard, R.A., Grossan, B., Bradt, H.V., Ohashi, T., Hayashida, K., Makino, F., Tanaka, Y.: Nature **350**, 589 (1991)

Romero, G.E., Surpi, G., Vucetich, H.: A& A **301**, 641 (1995)

Schindler, K., Birn, J., Hesse, M.: JGR **93**, 5547 (1988)

Schindler, K., Birn, J., Hesse, M.: ApJ **380**, 293 (1991)

Shapiro, V.D., Shevchenko, V.I., Cargill, P.J., Papadopoulos, K.: JGR **99**, 23.735 (1994)

Sotnikov, V.I., Shapiro, V.D., Shevchenko, V.I.: Sov. J. Plasma Phys. **4**, 252 (1978)

Spicer, D.S., Mariska, J.T., Boris, J.P.: in *Physics of the Sun*, Eds. P.A. Sturrock, T.E. Holzer, D.M. Mihalas, R.K. Ulrich, Reidel, Dordrecht, p. 181

Wagner, S., Sanchez-Pons, F., Quirrenbach, A., Witzel, A.: A& A **235**, L1 (1990)

Wagner, S., Witzel, A.: Ann. Rev. Astron. Astrophys. (in press) (1995)

Weatherhall, J.C.: Phys. Rev. Lett **60**, 1302 (1988)

Weatherhall, J.C., Benford, G.: ApJ **378**, 543 (1991)

Windsor, R.A., Kellogg, P.J.: ApJ **190**, 167 (1974)

Zeldovich, Y.B., Levich, E.V., Sunyaev, R.A.: JETP **35**, 733 (1972)

Particle Acceleration in Extended Radio Sources – A Critical Review

Klaus Meisenheimer

Max-Planck-Institut für Astronomie
Königstuhl 17, D–69117 Heidelberg, Germany

The fact that the high-frequency spectra of some extended extragalactic radio sources reach out to optical frequencies was recognized already in 1956 when Baade established the high polarization of the optical jet in M 87. Immediately afterwards Burbidge (1956) pointed out that the observation of synchrotron light from the outer parts of the jet is hardly compatible with the ultra-relativistic electrons being supplied from the core since their synchrotron loss time is much smaller than the light travel time along the jet. The most thorough discussion of this dilemma is due to Felten (1968) who demonstrated that simple ways to solve the transport problem (*e.g.* a special field geometry) are in conflict with the stability of the plasma flow in the jet. He therefore favoured either a relativistic jet flow ($v_{jet} \simeq c$) or an *"in situ"* re-acceleration based on an – at that time – unknown process of high efficiency. Although in the meantime an effective way of particle acceleration has been proposed (first-order Fermi acceleration at strong shocks, Bell 1978), recent observations of some radio sources reveal that high-frequency synchrotron emission is too widespread to be easily explained by standard shock acceleration theory. So Felten's dilemma might be refined but remains still unsolved: We need either a way to transport ultra-relativistic electrons over large distances ($\gg 1\,\mathrm{kpc}$) by avoiding severe synchrotron losses or a re-acceleration process which is not restricted to knots or hot spots.

Our workshop organizer, Wolfgang Kundt, has always favoured the former alternative. Here I will give my view of the *pros* and *cons* of both alternatives.

1 The problem

The essential point of our problem is a straightforward result from standard synchrotron theory. Optically radiating electrons in a magnetic field of 30 nT (typical for the brighter parts of extragalactic radio sources) have an energy $\gamma \equiv E/m_e c^2 \simeq 10^6$. They lose half of their energy on a time scale

$$\tau_{1/2} = 185 \left(\frac{B\sin\theta}{30\,\mathrm{nT}}\right)^{-2} \left(\frac{\gamma}{10^6}\right)^{-1} \text{ [years]} \tag{1}$$

where θ is the angle between the magnetic field and the line of sight (∥ to the momentum vector of the radiating particle). Even if we assume that the particles stream freely from one place to another (which is very optimistic since the particles have to gyrate around the field lines) we find that the maximum range is $c\tau_{1/2} \lesssim 60\,\mathrm{pc}$ for the above fiducial values of $\mathrm{B}\sin\theta$ and γ.

The standard argument for *in situ* acceleration assumes that the field strength in the emission region is similar to that in the region between the place of acceleration and the emission region. In this case the discrepancy between extent of the source L and synchrotron loss scale would be as high as $L/c\tau_{1/2} > 10^3$ (for the high-frequency radio emission of the hot spots in Cygnus A) or even $L/c\tau_{1/2} \gtrsim 10^4$ (for the optical synchrotron radiation from the western hot spot of Pictor A, Röser & Meisenheimer 1987). This argumentation clearly overstates the problem. In the standard jet model the hot spot field is strongly enhanced above the jet value: $\mathrm{B}_{HS} \simeq 4\,\mathrm{B}_{jet}$. But even assuming the lower jet field we derive $L/c\tau_{1/2} > 100$ for the largest radio sources which emit synchrotron light from their hot spots. In the jets of M 87 and 3C 273 where we can directly measure the energy loss scale $\lambda_{1/2} \equiv \frac{\gamma}{d\gamma/dz} \simeq 5\,\mathrm{kpc}$ (see Meisenheimer *et al.* these proceedings), we find $\lambda_{1/2}/c\tau_{1/2} \gtrsim 80$. So one of the following general alternatives has to be realized by nature:

(1) the energy losses due to the – observed – synchrotron radiation are continuously substituted by *in situ* re-acceleration.
(2) Ultra-relativistic particles (electrons, positrons, protons?) are transported almost loss-free from their place of acceleration (core, hot spots) into the extended synchrotron emission regions.

I will show in the following that both alternatives face severe problems in explaining the observations.

2 The case for *in situ* particle acceleration

Radio hot spots

Let me start with the simplest case: Optical synchrotron emission from radio hot spots. As we have shown (Meisenheimer *et al.* 1989) the emission of synchrotron light from radio hot spots[1] can comfortably be explained by 1st-order Fermi acceleration at the strong shock which should be present in radio hot spots. Essentially, no more than moderate acceleration from $\gamma_{max} \lesssim 10^4$ to $\gamma_{max} \simeq 10^6$ is required. The loss scale for $\gamma = 10^4$ electrons along the radio jet ($\mathrm{B}\sin\theta < 10\,\mathrm{nT}$) is at least 60 kpc that is close to the distance from the core. The time scale for Fermi acceleration is a few hundred to thousand years. For that time a quasi-steady shock structure should be maintained. So Fermi acceleration could work in hot spots!

Optical jets

The synchrotron loss scale $c\tau_{1/2}$ of the most energetic electrons in the jet of M 87 ($\gamma_c \simeq 10^6$) is $\lesssim 60\,\mathrm{pc}$ and even smaller for the jet in 3C 273 ($\gamma_c \simeq 10^7$,

---

[1] Only a small fraction of radio hot spots do emit optical synchrotron radiation. So we are discussing an extreme subsample.

see Meisenheimer *et al.* these proceedings). Nevertheless, the *observed* energy loss scale is $\lambda_{1/2} \gtrsim 5\,\mathrm{kpc} > 80 \times c\tau_{1/2}$. In principle, the *global* losses could be balanced by effective acceleration in the knots. However, since the changes of ν_c between knot and inter-knot regions of M 87 never exceed the factor of 2 which is most naturally explained by variations of the local field strength $B_\perp$, we are forced to conclude that $\lambda_{1/2} = 5\,\mathrm{kpc}$ is maintained also *locally*. In order to keep γ_c constant to such a degree, one would require ~ 100 strong shocks between each two of the prominent knots in the jet. This would slow down the jet plasma very rapidly. Therefore, 1st-order Fermi is virtually ruled out and we have to invoke an ubiquitous and highly efficient – but totally unknown - re-acceleration process to explain the spectral evolution along optical jets.

X-ray emission from jets

The jets of M 87, 3C 273 and Centaurus A seem to be detected in the X-ray band. Although neither the synchrotron origin of these X-rays is established beyond doubts nor their source can be clearly identified, we have to note that the presence of synchrotron X-rays would limit $c\tau_{1/2}$ to $\leq 2\,\mathrm{pc}$, putting extremely tight constrains on possible re-acceleration mechanisms. It is not clear whether Fermi acceleration could produce such high electron energies at all.

Radio lobes

K-band images of the inner lobes of M 87 clearly show that the brightest parts of both the western and the eastern lobe (in which no jet is visible) emit near-infrared light, which is most likely of synchrotron origin (Neumann *et al.* 1995). Using the standard minimum energy estimate for the field strength in these brightest patches of the lobes one finds that γ_c does not differ much from the value in the jet ($\gamma_c = 10^6$). Similar extended synchrotron light emission is detected on a deep K-band image of 3C 273 (Neumann 1995) and on an ultra-deep red image of the filament near the western hot spot in Pictor A (recent data taken by H.-J. Röser). Thus we conclude that the regions of optical synchrotron emission extend well beyond the knots or hot spots of some extended radio sources. Obviously this finding cannot be understood in terms of a localized particle acceleration region (strong shock).

Detailed maps of the radio spectrum in the lobes around the hot spots of Cygnus A (Carilli *et al.* 1991) point to a similar phenomenon: By no means can their radio spectrum be explained by those electrons which left the hot spot region downstream (see Fig. 1). Thus we have to demand distributed acceleration processes not only in the jets but also in the lobes – making acceleration at strong shocks an even less attractive model.

In summary, I conclude that shock acceleration may well account for the high-frequency radiation of hot spots. But it can hardly explain the distributed occurrence of optical synchrotron emission found in some other radio sources. Thus we have to invoke an additional, unknown acceleration process if we want keep *in situ* acceleration as a general concept for extended extragalactic synchrotron

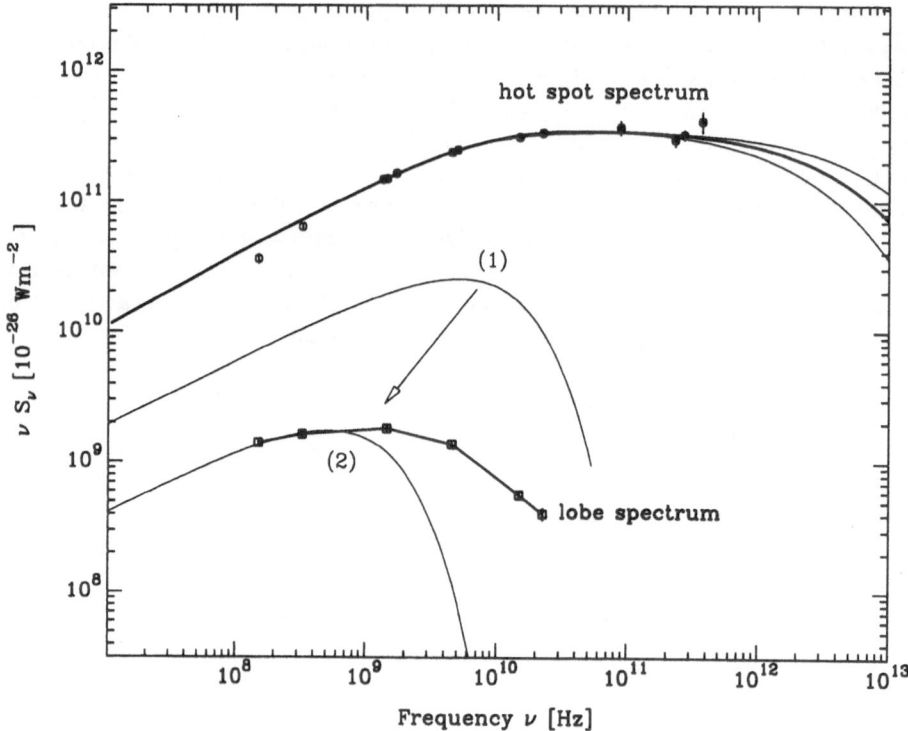

Fig. 1. Comparison between the hot spot and lobe spectrum in the eastern lobe of Cygnus A. Due to adiabatic losses in the expanding flow from the hot spot into the lobe, one expects that the spectrum (1) (of electrons leaving the hot spot, from Meisenheimer *et al.* 1996) is modified into spectrum (2). Obviously the observed lobe spectrum (□, from Carilli *et al.* 1991) is much flatter and reaches to higher energies. This seems to demand additional particle acceleration. Note that even the un-shifted spectrum (1) is much steeper than the observed lobe spectrum at $\nu > 10\,\mathrm{GHz}$.

sources. Although it is possible to weaken this conclusion considerably for the case of optical jets (in which relativistic beaming could mislead our estimate of $c\tau_{1/2}$), I see no way to put aside the equally strong evidence based on the near-infrared/optical light from extended radio lobes.

3 The case for a "loss-free transport" of particles

As already outlined by Felten (1968), one has to distinguish between two very different models:

(a) The relevant particles are ultra-relativistic protons. In such a scenario the *transport* problem would disappear: $c\tau_{1/2}(protons) = 1836 \times c\tau_{1/2}(e^-)$, but the proton to electron/positron conversion rate by collisions enters as a limiting factor.

(b) The relevant particles are the ultra-relativistic electrons (positrons) themselves which emit the synchrotron light. In this case we face the transport problem outlined in section 1.

3.1 Case (a): Radio jets contain ultra-relativistic protons

One might speculate whether ultra-relativistic protons could be the main component carrying very high energies. Their production is unavoidable in many particle acceleration processes, and very energetic protons are a major constituent of the cosmic ray flux outside the earth's atmosphere. Such a scenario would be very attractive since it would solve two problems:

- The synchrotron loss scale $c\tau_{1/2}$ of the required protons is comfortably below the observed loss scale $\lambda_{1/2}$.
- Any conversion of protons into secondary leptons (electrons, positrons, etc.) would essentially mirror the proton energy distribution and thus provide a natural explanation for the constancy of the particle spectrum along radio jets and even into the lobes.

However there is a major problem: The protons have to be converted into secondary particles by collisions with other particles. Since the standard way (proton-nucleon collisions) leads either to an unreasonably high proton flux or nucleon density in the jet, protons have been ignored in the last 25 years. However, as proton-photon interaction might well play a role in quasar cores (Mannheim & Biermann 1992, Mastichiadis & Kirk 1995) one should perhaps look more closely into the possibility that protons also could be important in extended radio sources. Together with Heinz Völk (MPI für Kernphysik, Heidelberg) I am currently exploring this possibility.

3.2 Case (b): Transport of ultra-relativistic electrons/positrons

There are at least three possibilities how one could realize a jet which is able to transport ultra-relativistic electrons (positrons) over large distances:

A highly relativistic e^+e^- jet: This model has been favoured by Wolfgang Kundt over more than a decade (Kundt & Gopal-Krishna 1980): If radio jets are composed of a highly relativistic pair plasma ($\Gamma_{jet} \equiv \gamma_{e^+ + e^-} \simeq 10^4$) one needs a special field geometry which avoids synchrotron losses. Wolfgang's favorite model is an $\mathbf{E} \times \mathbf{B}$ drift (that is a situation in which the relativistic particles see $B \equiv 0$ in their frame of reference), but any model in which $B\sin\theta < 3\,\text{nT}$ would do. In such a model strong energy losses (which produce the observed radiation) occur where the e^+e^- beam hits "obstacles" in which the field geometry diverges from the regular pattern. I have three problems with this scenario:

- The $\gamma = 10^4$ e^+e^- beam will propagate well in vacuum. But what about its stability in a plasma environment (which is necessary to anchor the "obstacles") ? From discussions with people who understand more plasma physics than I do, I got the impression that the very "obstacles" which produce the radiation would drive strong plasma waves into the beam and destroy the regular field pattern which is required for the loss-free transport.

- How to get rid of the "used" electrons and positrons? The electrons which strongly radiate should be trapped in the magnetic irregularities and lose their energy progressively fast. This would cause that more and more low-energy particles are stored within the jet. But this should significantly modify the radio-optical spectral index along the jet. Such changes are ruled out by our observations of M 87 and 3C 273 !

- The ordered morphological pattern of the jet in M 87 becomes disturbed beyond knot C and is certainly destroyed beyond knot G. Any highly relativistic e^+e^- beam can hardly pass this turbulent region without losing most of its energy. Thus the NIR/optical light from the turbulent end of the jet and from the lobes has to be explained by additional re-acceleration. But if re-acceleration is required there – why not admitting it at other places?

A "loss-free" channel within the jet: In order to explain the low synchrotron losses along the jet in M 87, Owen *et al.* (1989) proposed a jet model which consists of a loss-free central channel (with very low $B \sin \theta$) plus a boundary layer in which the field has about the equipartition value. Electrons with the required $\gamma_{max} \simeq 10^6$ could be advected without losses in the channel and would only radiate when they diffuse into the boundary layer. My objections to this scenario are:

- The electrons which diffuse into the boundary would create strong plasma waves which are bound to back-react violently on the channel. Thus the stability considerations along Felten's lines fully apply. They show that such a jet would be highly unstable and disrupted after 100 pc or so.

- The rapid synchrotron losses in the boundary layer will strongly affect the observed power-law index (the situation is equivalent to continuous injection into an inhomogeneous field). Thus a flat, constant $\alpha_{PL} = -0.66$ is totally unexpected in this model.

- With regard to the lobe emission the same argument as above applies: Morphology and polarization structure of the M 87 jet indicate that the ordered structure of the jet breaks down around knots A and B and definitely disappears beyond knots C or G. Dramatic changes of the energy spectrum would be expected downstream of knot B or C.

Highly magnetized filaments in a low-field jet: Together with Alan Heavens (University of Edinburgh) I looked into the possibility that the jet consists of an electron-proton plasma with a rather low magnetic field ($\lesssim 0.1 \times B_{m.e.}$) and relativistic electrons with $\gamma_{max} \simeq 3 \times 10^5$. This obviously would enhance the synchrotron loss scale $c\tau_{1/2}$ to the required values. The observed synchrotron radiation could be produced in transient, highly magnetized filaments in which the field is enhanced by a factor of ~ 100 (*i.e.* $\sim 10 \times B_{m.e.}$). This model would avoid the stability problem of the above models since we assume that the shrinking magnetic instabilities could rapidly decouple from the jet plasma. However the occurrence of a constant α_{PL} remains as mysterious as in the model by Owen *et al.* In addition, neither the timescale for filamentation (by radiative synchrotron instabilities) nor the time during which these filaments would produce high-frequency radiation seem correct to meet the requirements inferred from the observations. So we abandoned the model.

Finally, I would like to point out a general problem which is inherent in every model trying to explain the astonishing constant synchrotron spectra along radio jets with relativistic particles from the core: In any such scenario, the particles in the outer parts of the jet have left the core at least a light travel time earlier than those in the inner parts. That implies that the problem of a *spatially* constant spectrum is transformed into the problem of the lack of *temporal* variability in the core. Since we know that rapid and strong variability is common in compact synchrotron sources it is by no means clear how the extended jets could be supplied with a constant energy spectrum over 10^4 to 10^5 years.

4 Conclusions

There are now several detailed multi-frequency observations of the synchrotron emission from extended radio sources available. Many of them reveal that the energy spectra of the underlying electrons (positrons) do not suffer the rapid losses expected from standard synchrotron theory. This implies either that losses are balanced "in situ" by a *permanent* re-acceleration process or that the observed losses do not affect the particle spectrum injected into various places of the radio source.

Quite uncommon for investigations of extragalactic radio sources, we presently face a situation in which clear-cut observational evidence conflicts with all proposed theoretical concepts: None of the above models is able to provide a convincing explanation how radio sources manage to emit synchrotron light over an extent of many kpc. So there is an urgent demand for employing more theoretical vision. In my opinion two alternatives are most worthwhile being explored:

(I) The role of ultra-relativistic protons in large-scale extragalactic radio sources. We have strong indications from the cosmic ray background and current models of the γ-ray emission from quasars that these protons are produced. So we should try harder to make use of them for the solution of the particle acceleration dilemma.

(II) Permanent re-acceleration processes in magnetized plasma. If one could find a reason why a strongly magnetized plasma (as is present in radio jets) tends to keep the relativistic particle spectrum constant, one would have the most natural explanation for the observations. One might speculate that a self-regulation process channels field energy into particles whenever losses occur and also keeps the maximum energy limited. Since the process has to be more efficient in regions of high magnetic field (where losses are more rapid), I suspect that one has to look into electromagnetic acceleration processes which provide a straightforward explanation why the acceleration gain should be coupled to the local field-energy density.

References

Baade, W. 1956, *Astrophys. J.* **123**, 550

Bell, A.R. 1978, *Mon. Not. R. astr. Soc.* **182**, 147

Burbidge, G.R. 1956, *Astrophys. J.* **124**, 416

Carilli, C.L., Perley, R.A., Dreher, J.W. and Leahy, J.P. 1991, *Astrophys. J.* **383**, 554

Felten, J.E. 1968, *Astrophys. J.* **151**, 861

Kundt, W. and Gopal-Krishna 1980, *Nature* **288**, 149

Mannheim, K. and Biermann, P. 1992, *Astron. Astrophys.* **221**, 211

Mastichiadis, A. and Kirk, J.G. 1995, *Astron. Astrophys.* **295**, 613

Meisenheimer, K., Röser, H-J., Hiltner, P., Yates, M.G., Longair, M.S., Chini, R. and Perley, R.A. 1989a, *Astron. Astrophys.* **219**, 63

Meisenheimer, K. Yates, M.G. and Röser, H.-J. 1996, submitted to *Astron. Astrophys.*

Neumann, M. 1995, PhD Thesis, Universität Heidelberg.

Neumann, M., Meisenheimer, K., Röser, H.-J. and Stickel, M. 1995, *Astron. Astrophys.* **296**, 662

Owen, F.N., Hardee, P.E. and Cornwell, T.J. 1989, *Astrophys. J.* **340**, 698

Röser, H.-J. and Meisenheimer, K. 1987, *Astrophys. J.* **314**, 70

Plasma Acceleration and Jet Formation by a Magnetized Rotator

S.V.Bogovalov

Moscow Engineering Physics Institute, Kashirskoje shosse 31, Moscow, 115409, Russia

Abstract: The results of numerical simulations of the stationary plasma flow in the magnetosphere of a magnetized axisymmetrical rotator are presented. The nonstationary problem was solved to obtain the stationary solution. It is shown that strong collimation of the plasma flow to the axis of rotation by toroidal magnetic field leads to the formation of jets directed along the axis of rotation at large distance from the central object. Estimates of the transversal size of jets for nonrelativistic and relativistic flows are given. In the nonrelativistic case the effectivity of the transformation of the Poynting flux into the kinetic energy of plasma due to centrifugal acceleration achieves 46%.

Keywords: plasma - magnetohydrodynamics - jets - pulsars - acceleration

1 Introduction

Jet formation and acceleration of plasma by rotating magnetized objects such as neutron stars, YSO, AGN's and others are among the most intriguing problems of astrophysics. The model of a magnetized axisymmetrical rotator gives the opportunity to study these phenomena in the simplest form when the flow is axisymmetric and stationary. There are hopes that these phenomena can be understood in this model from first principles.

The MHD flow of magnetized plasma is described by nonlinear equations [1],[2],[3],[4]. There is a range of self-similar solutions of the problem [5,6,7,8]. But the solution of the problem for the most interesting parameters can be obtained only numerically. The results of numerical solution of the problem for the nonrelativistic case are presented in this paper.

2 Stationary problem

Equations describing stationary ejection of plasma by an axisymmetric magnetized rotator were widely studied starting with the work [9]. Here we follow the
nonrelativistic formalism presented in paper [10]. An equivalent version of these
equations was considered in [11]. The relativistic version of the equations was
investigated in [4,12,13]. Equations in general-relativity formalism were considered in [1,2,3]. In the axisymmetric flow the magnetic field $\mathbf{H}$ can be presented
as a sum of the poloidal magnetic field $\mathbf{H_p}$ and azimuthal magnetic field $\mathbf{H_\varphi}$.
The poloidal magnetic field can be expressed as

$$\mathbf{H_p} = \frac{\nabla\psi \times \mathbf{e}_\varphi}{\rho}, \tag{1}$$

where ρ is the distance from the axis of rotation. $\mathbf{e}_\varphi$ is the unit vector corresponding to the rotation around the axis z. The function ψ is proportional to
the full flux of the poloidal magnetic field through a surface at radius ρ. In
frozen-in approximation the relationship between the electric field $\mathbf{E}$ and the
poloidal magnetic field is $\mathbf{E} = \frac{\rho\Omega}{c}\mathbf{q}(\psi)\mathbf{H_p} \times \mathbf{e}_\varphi$ [9], Ω is the angular velocity of
the central object. The function $q(\psi)$ is constant along the poloidal field force
line and describes differential rotation of force lines.

The first equation defining the dynamics of the plasma along the lines of the
poloidal field is the conserved energy flux

$$\frac{u^2}{2} + \mu + \phi - f(\psi)\rho\Omega q(\psi)H_\varphi = W(\psi). \tag{2}$$

The second equation is the conserved angular momentum flux

$$\rho\Omega u_\varphi - f(\psi)\rho\Omega H_\varphi = M(\psi). \tag{3}$$

Projection of the frozen-in condition on the electric field gives

$$\rho\Omega q H_p + u_p H_\varphi = u_\varphi H_p. \tag{4}$$

Here $u^2 = u_p^2 + u_\varphi^2$, u_p is the velocity of plasma along a field line, u_φ is the
azimuthal velocity of plasma, μ is the enthalpy of plasma per particle, ϕ is the
gravitational potential of the star, $f = \frac{H_p}{4\pi mnu_p}$. The functions $W(\psi)$ and $M(\psi)$
are proportional to the energy and angular momentum flux per particle. These
functions are constant along field lines. Therefore they depend only on ψ. The
last equation is the adiabatic condition

$$s = s(\psi), \tag{5}$$

where s is the entropy per particle.

The transfield equation for ψ in the nonrelativistic limit is as follows [10]

$$(u_p^2 - u_a^2)[(u_p H_\varphi^2 - H_p^2 U_m)\psi_{\rho\rho} + 2H_\rho H_z u_p \psi_{\rho z} + (u_p H_z^2 - H_p^2 U_m)\psi_{zz} + H_z H_p^2 \times$$

$$\times \{u_m + \frac{u_\varphi^2}{u_p}\} + \frac{H_p^4 k}{u_p}\rho^2(u_p^2(\ln f)_\psi' + \frac{v_\psi' u_\varphi}{\rho\Omega q}) - \frac{H_p^2 c}{\Omega q u_p}(\mathbf{E}\nabla\phi) + \frac{\rho^2 k H_p^4 u_\varphi^2}{u_p}(\ln q)_\psi'] =$$

$$-u_a^2 H_p \rho\Omega q H_\varphi \{2H_z - \frac{1}{(q\Omega)^2}(\frac{RH_p U_m}{f} -$$

$$-\frac{V_\psi H_p}{u_p}\mathbf{U}\cdot\mathbf{H}) + (\ln q)_\psi'\frac{\rho H_p}{q\Omega}(cE + \frac{u_\varphi}{u_p}\mathbf{U}\cdot\mathbf{H})\}, \tag{6}$$

where $V = W - Mq$, $k = (\frac{C_s}{u_p})^2$, $U_m = k\frac{(u_p^2 - u_a^2)}{u_p} + f\frac{H^2}{H_p}$, $R = W_\psi' + \rho\Omega H_\varphi(fq)_\psi'$, $\mathbf{U}\cdot\mathbf{H} = u_p H_p + u_\varphi H_p$. Symbol $()'$ denotes the derivative with respect to ψ.

Equation (6) is the mixed elliptic-hyperbolic equation. In the regions $0 < u_p < u_c$ and $u_{sl} < u_p < u_f$, it is an elliptic equation. In the regions $u_c < u_p < u_{sl}$ and $u_p > u_f$ equation (6) is hyperbolic. u_c is the cusp velocity [14], u_{sl} and u_f are the slow and fast magnetosonic velocities, C_s is the (adiabatic) sound velocity.

3 Time dependent model

To obtain a stationary solution of the plasma flow, a numerical simulation of the time dependent problem was performed. Apparently at present it is the only regular method of the solution of the stationary problem on the reasons discussed in [10]. Following [11] we assume that the poloidal magnetic field of the nonrotating star has monopole like structure. We consider the cold nonrelativistic wind ejected from the surface with the velocity u^0 exceeding the slow magnetosonic velocity. This model problem is the nonrelativistic analog of the magnetosphere of pulsars.

We assume that the poloidal magnetic field, the plasma velocity and the plasma density of the nonrotating star are spherically symmetrical. The Alfven surface (AS) coincides with the fast magnetosonic surface (FMS) and is placed at the constant distance R_a from the star center. This distance is defined by the relationship $u_0 = H_p(R_a)/\sqrt{4\pi m n(R_a)}$. Let the magnetic field and the plasma density on the AS of the nonrotating star be H_a, n_a. The magnetic field can be presented as a sum of poloidal and azimuthal ones. The poloidal magnetic field is expressed through the function ψ according to formula (1). Now ψ depends not only coordinates but time also. We introduce dimensionless variables $\tilde\psi = \frac{\psi}{H_a R_a^2}, U_x = \frac{u_x}{u_0}, U_z = \frac{v_z}{u_0}, U_\varphi = \frac{u_\varphi}{u_0}, \tilde n = \frac{n}{n_a}, x = \frac{\rho}{R_a}, \tilde z = \frac{z}{R_a}, \tilde H_p = \frac{H_p}{H_a}, \tilde H_\varphi = \frac{H_\varphi}{H_a}, \tau = \frac{t u_0}{R_a}$. The system of equations defining the dynamics of plasma in a cylindrical system of coordinates in the new variables is

$$\frac{\partial\tilde\psi}{\partial\tau} = -U_x\frac{\partial\tilde\psi}{\partial x} - U_z\frac{\partial\tilde\psi}{\partial\tilde z}, \tag{7}$$

$$\frac{\partial\tilde H_\varphi}{\partial\tau} = \frac{\partial(U_\varphi H_z - U_z\tilde H_\varphi)}{\partial\tilde z} - \frac{\partial(U_x\tilde H_\varphi - U_\varphi H_x)}{\partial x}, \tag{8}$$

$$\frac{\partial U_\varphi}{\partial\tau} = -U_x\frac{\partial x U_\varphi}{x\partial x} - U_z\frac{\partial U_\varphi}{\partial\tilde z} + \frac{1}{\tilde n}(H_x\frac{\partial x H_\varphi}{x\partial x} + H_z\frac{\partial H_\varphi}{\partial\tilde z}), \tag{9}$$

$$\frac{\partial U_x}{\partial \tau} = -U_x \frac{\partial U_x}{\partial x} - U_z \frac{\partial U_x}{\partial \tilde{z}} - \frac{1}{2x^2 \tilde{n}} \frac{\partial (x H_\varphi)^2}{\partial x} + \frac{U_\varphi^2}{x} + \frac{H_z}{\tilde{n}} \left(\frac{\partial H_x}{\partial \tilde{z}} - \frac{\partial H_z}{\partial x} \right), \quad (10)$$

$$\frac{\partial U_z}{\partial \tau} = -U_x \frac{\partial U_z}{\partial x} - U_z \frac{\partial U_z}{\partial \tilde{z}} - \frac{1}{2x^2 \tilde{n}} \frac{\partial (x H_\varphi)^2}{\partial \tilde{z}} - \frac{H_x}{\tilde{n}} \left(\frac{\partial H_x}{\partial \tilde{z}} - \frac{\partial H_z}{\partial x} \right), \quad (11)$$

$$\frac{\partial \tilde{n}}{\partial \tau} = -\frac{\partial \tilde{n} x U_x}{x \partial x} - \frac{\partial \tilde{n} U_z}{\partial \tilde{z}}. \quad (12)$$

Equations (7,8) are the frozen-in conditions. These equations can be obtained from the frozen-in equation $\frac{\partial H}{\partial t} = u \times H$. Direct application of this equation to the calculation of the poloidal magnetic field usually creates problems with non-conservation of the flux of this field due to nonzero ∇H_p generated in any known numerical scheme [15]. Equation (7) ensures the conservation of the poloidal magnetic flux. This is the important feature of the used numerical scheme.

A two step Lax-Wendroff differencing scheme was used for the simulation [16]. It was performed in a quarter of the total box of simulation. It was assumed that the solution is symmetrical in relation to the equator and to the axis of rotation. The radius of the star was taken equal to 0.5. The outer boundary of the box of simulation was placed far enough beyond the FMS. No signal can propagate from this boundary into the internal part of the box of simulation. In principle it is not important what boundary conditions are specified on this boundary. To avoid the formation of shock waves near the outer boundaries we accepted continuous derivatives of all physical variables on the outer boundary. Boundary conditions on the axis of rotation and on the equator follow from the system of equations (7-12) and from the symmetry of the problem in relation to the equator.

At the stellar surface we accept the following boundary conditions. 1. Function ψ is specified a priori on the star surface and does not depend on time. This boundary condition specifies the normal component of the poloidal magnetic field on the star surface. The tangential component is free.

2. The tangential component of the electric field is continuous at the star surface. This condition specifies the law of rotation of field lines $q(\psi)$. For the azimuthal magnetic field we have used so called floating conditions. At every step of time the azimuthal magnetic field was calculated from equation (8) under the assumption that the electric field in the star is known from the frozen-in condition $E + u \times H/c = 0$.

3. The plasma velocity at the star surface in the frame of the star is equal to the plasma velocity of the nonrotating star. The relationship $U_x^2 + U_z^2 + (U_\varphi - \alpha x)^2 = 1$ was accepted on the surface. Here $\alpha = \frac{\Omega R_a}{u_0}$ is the dimensionless angular velocity of the star.

4. The plasma density is specified a priori at the star surface and is kept constant in time.

The simulation was performed in several dimensionless boxes with sizes 4×4 (lattice 100×100) and 2×8 (lattice 50×200). The simulation was terminated when a stationary solution was achieved.

4 The structure of stationary solutions

Figure 1 shows the structure of one quarter of the magnetosphere. The star is placed in the left-hand lower corner of the figure. Scales in X and Z directions are different. That is why the star has nonspherical shape. Circles show the Alfvenic surface where the velocity of plasma equals the local Alfven velocity. Stars show the fast magnetosonic surface where the plasma velocity equals the local fast magnetosonic velocity.

A strong deflection of the flow towards the axis of rotation due to the compression by the azimuthal magnetic field is the most remarkable result. Earlier similar results in the numerical calculations were pointed out in the works [17,18].

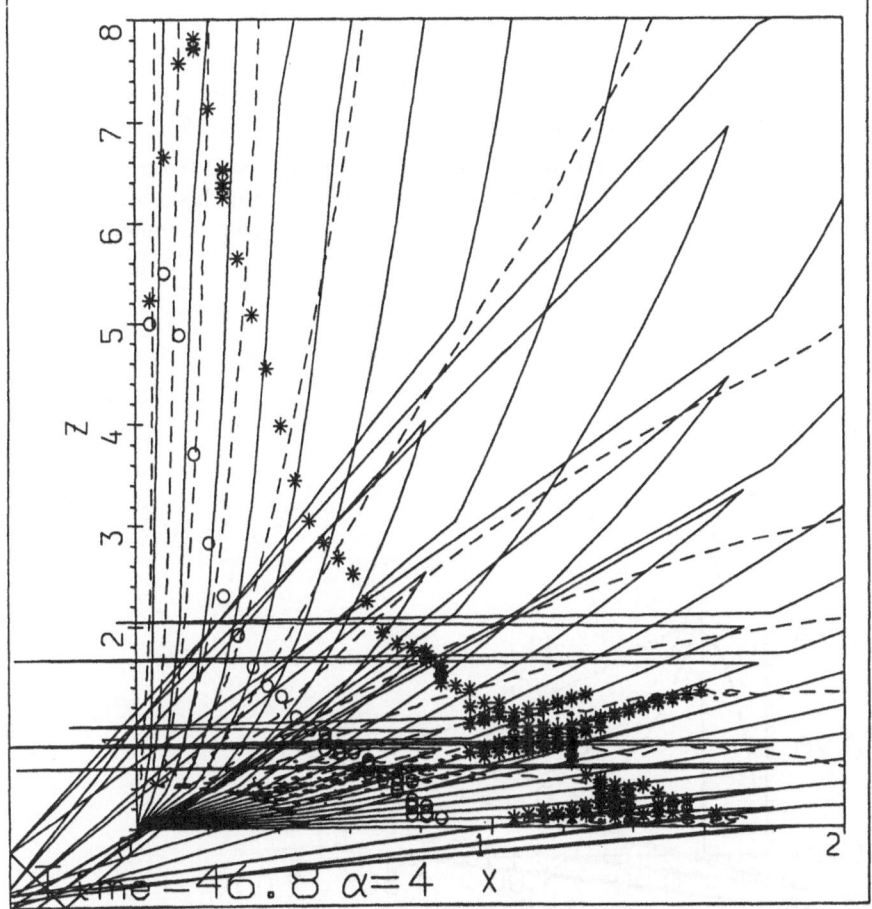

Fig. 1. The structure of the stationary magnetosphere of the axisymmetric rotator at dimension-less angular velocity $\alpha = 4$.. The scales in Z and X directions are different. Solid lines are the field lines of the poloidal magnetic field. Dashed lines are the lines of the poloidal electric currents. Circles o mark the points of the Alfven surface. Stars ⋆ mark the points of the fast magnetosonic surface.

Figure 1 clearly shows that the rotation of the magnetized object ejecting plasma leads to the collimation of the flow along the axis of rotation. This is valid not only in the nonrelativistic case. It was shown in [19] that in the relativistic case at very low angular rotation of the star leads to a similar collimation of the plasma flow along the axis of rotation. These results give strong evidence that this is a general property of magnetized plasma ejected by a rotating object. This conclusion is confirmed by a general analysis of the plasma flow at large distances from the star [20,21]. It shows that this collimated flow forms a jet at large distances from the star. It is interesting that there is a simple relationship between the parameters of jets and parameters of the star ejecting plasma.

The transversal dimension of the densest and apparently brightest part of the jet R_j (core of the jet) is defined as follows [21]

$$R_j = \sqrt{(1+\beta)}\frac{\gamma_j v_j}{\Omega},\tag{13}$$

where Ω is the angular velocity of the star's rotation, γ_j and v_j are the characteristic Lorentz-factor and the plasma velocity in the jet, $\beta = \frac{4\pi\delta P}{(\delta-1)H_0^2}$. δ is the adiabatic index of the plasma, P is the thermal plasma pressure and H_0 is the magnetic field on the axis of the jet. This relationship becomes especially simple when the plasma is so cold that the thermal pressure is much less than

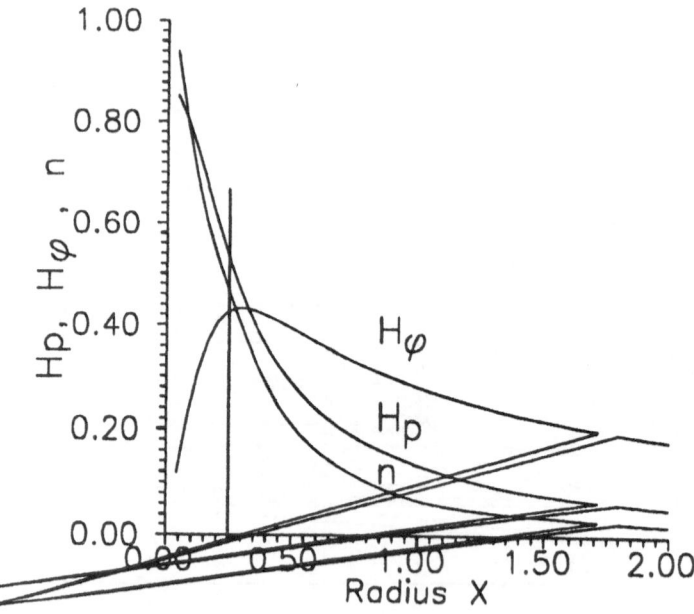

Fig. 2. The distribution of density, poloidal and azimuthal magnetic field in the plane of constant z at z=8. The vertical line at x=0.25 shows the theoretical prediction [21] of the transversal size of the jet.

the pressure of the magnetic field on the axis of the jet. For cold plasma, R_j is given by the expression

$$R_j = \frac{\gamma_j v_j}{\Omega}. \tag{14}$$

It is interesting that a similar estimate was obtained in [22] for cold relativistic plasma in massless approximation. These relationships can be used to define the parameters of real astrophysical objects from observational data. The results of numerical simulations allow to verify the validity of these estimates and to clarify their physical meaning. Figure 2 presents the distribution of a range of characteristics of the jet obtained in numerical simulations in dependence on the distance x from the axis of rotation at constant z equal to 8 for $\alpha = 4$. The vertical line at x=0.25 shows the estimate (14) of transversal size of the cold nonrelativistic jet. It is seen that the analytical estimate agrees with the numerical simulation. The transversal size of the jet is the size of the core of the jet. The poloidal magnetic field decreases approximately by a factor two and the azimuthal magnetic field achieves the maximum at the boundary of the core.

The jet discussed here can be unstable in relation to the helical instability [23]. In the result of the evolution of the jet due to this instability the jet can finally have helical structure as discussed in [24].

5 Centrifugal acceleration of plasma

Michel [25] was apparently the first who applied the ideal MHD to the investigation of centrifugal acceleration of plasma by rotation- powered pulsars in the model of an axisymmetrical rotator. He considered the dynamics of cold plasma in a monopole-like poloidal magnetic field. Deformation of the poloidal field by the moving plasma was not taken into account. The FMS was placed at infinity in his solution. As a result, the effectivity of transformation of Poynting flux carrying the main part of the rotational losses of the star to the kinetic energy of plasma appeared very low. This conclusion does not depend on the angular velocity of the star and the energy of the plasma.

The model used in this work is analogous to Michel's model. But here the full problem was solved. Plasma moves in the poloidal magnetic field affected by the plasma. The result drastically differs from that obtained in [25]. First of all it is evident from Fig. 1 that the FMS is placed at a finite distance from the star. The transformation of the Poynting flux to the kinetic energy of plasma appears very effective. The Poynting flux is proportional to the value $x H_\varphi$. The effectivity of the transformation can be characterized by the coefficient $\mu :=$ $\frac{((x H_\varphi)_s - (x H_\varphi)_{x=2})}{(x H_\varphi)_s}$. The index s marks the value $x H_\varphi$ at the surface of the star, the index x=2 marks the same value on the boundary of the box of simulation. This coefficient shows what part of the Poynting flux is transformed to the kinetic energy of the plasma. The dependence of μ on the angular velocity of the star is presented in Fig. 3. This coefficient was calculated for an equatorial field line. It increases with increasing α and achieves 46% at $\alpha = 4$.

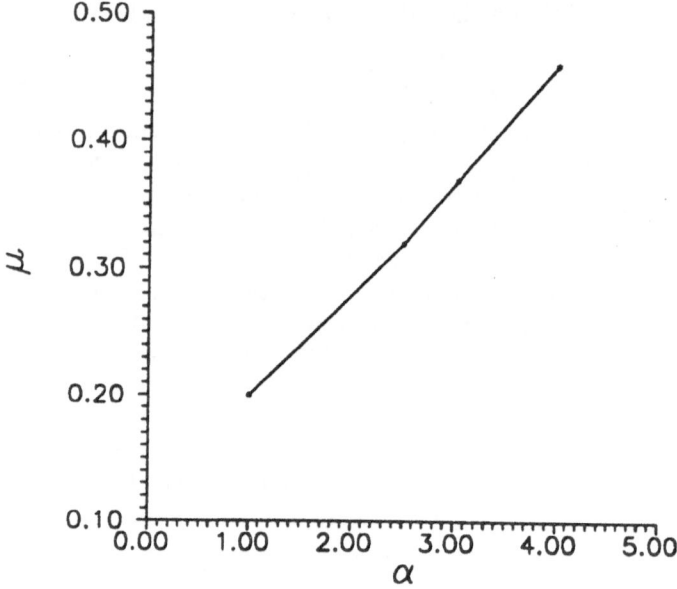

Fig. 3. The dependence on α of the coefficient of transformation of the Poynting flux to the kinetic energy of the plasma.

This result can be of importance for the physics of radio pulsars if extrapolate to the case of relativistic plasma ejected by a fast-rotating star. Existing models of electrostatic acceleration of plasma in inner [26,27,28,29] or outer gaps [30] ensure the effectivity of transformation of the rotational energy of the neutron star to the kinetic energy of the plasma of order 1%. At the same time, observations of the Crab Nebula [31,32,33] and pulsars in binary systems show that at least half of this energy is transformed to the kinetic energy of plasma. It follows from observations that some basic very effective mechanism of plasma acceleration in pulsars exists. According to the results of numerical simulations, this mechanism can be magnetocentrifugal acceleration.

It is important, however, to point out that it is impossible to transform 100% of Poynting flux to the kinetic energy of plasma as it was concluded in [31]. Our analysis of asymptotical behavior of the stationary MHD solutions shows that the net poloidal electric current flowing to infinity is never equal to zero [34]. According to the energy conservation law (2) it means that some possibly small part of the energy flux is always concentrated in the electromagnetic field.

73

References

1. M. Camenzind: Astronomy and Astrophysics **162** 32 (1986)
2. V.S. Beskin, V.I. Par'ev: Uspechi Phys. Nauk. **163** 96 (1993)
3. C.M. Mobarry, R.V.E. Lovlace: Astrophysical Journal **309** 455 (1986)
4. S.V. Bogovalov: Sov. Astron. Zhurnal **68** 1227 (1991)
5. R.D. Blandford, D.G. Payne: MNRAS **199** 88 (1982)
6. K. Tsinganos, E. Trussoni: Astronomy and Astrophysics **249** 156 (1991)
7. J. Contopoulos: Astrophysical Journal **432** 508 (1994)
8. Z. Li, T. Chiuen, M.C. Begelman: Astrophysical Journal **394** 459 (1991)
9. E.J. Weber, L.Davis: Astrophysical Journal **148** 217 (1967)
10. S.V. Bogovalov: MNRAS **270** 721 (1994)
11. T. Sakurai: Astronomy and Astrophysics **152** 121 (1985)
12. C.F. Kennel, F.S. Fujimura, I. Okamoto Geophys. Fluid. Dynamics **26** 147 (1983)
13. H. Ardavan: MNRAS **189** 397 (1979)
14. R.V. Polovin, V.P. Demutskii: Fundamentals of Magnetohydrodynamics, Consultants Buren, New York (1990)
15. J.U. Brackbill, D.C. Barness: J. Comput. Phys. **35** 426 (1980)
16. W.H. Press, B.P. Flannery, S.A. Teukolsky, W.T. Vetterling: Numerical Recipes, Cambridge University Press, Cambridge (1988)
17. H. Washimi, T. Sakurai: Solar Phys. **143** 173 (1993)
18. H. Washimi, S. Shibata: MNRAS **262** 936 (1993)
19. S.V. Bogovalov: Pis'ma Astron. Zh. **18** 832 (1992)
20. J. Heyvaerts, C. Norman: Astrophysical Journal **347** 1055 (1989)
21. S.V. Bogovalov: Pis'ma v Astron. Zh. **21** 633 (1995)
22. S. Appl, M. Camenzind: Astronomy and Astrophysics **274** 699 (1993)
23. S.V. Bogovalov: MNRAS in press (1995)
24. M. Villata, A.Ferrari: Astronomy and Astrophysics **293** 626 (1994)
25. F.C. Michel: 1969, Astrophysical Journal **158** 727 (1969)
26. M. Ruderman, P.C. Sutherland: Astrophysical Journal **196** 51 (1975)
27. J. Arons: Astrophysical Journal **266** 215 (1981)
28. V.S. Beskin, A.V. Gurevich, Ya. N. Istomin: Sov. Phys. JETP **58** 235 (1983)
29. W. Kundt, R. Schaaf: Astroph. Sp. Sci. **200** 251 (1993)
30. K.S. Cheng, C. Ho, M.A. Ruderman: Astrophysical Journal **300** 500 (1986)
31. W. Kundt, E. Krotscheck: Astronomy and Astrophysics **83** 1 (1980)
32. C.F. Kennel, F.V. Coroniti: Astrophysical Journal **283** 710 (1984)
33. R.T. Emmering, R.A. Chevalier: Astrophysical Journal **321** 334 (1987)
34. T. Chiueh, Z. Li, M.C. Begelman: Astrophysical Journal **377** 462 (1991)

The Astrophysical Plasma Gun

John Contopoulos

NAS/NRC Resident Research Associate
NASA/Goddard Space Flight Center, Greenbelt, MD 20771, USA

Abstract. In direct analogy to the focusing stage of a typical laboratory magnetized coaxial plasma gun, we suggest that the explosive release of the toroidal magnetic field generated in the interior of a differentially rotating accretion disk around a central black hole, might account naturally for the sequences of fast and slow moving blobs observed in galactic and extragalactic jets.

1 Introduction

Magnetic fields have been proposed in the past as a very efficient mechanism in accelerating and collimating outflows from accretion disks around stars and active galactic nuclei (see Blandford 1993 for a review). The most popular mechanism for magnetohydrodynamic plasma acceleration has been the so called 'centrifugal' driving mechanism of Blandford and Payne (1982).

The above mechanism works as follows: Plasma is frozen in the strong magnetic field which threads the accretion disk, and moves along the field like 'beads on a wire'. It is obvious that if the field geometry opens up 'sufficiently', and if the disk rotates fast enough, the field lines act as a slingshot, and accelerate plasma centrifugally away from the disk. There are however some rather special physical conditions which need to be satisfied in order for this mechanism to work in the case of an ionized accretion disk: (a) the field cannot be too strong since that would stop the rotation of the disk, and cannot be too weak since it is required to 'hold' the jet; (b) only a small percentage of the disk material is frozen onto the field and consequently outflows into the jet, the rest accretes by slipping through the magnetic field. We thus see that although this scenario might work very naturally in the magnetosphere of a rotating neutron star (where a strong field with a suitable geometry is provided by the star, and where frozen–in plasma is provided by electron–positron pair injection at the surface), there are questions about its applicability in the accretion disk scenario.

The above questions concern the region near the origin of the outflow. At large distances downstream (i.e. beyond the Alfvèn point), the situation simplifies very much. One can easily check that in most solutions of the centrifugal driving problem, the toroidal component of the magnetic field dominates strongly over the poloidal component, i.e. $B_p/B_\phi \to 0$ down the jet. In that limit, the axial flux density of electromagnetic (Poynting) energy is equal to $v_z B_\phi^2/4\pi$ to a very good approximation, and this suggests clearly that the energy density of the magnetic field ($B_\phi^2/4\pi$) is advected by the axial flow (v_z). Rotation of poloidal field lines, which has been the basis of centrifugal acceleration near the origin of the outflow, is not physically important at large distances down the jet.

The above considerations, together with the mathematical complications associated with the presence of a rotating poloidal magnetic field component, led us to consider an idealized magnetic field geometry where the poloidal component is zero everywhere, and the acceleration and collimation of the jet is due to a purely toroidal magnetic field. It should be obvious that, based only on large scale observations, one cannot differentiate between jets driven by a purely toroidal magnetic field, and jets driven centrifugally. As we will see, the generation and subsequent escape of toroidal magnetic field from a differentially rotating ionized accretion disk might very naturally produce jet-like outflows, and therefore, our present idealized discussion might not be too unphysical.

2   Steady–State Outflows

We will now discuss briefly the case of a steady–state axisymmetric ideal magnetohydrodynamic flow around a central gravitating compact object, when the magnetic field has only an azimuthal (toroidal) component. This simple exercise will convince the reader of the statement made in the Introduction, namely that jets driven by a purely toroidal magnetic field and jets driven centrifugally look very similar at large distances. Therefore, one should trust the present scenario as much as previous scenaria based on the mechanism of Blandford and Payne (e.g. Königl 1989; Sauty and Tsinganos 1994; Contopoulos and Lovelace 1994; Ferreira and Pelletier 1995; Li 1995). It is interesting however that, in the context of our picture, we can easily account also for the *time dependent* evolution of sequences of fast and slow moving blobs, such as observed in galactic and extragalactic jets.

We will work in the cylindrical coordinates r and z, with $\partial/\partial\phi \equiv 0$. The central compact object (black hole) is centered at $r = z = 0$, and the accretion disk lies at $z = 0$. Let us repeat our working hypothesis:

$$\mathbf{B}(r, z) \equiv B_\phi(r, z)\hat{\phi} , \tag{1}$$

(see Fig. 1). Following Contopoulos and Lovelace (1994), the basic equation of the problem is the usual force–balance (Euler's) equation,

$$\rho(\mathbf{v} \cdot \nabla)\mathbf{v} = -\nabla P + \rho\mathbf{g} + \frac{1}{c}\mathbf{J} \times \mathbf{B} , \tag{2}$$

supplemented by the continuity equation, Maxwell's equations, the ideal MHD condition, and an equation of state to relate the pressure and the density. Here, $\mathbf{v}$ is the flow velocity; $\mathbf{E}$ and $\mathbf{B}$ the electric and magnetic field respectively; $\mathbf{J}$ the electric current; P the pressure; ρ the matter density; and $\mathbf{g}$ the gravitational acceleration. Owing to axisymmetry and mass conservation, the velocity field can be separated into poloidal and toroidal parts ($\mathbf{v} = \mathbf{v}_p + \mathbf{v}_\phi$) with

$$\mathbf{v}_p = \frac{1}{4\pi\rho r} \left(-\frac{\partial\Phi}{\partial z}\hat{r} + \frac{\partial\Phi}{\partial r}\hat{z} \right) , \tag{3}$$

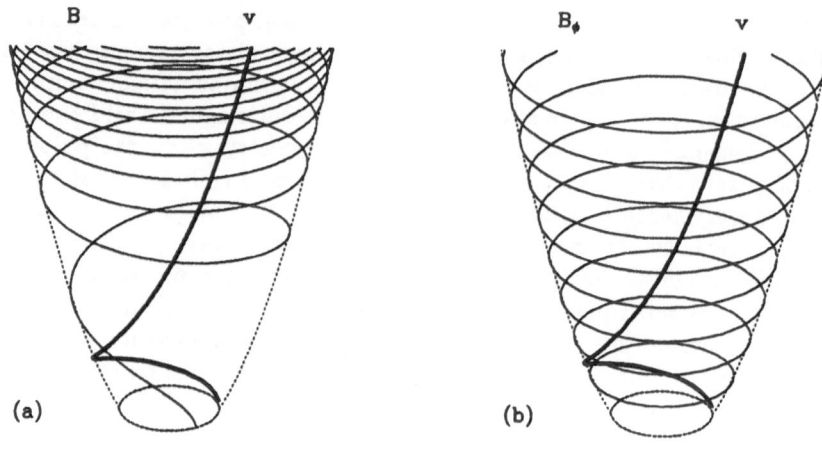

Fig. 1. Flux/flow surfaces ($z \geq 0$) inclined at some anlgle away from the viewer, for the present (b) and previous (a) field/flow geometries. Thin/thick lines represent field/flow lines respectively.

where $\Phi(r, z)$ is the Stokes stream function for the fluid motion (the total mass flow inside some radius r is equal to $\Phi(r, z)/2$). The lines of constant $\Phi(r, z)$ in the $r - z$ plane are the poloidal projections of the flow lines.

It is easy (although tedious) to show that the poloidal component of equation (2) perpendicular to the flow surfaces (i.e. parallel to $\nabla\Phi$), gives the following equation for $\Phi(r, z)$:

$$\Delta^*\Phi - \frac{1}{\rho}\nabla\rho \cdot \nabla\Phi = (4\pi\rho r)^2 \left(J' - \frac{LL'}{r^2} + 4\pi\rho r^2 \frac{N'}{N^3} - \frac{P}{\rho k_B}S'\right) , \qquad (4)$$

where, $\Delta^* \equiv r(\partial/\partial r)(1/r)(\partial/\partial r) + (\partial^2/\partial z^2)$, and

$$N(\Phi) = \frac{4\pi\rho r}{B_\phi} , \qquad (5)$$

$$L(\Phi) = v_\phi r , \qquad (6)$$

$$J(\Phi) = \frac{1}{2}|\mathbf{v}|^2 + v_A^2 + \int (dP/\rho)\Big|_{\Phi=\text{const.}} + \Phi_g , \qquad (7)$$

S is the entropy per unit volume, which is conserved along flow lines for an adiabatic flow, and primes denote differentiation with respect to Φ. Equation (7) expresses the condition for energy flow conservation (the Bernoulli equation), which includes matter flow, Poynting flux $(c/4\pi\rho v_p)E_p B_\phi \equiv B_\phi^2/(4\pi\rho) \equiv v_A^2$, enthalpy, and gravity $\Phi_g \equiv -GM(r^2 + z^2)^{-1/2}$ (M is the mass of the central compact object, and v_A is the Alfvèn velocity).

We emphasize that, as in the general case when $B_p \neq 0$ the general magnetohydrodynamic flow problem is reduced to the solution of one equation, the generalized Grad–Shafranov equation (see Lovelace et al. 1986), in the present case too, the problem is reduced to the solution of equation (4) for a given set of functions L, N, J, and S (which are clearly connected to the boundary conditions on the disk). It can be shown, by grouping together the second order derivative terms, that equation (4) is a second order partial differential equation of mixed type which is elliptic for $v_p^2 < c_s^2 + v_A^2$, and hyperbolic for $v_p^2 > c_s^2 + v_A^2$. Here, $v_p^2 \equiv v_r^2 + v_z^2$; and $c_s^2 \equiv \partial P/\partial \rho|_{\Phi=\text{const}}$. We see that the complications associated with the existence of three critical points in the Grad–Shafranov analysis (the slow/fast magnetosonic, and Alfvèn points) are not present in our formulation, since the flow field is separated by only one critical surface, the fast magnetosonic surface (where $v_p^2 = c_s^2 + v_A^2$).

As an interesting exercise and as a check of the above expressions, one can recover the present formalism as a limiting case of the general Grad–Shafranov formalism, in the limit $B_p/B_\phi \to 0$ (see Contopoulos 1995 for details). It is however important to stress here once again the physical difference between the present formulation and the Grad–Shafranov formulation: When $B_p = 0$ and $B_\phi \neq 0$, the magnetic field does not carry any angular momentum, and the element of rotation of magnetic field lines is not present.

Self–similar solutions of equation (4) which describe collimated plasma ouflows (jets) have been obtained, and we refer the reader to Contopoulos (1995) for details. At large axial distances, these solutions become almost indistinguishable from the respective self–similar solutions of the Grad–Shafranov equation (Contopoulos and Lovelace 1994). The difference between the two solutions can be found only near the origin of the outflow. As we have already said in the introduction, the centrifugal driving mechanism can account for the initial acceleration of the outflow, by converting rotational into axial flow velocity. On the other hand, the present mechanism requires an initial axial advection with a velocity comparable to the rotational velocity at the base of the flow.

We will argue that this is naturally expected in the case of a fully ionized thick accretion disk: Any weak radial magnetic field component inside a differentially rotating ionized disk will be 'wound' into a primarily toroidal field. It is generally believed that in most cases, this toroidal field is limited by buoyant escape to remain below equipartition with the ambient gas pressure of the disk material. Several people have studied the evolution of this field and have concluded that toroidal magnetic flux not only rises with velocities comparable to the dynamical velocity (in the case of a thick disk), but accumulates also in the region around the axis (e.g. Coroniti 1981; Chakrabarti and D'Silva 1994). Such a steady generation of advected toroidal magnetic flux might give rise to steady outflows similar to the ones described by the present steady–state formalism.

3   Explosive Release of the Toroidal Field

Another possibility however is discussed in Shibata, Tajima and Matsumoto (1990), namely that under certain conditions, buoyant escape appears to be suppressed, and does not prevent the toroidal field from growing to equipartition with the kinetic energy of the ambient disk material. To quote the above authors, *'once this upper bound on B_ϕ is reached, we expect explosive expulsion of the toroidal field; after this explosion, similar processes will occur again, so that the system shows quasi–periodic explosive energy release'*. We would like to emphasize at this point that the explosive evolution of such a strong toroidal magnetic field configuration will proceed in direct analogy to the focusing stage of a typical laboratory magnetized coaxial plasma gun (Mayo et al., 1995): the gradients of the toroidal field are associated with poloidal currents $\mathbf{J}_p \perp \mathbf{B}_\phi$, and the associated $\mathbf{J}_p \times \mathbf{B}_\phi$ force will expand the whole plasma–field configuration outwards in all directions. The central parts of this expanding 'toroidal solenoid' will clearly form an axially moving cylindrical pinch where the tension of the toroidal field will be held by rotation and any axial magnetic field present. We propose to call this jet acceleration mechanism, *the Astrophysical Plasma Gun*. Note that Shibata, Tajima and Matsumoto (1990) talk only about the throat collapse, and not about the general outward expansion of the configuration.

There is one more recently investigated interesting possibility that might also lead to such an explosive release of toroidal field and the formation of a fast moving cylindrical pinch: It has been observed repeatedly in global magneto-hydrodynamic numerical simulations, that shearing instabilities cause the entire surface of a magnetized unstable thick disk to lose most of its angular momentum and collapse dynamically towards the origin (Christodoulou and Hawley 1993; Matsumoto et. al 1995). Such a dynamical collapse of the surface layers will carry along any toroidal magnetic field which has been previously generated by differential rotation (Contopoulos et al. 1995). The radial collapse will at some point be held by rotation and axial magnetic fields, there is however nothing to hold the toroidal field from expanding in the axial direction. Thus, a significant percentage ($\sim$10%) of the inner disk might be expelled into an outflow, by the strongly amplified toroidal magnetic field that it contains. It is interesting that in this scenario, the ejection of fast moving 'blobs' is associated with the final stage of a period of dynamical accretion, as is suggested by hard X–ray outbursts which die out each time a radio component is generated in galactic superluminal sources (Mirabel et al. 1992; Tingay et al. 1995). It is also expected that the above scenario will repeat itself, and that this system will also show quasi–periodic explosive energy release.

The study of such time–dependent 'exploding' axisymmetric magnetospheres is based on the generalized form of eq. 2,

$$\rho \frac{d\mathbf{v}}{dt} = -\nabla P + \rho \mathbf{g} + \frac{1}{c}\mathbf{J} \times \mathbf{B} , \qquad (8)$$

where, $d/dt \equiv \partial/\partial t + (\mathbf{v} \cdot \nabla)$. The situation is greatly simplified when $\mathbf{B} = B_\phi \hat{\phi}$ for the following reason: the main problem of time dependent numerical MHD solutions has been the conservation of the condition $\nabla \cdot \mathbf{B} = 0$ along the numerical integration; in the case of a purely toroidal magnetic field, this condition is identically satisfied! As a result, the time–dependent equations are amenable to a straightforward Lagrangian integration scheme. We can get a feeling for the time evolution of such an explosion by considering the following simple idealized case:

$$\mathbf{v}(r, z; t) = v_z(z; t)\hat{z} \tag{9}$$

$$\mathbf{B}(r, z; t) = B_\phi(z; t)\frac{r_o}{r}\hat{\phi} \tag{10}$$

$$\rho(r, z; t) = \rho(z; t)\left(\frac{r_o}{r}\right)^2 \ . \tag{11}$$

The flow consists of concentric cylinders with matter density decreasing outwards as $1/r^2$. Electric currents run horizontally (radially) across the flow and produce azimuthal magnetic fields with a $1/r$ radial dependence. Clearly, the electric circuit closes along the outer and inner surfaces of the flow (the inner surface can very well be the symmetry axis of the flow), i.e. it forms a toroidal solenoid. From the above it is obvious that $\mathbf{J} \perp \mathbf{B}$, hence we are in a clearly non force–free situation in the axial direction. The gradient of the azimuthal magnetic field pressure will produce an outflow in very much the same way as plasma acceleration in a laboratory magnetized coaxial plasma gun. We have neglected pressure and gravity in order to concentrate on the effect of the magnetic field in accelerating the outflow.

One can easily check that the equations of time–dependent nonrelativistic ideal magnetohydrodynamics give

$$\frac{\mathrm{d}v_z(z; t)}{\mathrm{d}t} = -\left(\frac{B_\phi}{\rho}(z; t)\right)\frac{\partial B_\phi(z; t)}{\partial z} \tag{12}$$

$$\frac{\mathrm{d}B_\phi(z; t)}{\mathrm{d}t} = -B_\phi(z; t)\frac{\partial v_z(z; t)}{\partial z} \tag{13}$$

$$\frac{\mathrm{d}\rho(z; t)}{\mathrm{d}t} = -\rho(z; t)\frac{\partial v_z(z; t)}{\partial z} \ . \tag{14}$$

We start our simple Lagrangian numerical integration with zero flow velocity, and with a field and density distribution decreasing towards higher z, in order to see the accelerating power of the vertical magnetic pressure gradient at work (see Fig. 2). The first thing we notice is that, irrespective of the initial magnetic field profile, the flow evolves towards a *self–similar expansion* with a uniformly decreasing B_ϕ and a linear distribution of velocities in the interior of the plasma column. The second thing we notice is that field discontinuities (which are equivalent to horizontal current sheets) appear at the upper end of the configuration. Both these effects are very reminiscent of self–similar supernova explosions. The main difference however between the two types of explosions is that in the case of a supernova the explosion is (almost) isotropic (i.e. spherically symmetric), whereas in the present case, due to the presence of the azimuthal magnetic field B_ϕ, the explosion is very anisotropic, and leads to a strong cylindrical pinch around the axis of symmetry.

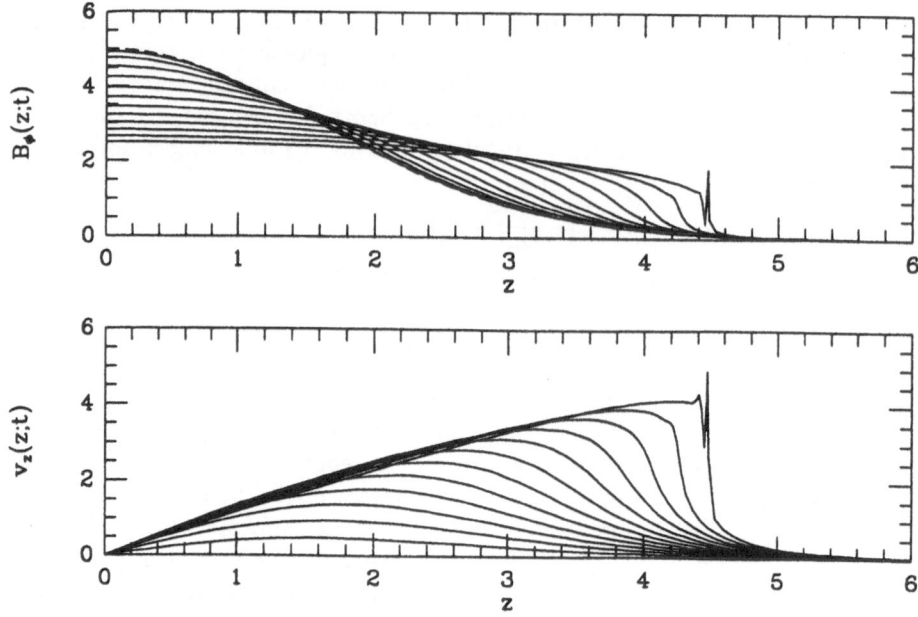

Fig. 2. Time dependent Lagrangian evolution of a cylindrical magnetic field configuration which is allowed to expand in the axial (z) direction (we plot $B_\phi(z;t)$, $v_z(z;t)$–defined by equations (9), (11)–in arbitrary units). We start with a smooth gradient in $B_\phi(z;0)$ (dashed line) and zero axial velocity, and integrate equations (12)—(14). $\rho(z;0) \propto B_\phi(z;0)$. We plot the configuration every 0.05 time units. We see that the field gradient indeed produces an axial acceleration of the plasma; at the same time however the plasma tends to self pinch into horizontal current sheets (this is very similar to the steepening of a sea wave). The system evolves towards a self–similar solution with a uniformly decreasing B_ϕ and a linear distribution of velocities in the interior of the plasma column. We stop the simulation when such a current sheet appears at the upper end of the configuration.

We propose to associate astrophysical jets with the above cylindrical pinch since it is there where matter densities are expected to be highest. Note the following characteristics, as expected naturally in this picture: (1) The jet is hollow, since matter is held against the tension force of the toroidal field by rotation and axial magnetic fields. (2) The cylindrical pinch (jet) lies along the axis of a much more extended region of expanding material which shields the jet from any possible irregularities in the ambient medium. (3) The explosion is expected to proceed in a quasi self–similar way, with velocities proportional to the distance from the origin. Highest velocities are expected at the top of the jet and at the surface of the expanding configuration. This might account for both the fast (relativistically) moving blobs at kpc scales, and the almost stationary blobs at pc scales, in the jets of M87 and 3C274 (Junor and Biretta 1994; Biretta, Zhou and Owen 1995). (4) Synchrotron emitting electrons might be energized in a turbulent layer on the surface of the cylindrical pinch. It is therefore expected that, as these energetic electrons diffuse in the region of strong field around the axis, their synchrotron spectrum will become steeper and steeper as we approach the axis. This is seen in VLBI observations of 4C39.25 (Alberdi et al. 1995).

More work needs to be done along the above lines. In particular, we need to understand quantitatively what percentage of the disk material accretes dynamically and what percentage goes into an outflow. We also need to obtain an estimate of the time intervals between ejections. Finally, detailed 2-D time–dependent numerical simulations of the explosive release of the toroidal magnetic field are needed.

We would like to thank Pr. Wolfgang Kundt for his invitation to attend the present workshop, and Heino Falcke, Jonathan Ferreira, Thomas Krichbaum, and Patrick Leahy for interesting discussions.

References

Alberdi et al., 1995, A & A, submitted

Biretta, J. A., Zhou, F. and Owen, F. N., 1995, ApJ, April 10

Blandford, R. D., 1993, in 'Astrophysical Jets', eds. D. Burgarella, M. Livio and C. P. O'Dea (Cambridge Univ. Press, Cambridge)

Blandford, R. D. and Payne, D. G., 1982, MNRAS, 199, 883

Chakrabarti, S. K. and D'Silva, S., 1994, ApJ, 424, 138

Christodoulou, D. M. and Hawley, J. F., 1993, BAAS, 25, 1475

Contopoulos, J. and Lovelace, R. V. E., 1994, ApJ, 429, 139

Contopoulos, J., 1995, ApJ, 450, 616

Contopoulos, J., Christodoulou, D. and Kazanas, D., 1995, ApJ, submitted

Coroniti, F. V., 1981, ApJ, 244, 587

Ferreira, J. and Pelletier, G., 1993, A & A, 295, 807

Junor, W. and Biretta, J. A., 1994, BAAS, 26, 1505

Königl, A., 1989, ApJ, 342, 208

Li, Z-Y, 1995, ApJ, 444, 848

Lovelace, R. V. E., Mehanian, C., Mobarry, C. M. and Sulkanen, M. E., 1986, ApJSS, 62, 1

Matsumoto, R. et al., 1995, ApJ, in press

Mayo, R. M., Bourham, M. A., Glover, M. E., Caress, R. W., Earnhart, J. R. D., and Black, D. C., 1995, Plasma Sources Science & Technology, in press

Mirabel, I. F. et al., 1992, Nat, 358, 215

Sauty, C. and Tsinganos, K., 1994, A & A, 287, 893

Shibata, K., Tajima, T. and Matsumoto, R., 1990, ApJ, 350, 295

Tingay, S. J. et al., 1995, Nat, 374, 141

Magnetized Accretion-Ejection Structures

Jonathan Ferreira

Landessternwarte, Königstuhl, D-69117 Heidelberg, Germany

Abstract: It is shown that powerful, self-collimated jets can be produced by keplerian accretion disks thread by a large-scale magnetic field. The general properties of the disk as well as the interplay between accretion and ejection processes are exposed, in the context of both active galactic nuclei and young stellar objects. Global self-similar solutions are presented for non-relativistic jets.

1 What Are MAES ?

In both galactic and extragalactic objects, powerful highly collimated jets are observed emanating from the innermost central region, whose luminosity is believed to be powered by accretion processes. It is then natural to seek for a common structure that tightly links ejection to accretion. Since the pioneering works of Chan & Henriksen (1980) and Blandford & Payne (1982, hereafter BP), it is now widely accepted that jets are magnetically self-collimated and probably driven (so-called cold jets).

There have been numerous studies of magnetized jets (e.g. Camenzind 1986, Heyvaerts & Norman 1989, Pelletier & Pudritz 1992, Contopoulos & Lovelace 1994, to cite a few), but they all suppose that the underlying disk would support the jets. However, if jets are to carry away the disk angular momentum, they must strongly influence the disk dynamics and one cannot rely anymore on the standard viscous accretion disk model (Shakura & Sunyaev 1973).

The investigation of Magnetized Accretion-Ejection Structures (MAES), where accretion and ejection are interdependent, requires a new theory of accretion disks. Despite some advances in magnetized disk physics (e.g. Wardle & Königl 1993, Li 1995), the following questions remained unanswered:
1) What are the disk physical conditions required to launch cold jets ?
2) What makes a vertical motion out of a purely accretion one ?
3) What type of jets (terminal velocity, degree of collimation) can be produced by keplerian accretion disks ?

This paper is organized as follows. We address in Section 2 the first two questions, give hints for the last one in Section 3 and show a non-relativistic, self-similar jet solution in Section 4. We finally conclude in Section 5.

2 Keplerian Accretion Disks Driving Jets

2.1 Magnetohydrodynamic Equations

An accretion disk is supposed to be thread by a large scale magnetic field of bipolar topology. Although this field has a profound dynamical influence on the disk, it is assumed weak enough not to significantly perturb the keplerian balance between the gravity of the central object (either a star or a compact object) and the centrifugal force. The magnetohydrodynamical equations describing stationary, axisymmetric MAES are then the following:

$$\nabla \cdot \rho \boldsymbol{u} = 0 \tag{1}$$

$$\rho \boldsymbol{u} \cdot \nabla \boldsymbol{u} = -\rho \nabla \Phi_G - \nabla P + \boldsymbol{J} \times \boldsymbol{B} + \nabla \cdot \boldsymbol{T}_{vis} \tag{2}$$

$$\boldsymbol{J} = \frac{1}{\mu_o} \nabla \times \boldsymbol{B} \tag{3}$$

$$\eta_m J_\phi = \boldsymbol{u}_p \times \boldsymbol{B}_p \tag{4}$$

$$\nabla \cdot (\frac{\nu'_m}{r^2} \nabla r B_\phi) = \nabla \cdot \frac{1}{r} (B_\phi \boldsymbol{u}_p - \boldsymbol{B}_p \Omega r) \tag{5}$$

where the subscript "p" refers to poloidal, $\Phi_G = -GM/(r^2 + z^2)^{1/2}$ is the gravitational potential of the central object (disk self-gravity neglected), $\boldsymbol{T}_{vis}$ is a turbulent viscous stress tensor, ν_m ($\eta_m = \mu_o \nu_m$) and ν'_m are turbulent magnetic diffusivities required for stationarity (see below), the other symbols keeping their usual meaning.

The poloidal magnetic structure is obtained from the magnetic flux function $a(r, z)$, such that $r\boldsymbol{B}_p = \nabla a \times \boldsymbol{e}_\phi$. A magnetic surface, defined by a constant magnetic flux, is directly labelled by $a = constant$.

In order to close the above system, along with an equation of state, one would need to solve the energy conservation equation. Since we suppose that the disk is turbulent in order to have the necessary transport coefficients (viscosity and magnetic diffusivity), we must self-consistently take into account a turbulent heat transport inside the disk (see e.g. Shakura et al. 1978). Such a term would greatly modify the disk thermal structure, since magnetized disks are probably convectively unstable (Ferreira & Pelletier 1995, hereafter FP95). Thus, as we lack knowledge of this heat transport in presence of magnetic fields and differential rotation, the neglect of this term could lead to serious problems in deriving the disk vertical structure. However, we will show below that most of the available energy is transferred into the jets, leaving thermal processes as an epiphenomenon.

Therefore, we will use a polytropic approximation $P = \kappa \rho^\gamma$ in the isothermal case ($\gamma = 1$) for the sake of simplicity. Since the jets studied here are not thermally driven, this last approximation does not affect the conclusions of this work.

2.2 How Is Accretion Achieved ?
2.2.1 Magnetic Diffusivity

Steady-state accretion is possible only if matter diffuses through the magnetic field lines. Such a diffusion process could be either due to ambipolar diffusion (e.g. Königl 1989) or a turbulent diffusivity (e.g. Ferreira & Pelletier 1993a). We favor this last possibility for two reasons: first, jets are produced in a wide variety of objects and thus do not seem to depend on the degree of disk ionisation (which must be small where neutral-ion collisions are relevant); second, magnetized accretion disks are known to display instabilities (e.g. Balbus & Hawley 1991, Tagger et al. 1992) that would probably saturate in the non-linear regime by providing turbulent transport coefficients. Thus, we choose a "poloidal" magnetic diffusivity that scales as

$$\nu_m = \alpha_m V_{Ao} h \tag{6}$$

where V_{Ao} is the Alfvén speed at the disk midplane, $h(r)$ is the disk half-thickness and α_m a phenomenological parameter describing the level of the turbulence. For $\alpha_m \sim 1$ resistive instabilities are saturated. Since magnetic instabilities would probably be enhanced with respect to the toroidal field, we allowed the turbulence to be anisotropic by using a "toroidal" diffusivity ν_m' in (5).

Equation (4) describes how the magnetic surfaces bend, with a characteristic scale height $l(r)$, in response to both advection and diffusion of the flow. This bending is then directly related to the magnetic Reynolds number

$$\mathcal{R}_m \equiv \frac{r u_{ro}}{\nu_o} = \frac{r^2}{\beta l^2} \tag{7}$$

where the subscript "o" refers to quantities evaluated at the disk midplane and $\beta \equiv \partial \ln a_o / \partial \ln r$. As BP showed, cold jets, that is jets that are not thermally driven, are possible only if the magnetic surfaces make an angle of more than 30° with the vertical axis at the disk surface. This constraint imposes $\mathcal{R}_m \gtrsim \varepsilon^{-1}$, where $\varepsilon \equiv h/r \ll 1$ is the disk aspect ratio. In such a parameter regime, all the dynamical terms become relevant at the disk surface. They must therefore be kept if a smooth transition between the quasi magnetohydrostatic disk and the high-velocity jet is desired (FP95).

2.2.2 Angular Momentum Transport

Once diffusion is allowed, accretion is possible if there is a means to transport the disk angular momentum. This can be done either by "internal" viscous stresses (transport through the disk) or by "external" magnetic stresses (transport by the jets). This magnetic torque arises from the unipolar induction effect: the rotation of the resistive disk induces an electromotive radial force $\Omega r B_z$ that drives a radial current J_r, responsible for the magnetic torque $J_r B_z$ (the disk behaves like a Barlow wheel). The relative importance of the two torques at the disk midplane is

$$\Lambda \equiv \frac{\text{mag. torque}}{\text{visc. torque}} \simeq \frac{\nu_m}{\nu_v} \mathcal{R}_m \gtrsim \frac{r}{h} \tag{8}$$

where ν_v is the Shakura-Sunyaev turbulent viscosity (which is expected to be comparable to the diffusivity). Hence, it is straightforward to realise that purely magnetically driven jets require a dominant magnetic torque on the disk (we can then neglect the viscous term in (2)). This has tremendous consequences on the energy budget (see Section 3.2).

2.3 From Accretion to Ejection

One has now achieved a situation where both angular momentum and mechanical energy are transferred to the magnetic field, allowing the matter to fall towards the central object. This accretion motion is also slightly converging towards the disk midplane through the combined action of the gravitational tidal force and the vertical magnetic pinching force. How does one go from this situation (u_r and u_z negative) to the one demanded for jets (both velocities positive) ?

2.3.1 The crucial role of the radial current profile

While accretion is characterized by a negative azimuthal component of the Lorentz force F_ϕ, magnetic acceleration occurring in jets requires a positive F_ϕ. Since $F_\phi = J_z B_r - J_r B_z$, the transition between these two situations depends mainly on the vertical profile of the radial current $J_r \equiv -\mu_o^{-1} \partial B_\phi / \partial z$, that is, on the rate of change of the magnetic shear with altitude. In order to switch from accretion to ejection, J_r must vertically decrease. This crucial issue is controlled by (5), that provides

$$\eta_m' J_r \simeq \eta_o' J_o + r \int_0^z dz \, \boldsymbol{B}_p \cdot \nabla \Omega - B_\phi u_z \qquad (9)$$

The first term on the right-handside of (9) describes the current due to the electromotive field, the second is the effect of the disk differential rotation and the third is the advection effect, only relevant at the disk surface layer. Thus, the vertical profile of J_r is mainly controlled by the ratio Γ of the differential rotation effect over the induced current. No jet would be produced without differential rotation ($\Gamma \ll 1$), for it is the only cause of the vertical decrease of J_r. However, the counter current due to the differential rotation cannot be much bigger than the induced current in the disk ($\Gamma \gg 1$), otherwise J_r would become strongly negative and lead to an unphysical positive toroidal field at the disk surface. Thus, steady state ejection is achieved only when $\Gamma \sim 1$ (Ferreira & Pelletier 1993b, FP95). Note that this implies a "toroidal" diffusivity such that $\nu_o' \simeq \nu_o 3 / \alpha_m^2 \Gamma$.

The radial current decreasing in a disk scale height has two important consequences, resulting in the ejection of a fraction of the matter: 1) the vertical Lorentz force decreases in such a way that the plasma pressure gradient slowly lifts up matter from the disk surface (see Figure 1); 2) angular momentum and mechanical energy are magnetically transferred back to matter, leading there to a radial centrifugal acceleration.

The global current topology is linked to the ejection efficiency (Ferreira 1994), through the ejection index defined by

$$\xi = \frac{d \ln \dot{M}_a}{d \ln r} . \tag{10}$$

This major parameter is a local measure of the jet formation efficiency ($\xi = 0$ in a standard disk without jet). As such, it is a link between radial and vertical structures in these intrinsically 2-D objects. The disk could a-priori build two different current topologies. For $\xi < 1/2$, a strong current flows down the jet axis, enters the disk at its inner edge and flows back along the jet. In this case, vertical motion is obtained while the Lorentz force still pinches the disk ($J_r > 0$ at the disk surface), allowing only a "tenuous" ejection. For $\xi > 1/2$, the current flows down the jet itself with the return current in an outer cocoon. Here, the gradient of toroidal magnetic pressure helps plasma pressure to lift up matter ($J_r < 0$), hence leading to "heavy" ejection.

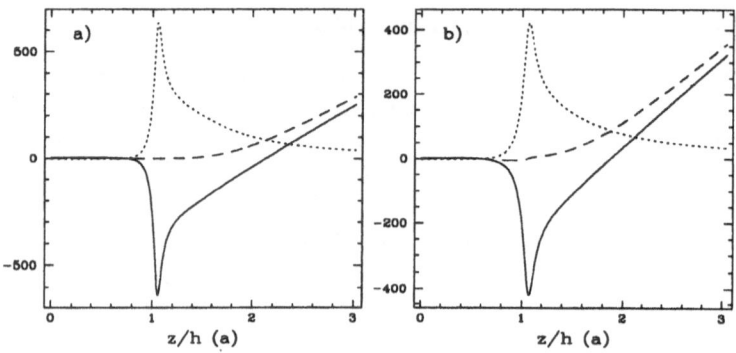

Fig. 1. Poloidal acceleration along the plasma motion $\boldsymbol{F} \cdot \boldsymbol{u_p}/\rho$, for the following forces (units are arbitrary): sum of gravitational, centrifugal and Lorentz forces (solid line), plasma pressure gradient (dotted line) and their sum (dashed line). Each curve is calculated vertically along a magnetic surface (constant magnetic flux a). The main acceleration occurs close to the disk surface, where the density decreases rapidly. In both cases (pannel a: $\xi = .01$; b: $\xi = .6$), the plasma pressure gradient develops an outward motion against the other forces.

2.3.2 From resistive disks to ideal MHD jets

As matter is expelled off the disk with an angle $\theta_{u_p} \equiv \arctan(u_r/u_z)$, magnetic stresses make it gradually flow along a magnetic surface (with $\boldsymbol{u_p} \parallel \boldsymbol{B_p}$). Indeed, (4) can be written

$$\eta_m J_\phi = u_r B_z \left(\frac{\tan \theta_{B_p}}{\tan \theta_{u_p}} - 1 \right) \tag{11}$$

where $\theta_{B_p} \equiv \arctan(B_r/B_z)$. As long as $\theta_{u_p} < \theta_{B_p}$, the toroidal current remains positive. This maintains a negative vertical Lorentz force (that decreases u_z) and a positive radial Lorentz force (that, along with the centrifugal term, increases u_r), thus increasing θ_{u_p}. If $\theta_{u_p} > \theta_{B_p}$, the toroidal current becomes negative, lowering both components of the poloidal Lorentz force and, hence, decreasing θ_{u_p}. Therefore, along with a vertical decrease of the magnetic diffusivity (for its origin lies in a turbulence triggered inside the disk), there is a natural mechanism that allows a smooth transition between resistive and ideal MHD regimes.

3 The Global Picture

3.1 MAES parameters

A complete structure is characterized by the following parameters, evaluated at the disk midplane: $\varepsilon = h/r$, $\mathcal{R}_{m_,}$ a measure of the disk magnetization $\mu = V_{Ao}^2/\Omega_k^2 h^2$, the disk accretion rate $\dot{M}_{ae}$ at its outer radius r_e, the inner radius r_i and the ejection index ξ. In the jets studied here, these parameters are constant throughout the disk radial extension. As a consequence, the fraction of mass ejected in the two jets is simply

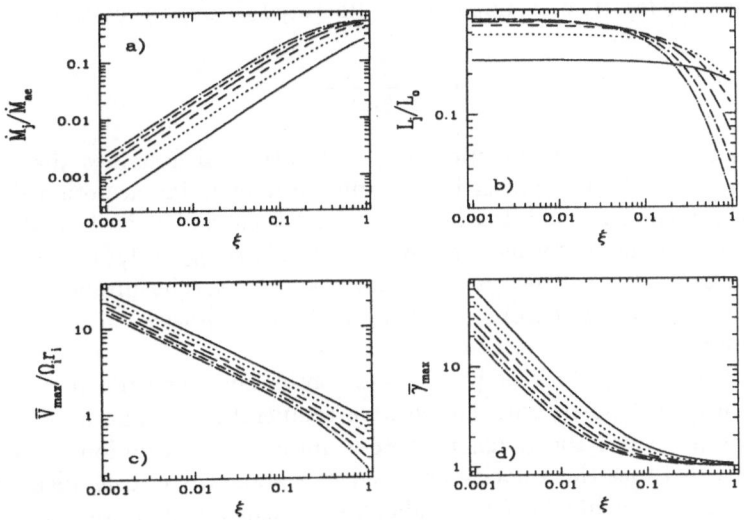

Fig. 2. Observational quantities evaluated with a disk radial extension r_e/r_i of 2 (solid line), 4, 10, 20, 45 and 100 (long dash-dotted line): a) fraction of ejected mass in one jet; b) power carried away in one jet in the form of both kinetic and MHD Poynting fluxes, normalized by $L_o = GM\dot{M}_{ae}/2r_i$; c) maximum mean jet velocity in the non-relativistic case, normalized by the keplerian velocity $\Omega_{ki}r_i$ at the inner disk radius r_i; d) maximum mean Lorentz factor for MAES settled around a black hole with $r_i = 3r_g$, r_g being the Schwarzschild radius. The exact velocities can be much lower than those displayed here, if jets radiate and/or the magnetic structure keeps stored a significant fraction of the power.

$$f \equiv \frac{2\dot{M}_j}{\dot{M}_{ae}} = 1 - \left(\frac{r_i}{r_e}\right)^\xi .$$ (12)

The requirement of stationarity will impose the values of μ and $\mathcal{R}_m$ for given ξ and ε, as regularity conditions at respectively the Slow-Magnetosonic and Alfvén critical surfaces encountered by the jet. Hence, the disk vertical equilibrium (related to the SM-critical point) imposes μ of order unity: MAES require magnetic fields close to equipartition with thermal energy (see parameter space in FP95). The other parameters, $\dot{M}_{ae}$, r_i and r_e, will be determined with observed global quantities, like total ejection rates or mean asymptotic jet velocities (Figure 2).

3.2 The Disk Luminosity

The energy conservation equation

$$P_{lib} = 2P_{rad} + 2P_{MHD}$$ (13)

tells that the available mechanical power P_{lib} is shared by radiation losses P_{rad} and an outward MHD Poynting flux P_{MHD} from each surface of the disk (Ferreira & Pelletier 1993a). The available power liberated through accretion is

$$P_{lib} = \eta_{lib} \frac{GM\dot{M}_{ae}}{2r_i}$$ (14)

where $\eta_{lib} < 1$ is a decreasing function of ξ and r_e/r_i and describes that the more mass is ejected, the less power is available. In a keplerian accretion disk, the fraction of the energy converted into heat (and then radiated) depends on the nature of the dominant torque (see (8)), with $2P_{rad}/P_{lib} \simeq 1/(1 + \Lambda)$ and $2P_{MHD}/P_{lib} \simeq \Lambda/(1 + \Lambda)$. Thus, a dominant magnetic torque implies a disk luminosity of a fraction, of the order of h/r, of the total power which is then carried away by the jets.

If the disk is optically thick, it will emit a continuum spectrum made of a superposition of black bodies, with an effective temperature scaling as $T_{eff} \propto r^{-3/4+\xi/4}$. However, since the magnetic torque imposes an accretion velocity of the order of r/h times the velocity achieved in a standard viscous disk, the corresponding density would be h/r smaller for an equivalent accretion rate. Thus, such weakly dissipative disks could already be optically thin for reasonable accretion rates (FP95).

3.3 Jet Energetics

If this MHD Poynting flux is completely converted into kinetic energy, the maximum asymptotic jet velocity is determined by

$$P_{MHD} \simeq P_{K,j} \equiv \frac{1}{2} \int_{r_i}^{r_e} d\dot{M}_j V_j^2 \equiv \frac{1}{2}\dot{M}_j \bar{V}_j^2$$ (15)

where V_j is the jet velocity along a particular magnetic surface and $\bar{V}_j$ the mean jet velocity. For non-relativistic jets, the maximum jet velocity scales as

$$V_j \simeq \Omega_{ko} r_o \xi^{-1/2} \qquad (16)$$

where the subscript "o" refers to the anchoring radius of the magnetic surface. Thus, very high velocities can be achieved for tiny ejection indices ξ. If MAES are settled around a compact object (e.g. in active galactic nuclei), relativistic speeds are possible if Compton drag can be avoided (Phinney 1987). In such a case, the maximum mean bulk Lorentz factor is given by

$$\bar{\gamma} \simeq 1 + \frac{\eta_{lib}}{2f} \frac{GM}{r_i c^2} \; . \qquad (17)$$

As an example, a MAES settled around a black hole with $\xi = .01$, $r_e/r_i = 10$ and $\dot{M}_{ae} = 1\ M_\odot . yr^{-1}$ would drive jets with some $10^{44}\ erg.s^{-1}$, a total ejection rate of order $10^{-2} M_\odot . yr^{-1}$ and a maximum bulk Lorentz factor of 4.

4 Non-Relativistic Self-Similar Jets

How much of the available MHD Poynting flux is transferred into kinetic energy depends on the jet model and is directly connected to the degree of collimation achieved. To address this question, we used a self-similar Ansatz (e.g. BP, Ferreira & Pelletier 1993a) and constructed global non-relativistic disk-jets solutions.

We obtain trans-Alfvénic jets for $\xi < 1/2$ only, exhibiting the following behaviour (see Figure 3): after a widening of the magnetic surfaces that leads to a very efficient acceleration, the jet reaches a maximum radius and starts to recollimate towards the jet axis. This refocusing stops where the flow meets the last MHD critical surface (Fast-Magnetosonic, FP95), whose treatment is beyond the scope of the present work. Whether or not this behaviour is imposed by self-similarity deserves further investigation (note however that this Ansatz is of no importance for the disk structure itself).

The jet acceleration efficiency is measured by the ratio of the MHD Poynting flux and the kinetic energy flux along a magnetic surface, namely

$$\sigma = \frac{-2\Omega_* r B_\phi B_p}{\mu_o \rho u^2 u_p} \qquad (18)$$

where Ω_* is the magnetic surface angular velocity. At the basis of the jet, identified here as the SM-surface, the disk provides $\sigma_{SM} \simeq \xi^{-1}$, showing that the ejection index acts like an injection parameter. Further on, σ decreases as magnetic acceleration takes place. Since it goes here to zero (see Figure 3), the final velocity can be rewritten as $V_j \simeq \sigma_{SM}^{1/2} \Omega_{ko} r_o$.

In order to make a link with the work done by BP, we translate their jet parameters into ours. They used ξ'_o, describing the angle of the magnetic field at the disk surface, the angular momentum parameter $\lambda_{BP} = \Omega_* r_A^2 / \Omega_o r_o^2$ and the mass load parameter $\kappa_{BP} = \sqrt{\mu_o \rho_A} \Omega_o r_o / B_o$. Here, r_A is the cylindrical radius where the poloidal velocity of the ejected matter reaches the poloidal Alfvén speed and ρ_A is the density at this Alfvén point. In our formalism, ξ_o' is directly related to $\mathcal{R}_m \varepsilon$, the magnetic lever arm writes

$$\lambda_{BP} \equiv \frac{r_A^2}{r_o^2} = 1 + \frac{1}{2\xi} \tag{19}$$

and the mass load is

$$\kappa_{BP} \equiv \left(\frac{\rho_A}{\rho_o} \right)^{1/2} \mu^{-1/2} \varepsilon^{-1} = 2\xi \left. \frac{B_\phi}{B_o} \right|_{SM}. \tag{20}$$

It can thus be seen that λ_{BP} and $\kappa_{BP} \simeq \xi$ are closely related to each other, both directly linked to the ejection index ξ. Because BP did not take into account the SM-surface, the parameter space for jets driven by keplerian accretion disks is smaller than the one they found. In particular, very high magnetic lever arms would require very small ejection indices, which is not consistent with the disk vertical equilibrium (FP95, Ferreira 1996).

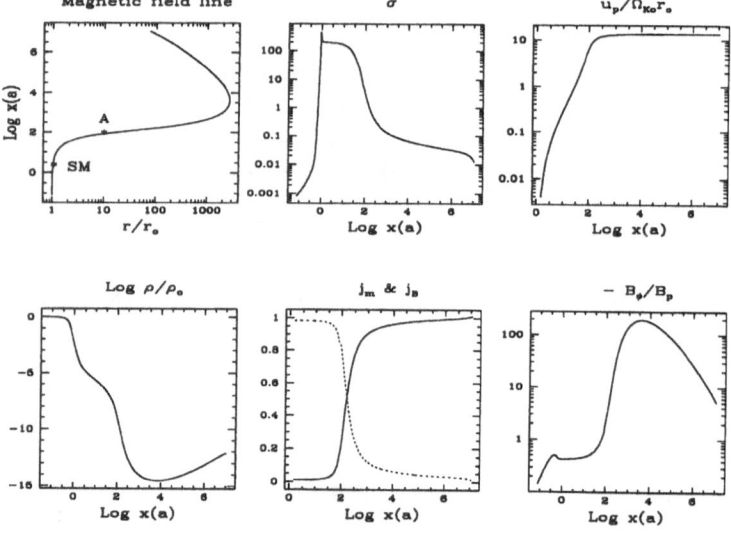

Fig. 3. Non-relativistic self-similar jet with $\xi = .005$ and $\varepsilon = .01$ ($\mu = .356$, $\mathcal{R}_m = 220$). All the quantities are calculated along a magnetic surface ($x = z/h$). Each surface opens up to 2800 times its initial anchoring radius r_o, before converging slowly towards the jet axis. The efficiency of energy transfer is very high (σ becomes much smaller than unity), allowing the flow to reach its maximum velocity. The total specific angular momentum carried away by the jet, $\Omega_* r_A^2$, stored in the magnetic field (j_B, dotted line) at the disk surface, is progressively transferred back to matter (j_m, solid line).

5 Conclusion

In order to steadily launch cold (i.e. non thermaly driven) jets, keplerian accretion disks must fulfill the following requirements:

1) a large-scale magnetic field of bipolar topology, at roughly equipartition with thermal energy inside the disk; such a field could be either advected from the interstellar medium or built in by dynamo processes.

2) an MHD turbulence, providing the required effective magnetic diffusivity; its origin and the level it can achieve remain to be worked out.

3) a current flowing in the disk and decreasing vertically on a disk scale height; the electric circuit controlling the kind of jets produced would probably depend on the local environment.

These results are general and do not depend on the self-similar Ansatz we used to obtain global disk-jets solutions.

It has been found that trans-Alfvénic jets are possible for "tenuous" ejection, feeding jets with almost all the available accretion power. Such powerful jets require a strong current flowing in the jet axis, a situation that probably favors a coupling between the central object and the surrounding accretion disk.

The fact that our non-relativistic jet solutions always seem to recollimate towards the jet axis is tightly linked with the very high acceleration efficiency. Such a behaviour could well be a consequence of self-similarity, since it imposes geometrical constraints that have a big influence on the transverse equilibrium of jets.

References

Blandford, R.D., Payne, D.G., 1982, MNRAS, 199,883

Camenzind, M., 1986, A&A, 156, 137

Chan, K.L., Henriksen, R.N., 1980, ApJ, 241, 534

Contopoulos, J., Lovelace, R.V.E., 1994, ApJ, 429, 139

Ferreira, J., Pelletier, G., 1993a, A&A, 276, 625

Ferreira, J., Pelletier, G., 1993b, A&A, 276, 637

Ferreira, J., 1994, PhD Thesis, Paris 7 University

Ferreira, J., Pelletier, G., 1995, A&A, 295, 807

Ferreira, J., 1996, A&A, to be submitted

Heyvaerts, J., Norman, C., 1989, ApJ, 347, 1055

Königl, A., 1989, ApJ, 342, 208

Li, Z-Y., 1995, ApJ, 444, 848

Pelletier, G., Pudritz, R.E., 1992, ApJ, 394, 117

Phinney, E.S., 1987, in Superluminal Radio Sources,eds. J.A. Zensus and T.J. Pearson, Cambridge Univ. Press

Shakura, N.I., Sunyaev, R.A., 1973, A&A, 24, 337

Shakura, N.I., Sunyaev, R.A., Zilintikevitch, S.S., 1978, A&A, 62, 179

Wardle, M., Königl, A., 1993, ApJ, 410, 218

Jet Formation in Astrophysical Converging Flows

V.V.Gvaramadze

Abastumani Astrophysical Observatory, Republic of Georgia

Abstract. We consider the so-called "cumulative effect" arising in converging flows of continuous media. Possible astrophysical applications of this effect are discussed. We also propose the mechanism of unsteady mass ejections from magnetospheres of accreting magnetized stars (classical T Tauri stars, white dwarfs and neutron stars).

1 Introduction: the cumulative effect

In converging spherical or axisymmetrical flows of continuous media, energy concentration arises, respectively, in the point or on the axis of symmetry. The energy concentration is inhomogeneous if the flow's symmetry differs from the simplest types of axial symmetry (spherical or cylindrical). In this case, an axial pressure gradient is built up which pushes matter along the symmetry axis. In this process, a significant part of the energy of the converging flow can be transferred to a small part of the matter which forms a narrow, jet-like ejection. Effects of such type are called *cumulative* (see, e.g., Lavrent'ev & Shabat, 1977).

The cumulative effect can be observed in a simple experiment. A test-tube with water vertically falls from the height of about $10 \div 20$ cm onto a table. Right after the impact a narrow jet, about one meter long, emanates from the test-tube. The qualitative explanation of this effect is the following. At the moment of the impact, a part of the water adjoined to the tube's wall got, due to the adherence, a velocity directed down. The closer to the tube's axis, the smaller this velocity. But the central part of the water (on the tube's axis) got velocity directed upward. Just this part of the water forms the jet. In other words, after the impact, the kinetic energy of the test-tube redistributed and concentrated on the tube's axis; that leads to the jet formation. This jet possesses a considerable part of the kinetic energy of the test-tube.

The cumulative effect can also be observed after the impact of a drop falling from a large height onto a water surface (see Fig. 2). A quasi-spherical cavity forms after collision, collapses, and produces a plume. The mass of this plume is a few times larger than the mass of the drop. The plume is driven by the axial pressure gradient arising due to the non-spherical collapse of the cavity. A similar effect arises when a bullet enters the water at an arbitrary angle. In this case a plume is directed opposite to the bullet motion and inclined at the same angle.

The most well-known example of the cumulative effect is the so-called shaped charge effect (Birkhoff *et al.*, 1948; Lavrent'ev & Shabat, 1965; see also Pfaehler,

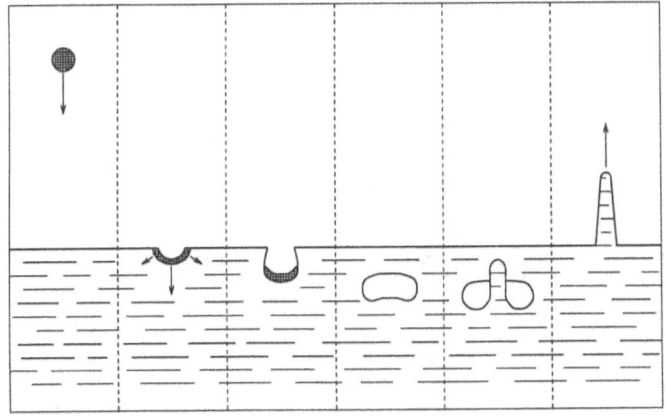

Fig. 1. Jet formation due to the non-spherical collapse of a cavity.

1993), which is being applied in mining and artillery. Within the explosive charge, there is a hollow conical metal insert (Fig. 3). This conical insert collapses under the influence of the detonation pressure. During the collapse two jets are formed: a wide and a narrow one (see Fig. 3).The detonation pressure is so high that the motion of the metal insert can be considered as fluid. For a small cone half-angle α, the velocity of the narrow (cumulative) jet is estimated as $v_j \simeq v_{\text{converg}}/\alpha$, and may be much larger than the velocity v_{converg} of the collapsing metal insert. The width and mass of the cumulative jet sharply decreases with decrease of the cone half-angle.

The theory of shaped charge effect is based on the theory of head-on collision of two hydrodynamical cylindrical jets with different radii ($r_1 > r_2$) (Lavrent'ev & Shabat, 1965; see also Batchelor, 1970). A hollow quasi-conical shroud arises after the collision of the jets (see Fig. 4). The cone half-angle $\alpha = \arccos[(r_1^2 - r_2^2)/(r_1^2 + r_2^2)]$, decreases with decreasing radius r_2 of the narrow jet. Using the principle of flow inversion, we can assert that a flow converging along the conical shroud produces two jets moving along the symmetry axis in opposite directions. The correlation of mass fluxes in these jets depends on the cone half-angle: $m_1/m_2 = (1 - \cos\alpha)/2$.

Cumulative jets also arise in converging conical shock waves (Birkhoff, 1950; Sokolov, 1988) (see Fig. 5). The matter in a jet is separated from the unperturbed medium by a Mach shock wave. Note that the effect of plasma ejection along the symmetry axis is absent in the case of a converging cylindrical shock wave. In this case a strong pressure increase occurs on the axis, but there is no axial pressure gradient. Therefore, the cylindrical shock wave is simply reflected from the axis.

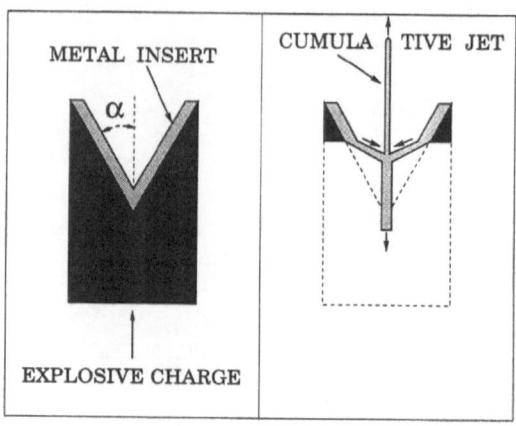

Fig. 2. The shaped charge effect

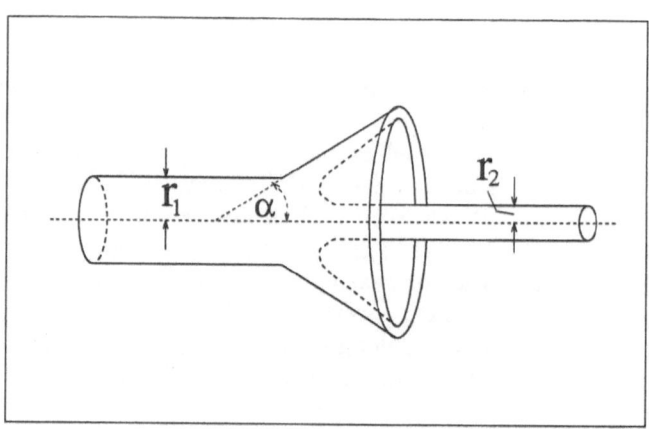

Fig. 3. Head-on collision of two cylindrical jets

Only in the case of a conical shock wave, a part of its energy is transferred to the plasma ejected along the cone axis. Its value depends on the cone half-angle.

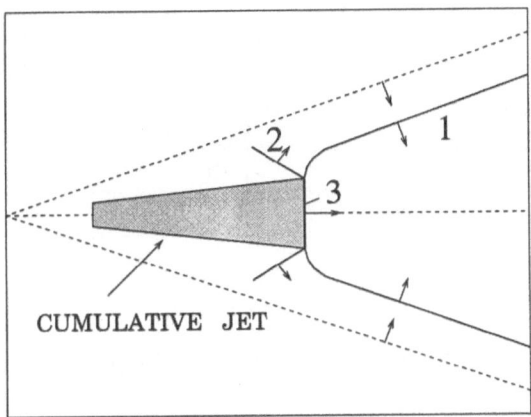

Fig. 4. Formation of a jet due to the cumulative effect in a converging conical shock wave: 1 – the converging shock wave; 2 – the reflected shock wave; 3 – the Mach shock wave.

The knowledge of the cumulative effect suggests a new interpretation of the well-known Landau-Squire solution for conically similar flows of an incompressible viscous fluid. In this class of motions, the velocity is inversely proportional to the distance from some origin (see, e.g., Landau & Lifshitz, 1979; Pillow & Paull, 1985; Goldshtik, Shtern & Yavorskaya, 1989): $\mathbf{v} = \nu\mathbf{U}(\varphi,\theta)/R$, where ν is the (constant) kinematic viscosity, R, φ, θ are spherical polar coordinates and $\mathbf{U}$ is a dimensionless vector function of φ and θ. The Landau-Squire solution describes a submerged jet – the flow induced in a fluid by a source of axial momentum (along the polar axis) at the origin. The central part of this flow represents a conical flow with a vertex at the origin (see Fig. 6). The jet becomes stronger and narrower as the axial momentum increases. For strong jets, the half-angle α of the jet is proportional to $F^{-1/2}$ (Landau & Lifshitz, 1979; Batchelor, 1970), where F is the axial momentum. Generalizations of the Landau-Squire solution have been applied to astrophysical jets by many authors (see, e.g., Lovelace & Scott (1982), Scott & Lovelace (1987), Henriksen (1986, 1987), and Henriksen & Valls-Gabaud (1994)). An important question in this problem is the origin of the required source of the axial momentum. It was supposed that this source might be either photon or Poynting flux absorbed by the matter at the jet origin.

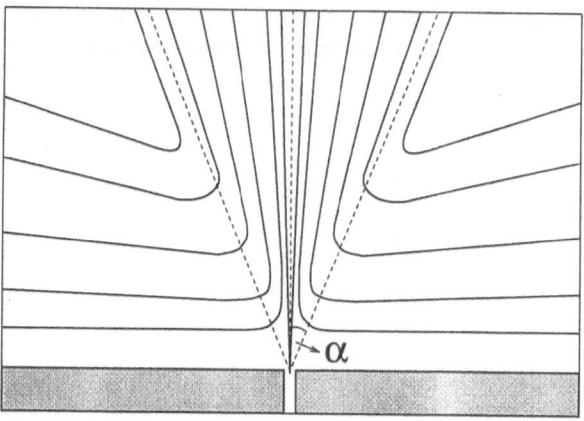

Fig. 5. The Landau-Squire jet

A more productive interpretation of the Landau-Squire solution has been done by Goldshtick & Shtern (1988) (see also Goldshtick, 1990). They consider unbounded space filled with a viscous incompressible fluid and suppose that there exists a plane (say an accretion disk), the material of which moves (because of gravitation) towards an origin (a black hole), so that the origin is a mass sink. Due to adherence, ambient fluid is driven to the symmetry axis (which is normal to the plane) and then flows away along the axis. As a result, near the axis an induced Landau-Squire jet arises. The half-angle of the jet is determined by the strength of a mass sink, i.e. ultimately by the accretion rate and by the mass of the black hole. (The same effect can be observed in a bath-room where the streaming water causes surrounding air (tinted by smoke) to form a jet just above the sink.) In this interpretation, the momentum concentration near the origin, which drives the jet, is only a consequence of the convergence of a conical flow, i.e., the axial pressure gradient is built up in the conical converging flow in a natural way due to the cumulative effect. This example is idealized and can hardly be applied to astrophysical objects, but it shows the importance of the cumulative effect.

The energy concentration in various types of the cumulative effect is connected with the axisymmetry of converging flows. It occurs (in the absence of a matter sink) in an impulsive manner, and the consequent mass ejection is unsteady. The above-mentioned cumulative jets are short-lived events because of the specific physical conditions inherent in laboratory experiments.

2 Some astrophysical applications of the cumulative effect

Apparently, the first example of an astrophysical application of the cumulative effect was given in the paper by Severny & Shaposhnikova (1960), where they have discussed the possibility of a cumulative origin of limb flares on the Sun.

The next example can be found in the book "Relativistic Astrophysics" by Zel'dovich & Novikov (1967) (note that this example is absent in the English version of the book!). They have proposed that a cumulative jet can be formed behind an attracting body moving through a continuous medium. It is known (Bondi & Hoyle (1944); Salpeter (1964)) that behind a supersonically moving compact object (with a velocity v), a standing quasi-conical shock wave develops with a cone angle $\alpha = \arcsin(M^{-1})$, where $M \equiv v/a$ is the Mach number (a is the sound speed). Gas particles with impact radii less than $r_{\text{accr}} \simeq GM/(v^2 + a^2)$ will be captured (see Fig. 7). In the case of a uniform medium (and low Mach numbers) this flow is stationary. But, the picture changes if the attracting body runs into a region with enhanced density (say in a gaseous cloud). The gas closes up behind the body and produces a cumulative jet.

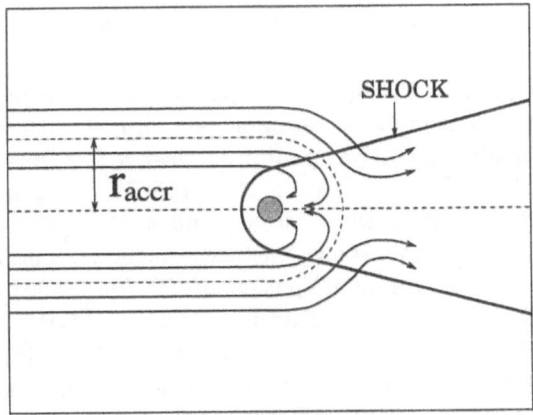

Fig. 6. Accretion onto a moving attracting body.

Formation of "bullets" in converging conical shocks was considered by Tenorio-Tagle & Rozyczka (1984a,b). According to their model, the converging conical shocks take place whenever strong shocks (for example, supernova or wind-driven interstellar shocks) overtake a dense cloudlets. Bullets result from cooling and condensation of matter swept up by the converging conical shocks. This effect was applied for an explanation of the origin of Herbig-Haro objects.

The possibility of jet formation, owing to the cumulative effect, in funnels of rapidly rotating protostars was discussed by Pfaehler (1993). According to this paper, two jets emanate in opposite directions along the rotation axis of the protostar if it pulsates due to some reason.

Cumulative formation of jets in funnels of magnetized accretion disks of active galactic nuclei was considered by Gvaramadze & Machabeli (1994). Accumulation and/or generation of strong large-scale magnetic fields in the inner part of an accretion disk leads to the conical geometry of the accretion flow and, correspondingly, to the formation of jet-like mass ejections.

It also seems attractive to apply the cumulative effect for an explanation of numerous dark lines and filaments emerged from gaseous disks of Our and some other galaxies (Heiles, 1976; Sofue, 1987, 1988). These filaments may arise due to the refilling of cavities produced by collisions of high-velocity ($100 \div 300$ km s^{-1}) molecular clouds with gaseous disks. The collapse of a quasi-spherical cavity may produce a pressure gradient directed opposite to a cloud motion, thereby pushes out a jet-like ejection from the disk. Non-spherical cavities and, correspondingly, cumulative jets also may arise due to the flare activity of stars.

3 Unsteady jets from magnetized stars

Another field of a possible astrophysical application of the cumulative effect is the problem of jet formation in accreting stellar objects with sufficiently strong regular magnetic fields (the Alfven radius at least a few times larger than the stellar radius), such as classical T Tauri stars (CTTSs; magnetic field $B \simeq 10^3$ G), white dwarfs (WDs; $B \simeq 10^5 \div 10^9$ G) and neutron stars (NSs; $B \simeq 10^{12}$ G). These objects often demonstrate unsteady outflows. The best examples of such ejections are the "superluminal" jets emanated from the two X-ray transient sources: GRS 1915+105 (Mirabel & Rodriges, 1994) and GRO J1655-40 (Tingay et al., 1995), which are probably associated with neutron stars. These sources change their appearance on the time-scale of a few days. Unsteady mass ejections are also observed in many other systems: star-forming regions (Herbig-Haro objects; a striking example is the HH 30 jet in the HL Tauri region consisting of a multitude of aligned knots (López et al., 1995)), planetary nebulae (mass ejections with point symmetry (Corradi, Schwarz & Stanghellini, 1993; López, Roth & Tapia, 1993); fast, low-ionization emission regions (Balick et al., 1994)), SS 433 (radio blobs (Vermeulen et al., 1987, 1993a)), symbiotic stars (CH Cyg (Taylor, Seaquist & Mattei, 1986); R Aqr (Sopka et al., 1982)).

The majority of current models of jet formation in stellar objects (see, e.g., Pudritz & Norman, 1983; Camenzind, 1990; Lovelace, Berk & Contopoulos, 1991) is based on accretion disk-driven magneto-centrifugal winds (Blandford

& Payne, 1982). The main shortcomings of these models are that they deal with stationary solutions of (ideal) MHD equations and can explain neither the source of initial momentum, which drives matter into jets, nor the unsteady motions typical for these jets. These models proceed from the assumption that jets are already existing and then describe their further motion and collimation. In a number of models the knotty structure of jets is explained in terms of internal shocks excited in continuous flows due to the *a priori* given time variability of the jet sources (see, e.g., Raga (1992) and Stone & Norman (1993)). But there exists another possibility to explain observational data: the optical variability of the jets in SS 433 may be explained in terms of optical "bullets" (Vermeulen *et al.*, 1993b; see, however, Kundt (1991)); the point symmetry in planetary nebulae (e.g. IC 4634) may be the result of multiple mass ejections from precessing central objects (Corradi *et al.*, 1993); the phenomenon of gamma-ray bursts may be the consequence of unsteady mass ejections (Rees & Mészáros, 1994; Gvaramadze, 1995a). The common feature in these proposals is the necessity of episodic matter outflow. At the same time, there is no doubt that mass ejections are closely connected with unsteady accretion and that the regular magnetic fields and stellar rotation play a crucial role in these two processes. Moreover, the observed large proper motions of outflows witness that they must originate in the very close vicinity to the central stars.

Proceeding from the above-mentioned facts, we propose a model of unsteady jet formation in stellar objects (Gvaramadze, 1995b,c). The main point of our model is the statement that jets are the result of unsteady mass ejections from the magnetospheres of accreting magnetized stars (CTTSs, WDs or NSs). The necessary ingredient of the model is the stellar regular magnetic field, which provides the conical geometry of accretion flows. We assume that the stellar magnetic field is dipolar and that its axis is aligned with the rotation axis. The source of accreting matter is a remnant or a newly acquired accretion disk. We also assume that the accretion rate episodically increases due to the accretion instability at the inner edge of the accretion disk. The cause of this instability is the centrifugal barrier (Illarionov & Sunyaev, 1975; for more details see Lipunov, 1992) or a sudden magnetic coupling of a ring of cold and dense matter to the stellar magnetosphere (Michel, 1991). Old neutron stars can episodically suffer accretion during their passages through interstellar gas clouds (Kundt & Chang, 1994). Provided that the above-mentioned assumptions are realized we can suggest the following scenario of formation of unsteady mass ejections.

The pressure of the stellar magnetic field disrupts the inner part of a circumstellar accretion disk, and the accreted plasma "freezes" onto the field lines (see, e.g., Lamb, 1989). Because only a part of stellar magnetic field lines threads the disk outside its inner edge, the frozen-in plasma falls towards the magnetic poles only along hollow accretion channels of a quasi-conical geometry.

At a low accretion rate $\dot{M} < \dot{M}_{cr} (\simeq 4 \times 10^{16} \mathrm{g\ s^{-1}}$ in the case of a NS with a mass $\simeq 1 M_\odot$) the plasma motion inside the accretion channels is magnetically controlled (the magnetic pressure is larger than the dynamical pressure of freely falling matter) and stationary (Basko & Sunyaev, 1976). At the ring-like base

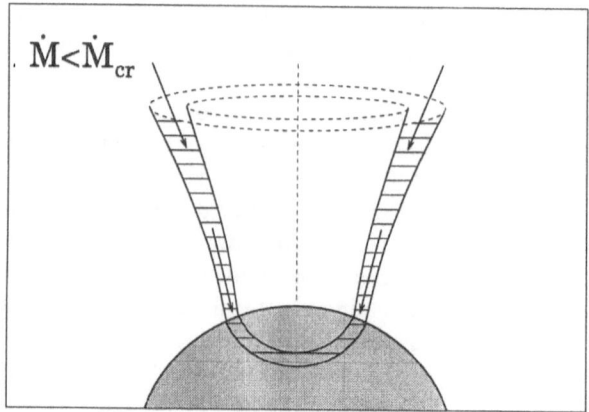

Fig. 7. The base of a hollow accretion channel. The accretion flow is shown by horizontal shading.

of the accretion channel, plasma is decelerated by interaction with particles of the stellar photosphere and its kinetic energy partially released in a standing radiative shock (a number of observed manifestations of accreting stars may be explained in terms of hot-spots arising from the accretion shocks). Accumulation of plasma at the base of the channel can violate magnetic confinement. This instability may explain fast quasi-periodic luminosity variations in accreting magnetized stars (Livio, 1984; see also Gvaramadze, 1995b) and (in the case of NSs) it may be a powerful high energy particle accelerator (Arons, 1988).

However, the situation drastically changes if the accretion rate episodically exceeds $\dot{M}_{cr}$ (Basko & Sunyaev, 1976; Wang & Frank, 1981). A sudden drop of the redundant portion of matter into the accretion channel leads to the rapid rise of a radiative shock. The radiative pressure slows down and compresses the accretion flow. This obstacle on the way of initially freely falling plasma leads to the violation of local magnetic confinement, and the plasma spreads, due to the plasma-field interchange instability, over the stellar magnetosphere. Part of the plasma rushes into the hollow accretion channel (see Fig. 11). This axisymmetrical quasi-conical converging flow produces an axial pressure gradient, which pushes out from the stellar magnetosphere a jet-like blob of matter. The closer to the stellar surface this occurs, the higher the velocity of the blob, and the smaller its mass. The blob by itself is unstable, because the overpressure, created by the converging flow, tends to re-expand it. But radiative cooling and joint action of magnetic fields and stellar rotation may prolong its lifetime. Moreover, frequent and unidirected ejections of blobs may mimic quasi-steady outflows.

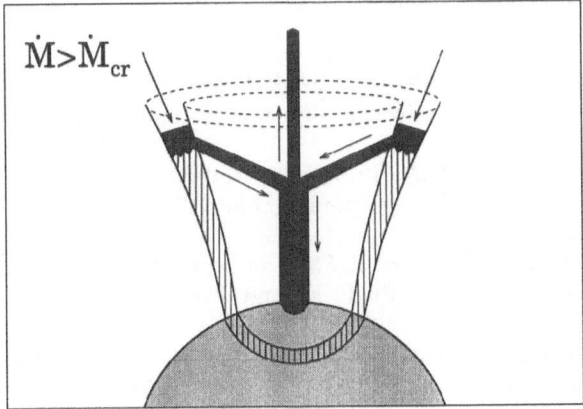

Fig. 8. Cumulative jet formation in a hollow accretion channel. The radiative shock is shown by vertical shading. The accretion flow and the jet-like blob of matter are drawn in black.

4 Conclusion

In conclusion we would like to stress the role of magnetic fields in jet formation. It seems quite plausible that regular magnetic fields are only the agent which provides the conical geometry for converging flows of plasma and helps to cumulate their energy on the symmetry axis. Thereby magnetic fields determine the orientation of outflows. Of course they can promote the further collimation of jets and can cause various MHD instabilities in jets. But, we think that the main source of momentum which drives unsteady outflows is the pressure gradient produced by the cumulative effect in converging conical and quasi-spherical flows.

Acknowledgements. I am grateful to Prof. W.Kundt for an invitation to the workshop and for critical remarks concerning the manuscript. I am also thankful to Dr. M.Camenzind for useful discussions during my stay at the Landessternwarte Heidelberg, where this work was partially carried out. The work was partially supported by the Deutsche Forschungsgemeinshaft (DFG).

References

Arons,J.: 1988, *Plasma Astrophysics*, **ESA SP-285**, 117

Balick, B., Perinotto, M., Maccioni, A., Terzian, Y., and Hajian, A.: 1994, *Astrophys. J.*, **424**, 800

Basko, M.M., and Sunyaev, R.A.: 1976, *Mon. Not. R. Astron. Soc.*, **175**, 395

Batchelor, G.K.: 1970, *An Introduction to Fluid Dynamics*, Cambridge University Press: Cambridge

Birkhoff, G.: 1950, *Hydrodynamics: A Study in Logic, Fact and Similitude*, Princeton

Birkhoff, G., MacDougall, D.P., Pugh, E., and Taylor, G.: 1948, *J. Appl. Phys.*, **19**, 563

Blandford, R.D. and Payne, D.G.: 1982, *Mon. Not. R. Astron. Soc.*, **199**, 883

Bondi, H., and Hoyle, F.: 1944, *Mon. Not. R. Astron. Soc.*, **104**, 273

Camenzind, M.: 1990, in G.Klare, ed., *Rev. Modern Astrophys.*, **3**, Springer-Verlag: Berlin, 234

Corrali, R.L.M., Schwarz, H.E. and Stanghellini, L.: 1993, in R.Weinberger, A.Acker, eds., *Planetary Nebulae*, Kluwer Academic Publishers: Dordrecht, 216

Ghosh, P., and Lamb, F.K.: 1978, *Astrophys. J.*, **223**, L183

Goldshtick, M.A.: 1990, *Annu. Rev. Fluid Mech.*, **22**, 441

Goldshtick, M.A., and Shtern, V.N.: 1988, *Proc. R. Soc. Lond.*, **A419**, 91

Goldshtick, M.A., Shtern, V.N., and Yavorskaya, N.I.: 1989, *Viscous Flows with Paradoxical Properties*, Nauka: Novosibirsk (in Russian)

Gvaramadze, V.V.: 1995a, *Astrophys. Space Sci.*, **231**, 411

Gvaramadze, V.V.: 1995b, in H.J.Staude, ed., *Disks and Outflows Around Young Stars*, Springer-Verlag: Heidelberg (in press)

Gvaramadze, V.V.: 1995c, *Astrophysical Letters & Communications* (submitted)

Gvaramadze, V.V., and Machabeli, G.Z.: 1994, in G.V.Bicknell, M.A.Dopita, and P.J.Quinn, eds., *The Physics of Active Galaxies*. ASP Conf. Ser., vol. 54, p. 85

Heiles,C., and Jenkins,E.B.: 1976, *Astron. Astrophys.*, **46**, 333

Henriksen, R.N.: 1986, *Canad. J. Phys.*, **64**, 403

Henriksen, R.N.: 1987, *Astrophys. J.*, **314**, 33

Henriksen, R.N., and Valls-Gabaud, D.: 1994, *Mon. Not. R. Astron. Soc.*, **266**, 681

Illarionov, A.F., and Sunyaev, R.A.: 1975, *Astron. Astrophys.*, **39**, 185

Kundt, W.: 1991, *Comm. Astrophys.*, **15**, 255

Kundt, W., and Chang, H.-K.: 1994, in G.J.Fishman, J.J.Brainerd and K.Hurley, eds., *Gamma-Ray Bursts*, AIP Press: New York, 596

Lamb, F.K.: 1989, in H.Ögelman, E.P.J. van Heuvel, eds., *Timing Neutron Stars*, Kluwer Academic Publishers: Dordrecht, 649

Lamb, F.K., Pethick, C.J., and Pines, D.: 1973, *Astrophys. J.*, **184**, 271

Landau, L.D., and Lifshitz, E.M.: 1979, *Fluid Mechanics*, Pergamon Press: Oxford

Lavrent'ev, M.A., and Shabat, B.V.: 1965, *Methods of the Theory of Complex Functions*, Nauka: Moscow (in Russian)

Lavrent'ev, M.A., and Shabat, B.V.: 1977, *Problems of Hydrodynamics and Their Matematical Models*, Nauka: Moscow (in Russian)

Lipunov, V.M.: 1992, *Astrophysics of Neutron Stars*, Springer-Verlag: Berlin

Livio, M.: 1984, *Astron. Astrophys.*, **141**, L4

Lopéz, J.A., Roth, M., and Tapia, M.: 1993, *Astron. Astrophys.*, **267**, 194

López, R., Raga, A., Riera, A., Anglada, G., and Estalella, R.: 1995, *Mon. Not. R.Astron. Soc.*, **274**, L19

Lovelace, R.V.E., and Scott, H.A.: 1981, *Plasma Astrophysics*, **ESA SP-161**, 215

Lovelace, R.V.E., Berk, H.L., and Contopoulos, J.: 1991, *Astrophys. J.*, **379**, 696

Mirabel, I.F., and Rodriguez, L.F.: 1994, *Nature*, **371**, 46

Michel, F.C.: 1991, *Theory of Neutron Star Magnetospheres*, The University of Chicago Press: Chicago

Pfaehler, J.: 1993, in L.Errico and A.A.Vittone, eds., *Stellar Jets and Bipolar Outflows*, Kluwer Academic Publishers: Dordrecht, 335

Pillow, A.F., and Paull, R.: 1985, *J. Fluid Mech.*, **155**, 327

Pudritz, R.E., and Norman, C.A.: 1983, *Astrophys. J.*, **274**, 677

Raga, A.C.: 1992, *Mon. Not. R. Astron. Soc.*, **258**, 301

Rees, M.J., and Mészáros, P.: 1994, *Astrophys. J.*, **430**, L93

Salpeter, E.E.: 1964, *Astrophys. J.*, **140**, 796

Scott, H.A., and Lovelace, R.V.E.: 1987, *Astrophys. Space Sci.*, **129**, 361

Severny, A.B., and Shaposhnikova, E.F.: 1960, *Izvestia Krimskoi Astrofizicheskoi Observatorii*, **24**, 235 (in Russian)

Sofue,Y.: 1987, *Publ. Astron. Soc. Japan*, **39**, 547

Sofue,Y.: 1988, *Publ. Astron. Soc. Japan*, **40**, 567

Sokolov, I.V.: 1988, *Teplofizika Visokikh Temperatur*, **26**, 560 (in Russian)

Sopka,R.J, Herbig,G., Kafatos,M., and Michalitsianos,A.G.: 1982, *Astrophys. J.*, **258**, L35

Stone, J.M., and Norman, M.L.: 1993, *Astrophys. J.*, **413**, 210

Taylor, A.R., Seaquist, E.R., and Mattei, J.A.: 1986, *Nature*, **319**, 38

Tenorio-Tagle,G., and Rozyczka,M.: 1984a, *Astron. Astrophys.*, **137**, 276

Tenorio-Tagle,G., and Rozyczka,M.: 1984b, *Astron. Astrophys.*, **141**, 351

Tingay, S.J., et al.: 1995, *Nature*, **374**, 141

Vermeulen, R.C., Schilizzi, R.T., Icke, V., Fejes, I., and Spencer, R.E.: 1987, *Nature*, **328**, 309

Vermeulen, R., et al.: 1993a, *Astron. Astrophys.*, **270**, 177

Vermeulen, R., et al.: 1993b, *Astron. Astrophys.*, **270**, 204

Wang, Y.M., and Frank, J.: 1981, *Astron. Astrophys.*, **93**, 255

Zel'dovich,Y.B., and Novikov,I.D.: 1967, *Relativistic Astrophysics*, Nauka: Moscow (in Russian)

Observational Properties of Jets from Young Stars

Jochen Eislöffel [1] [2]

[1]Laboratoire d'Astrophysique, Observatoire de Grenoble, B.P. 53X,
F-38041 Grenoble Cedex, France
[2]Thüringer Landessternwarte Tautenburg, Karl-Scharzschild-
Observatorium, D-07778 Tautenburg, Germany

Abstract: Observations of jets from young stars and of their bow shocks from the ultraviolet to centimetre wavelengths have revealed many of the characteristic properties of these objects. Here I summarize how optical imaging and spectrocopy have contributed to our understanding of the morphology, the kinematics, and the excitational state of the gas in these objects.

1 Introduction

Jets from young stars are certainly among the most spectacular sign-posts of star formation. They constitute the highly-collimated component of a bipolar outflow, which emanates from the immediate environment of a newly formed star. They consist of a more or less well-aligned chain of bright knots, often terminated by an arc-shaped Herbig-Haro (HH) object, which is considered as the bow shock where the flow of the jet impinges on the material ahead of it.

Here I shall give an overview of the observational properties of these jets. A wealth of data has been collected on them from the UV to cm-wavelengths over the past 15 years. I will, however, restrict myself to optical wavelengths. Results and problems from UV observations are discussed by Böhm (1990), some new observations of molecular outflows in the near-infrared molecular hydrogen lines can be found in this volume (Eislöffel & Davis), and an excellent overview over all aspects of radio observations of these outflows is given by Bachiller (1996).

This review is divided into two major parts. The first part summarizes the results of optical imaging. It describes the morphology, opening angle and collimation of the flow, proper motions and tangential velocities of the knots in jets and bow shocks, their pattern motions, and photometric variability of the knots. The second part then deals with results obtained by spectroscopic methods. It describes radial velocities, line profiles and velocity dispersion, line ratios and shock velocities derived from them, and the electron densities along the flow.

2 Optical imaging

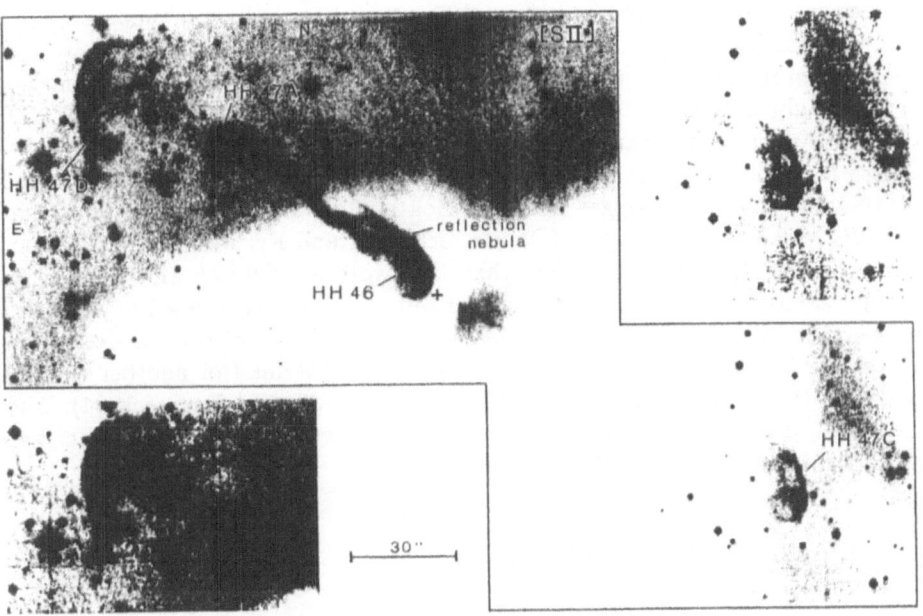

Fig. 1. [SII]$\lambda\lambda$6716, 6731 image of the HH 46/47 jet system. The bow shocks HH 47A and HH 47C have been marked. The position of the (invisible) source is also marked by a cross. The length of the scale bar corresponds to 0.05 pc for an assumed distance of 350 pc to HH 46/47.

Since the discovery of the first jets by Mundt & Fried (1983), more than 30 highly-collimated jets and a total of more than 280 HH objects have been found by many researchers. Optical imaging, in the Hα or the [SII]$\lambda\lambda$6716, 6731 emission lines, is usually employed to find these objects and study their striking morphologies (e.g., Mundt et al. 1984; Mundt 1988; Reipurth 1989a). Such images also reveal other characteristic properties of the jets, like their opening angles and the collimation of the flows. Comparison of images taken at various epochs let us measure the proper motions of the knots and hence their tangential velocities. Such series of images also allow us to study their photometric variability.

2.1 Morphology

In the optical, jets are usually observed as chains of bright emission-line knots that are more or less well-aligned. They emanate from the immediate surroundings of a young star and some of them terminate in bright bow-shaped objects, which were long known as Herbig-Haro (HH) objects. A fine example of such a jet system, HH 46/47 (Schwartz 1977), is shown in Fig. 1 and may serve to illustrate the constituents that typically make up such a system. HH 46/47 is a bipolar jet that emanates from an active young star, which is embedded in a Bok globule (Bok 1978). The north-east pointing jet is extremely wiggly and, over part of its length even splits up into two parallel strands. The jet terminates in the bright and rather compact bow-shaped Herbig-Haro object HH 47A. To the south-west, and separated from the bright jet by a dark lane in which the source is hidden, a short and much fainter counter jet is visible. Even further south-west, the flow emerges from the Bok globule and its counter bow shock HH 47C becomes visible again (Dopita, Schwartz and Evans 1982). The jet system, however, may extend further than the two bows HH 47A and HH 47C. To the north-east of HH 47A another fainter but much more extended bow HH 47D is seen (Hartigan, Raymond and Meaburn 1990). Series of two, or even more, successive bow shocks are not uncommon in jet systems (for another example see, e.g., HH 34: Reipurth and Heathcote 1992; Bally and Devine 1994). They have been attributed to episodes of varying outflow activity driven by increased accretion from a circumstellar disk (see also 2.5).

2.2 Opening angle and collimation

Some outflows show a length-to-diameter ratio of more than 30 (e.g., the HH 30 jet: see Fig. 2 and Mundt et al. 1990). This leads us to the interesting question of how the collimation of the jet is initially made and how it can be maintained over such a length. Does the collimation and so the opening angle change along the flow? And where does the initial collimation take place? Is the outflow ejected from the star or from a circumstellar accretion disk? The measurement of the width of a jet with distance from the source may provide us with some answers to these questions.

Mundt, Ray & Raga (1991), measured for 15 jets changes of the width of the knots as a function of distance from the source on high-quality CCD images taken under sub-arcsec seeing conditions. They found that jets have overall opening angles of 1 to 10 degrees. In almost all flows, however, the opening angle shows an approximately monotonic decrease with increasing distance from the source (see Fig. 2). Typical opening angles near the jet's "end" are of only $0.5 - 5$ degrees. Close to the source the collimation is much poorer with opening angles of up to several tens of degrees and the initial collimation must take place within less than 300 AU from the star. Observations by Hirth, Mundt and Solf (1994) and Hirth et al. (1994) showed that the jet exists already at a distance of only $15 - 30$ AU from the star. These results are confirmed by Ray and Mundt (1996,

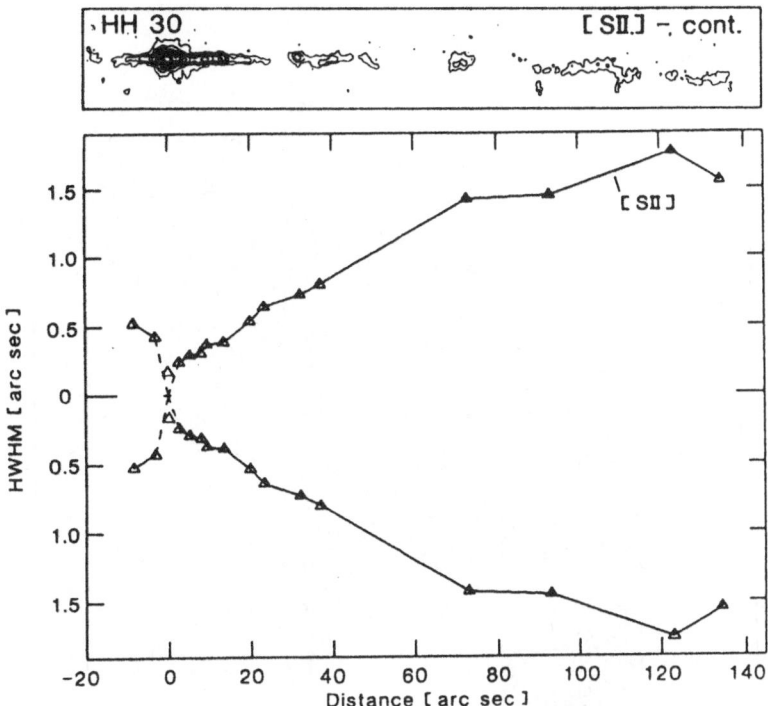

Fig. 2. Collimation of the HH 30 jet. The upper panel shows a continuum-subtracted contourplot of the jet in [SII]. The lower panel shows the width of this jet as a function of distance from the source (Mundt, Ray, and Raga 1991). A length of 140 arcsec corresponds to 0.10 pc for an assumed distance of 150 pc to HH 30.

in preparation) on the basis of HST observations. This calls for a very efficient mechanism that is capable of collimating the jet at such short length scales.

2.3 Proper motions and tangential velocities

A very important parameter for the kinematics of the outflows are proper motions and tangential velocities of the knots in the jets and the condensations in the bow shocks. Early work was carried out by Herbig and coworkers on photographic plates (Cudworth and Herbig 1979; Herbig and Jones 1981, 1983; Schwartz, Jones and Sirk 1984). They measured proper motions of HH objects, found several outflows to be bipolar, and determined the exciting sources of some HH objects.

With today's CCDs it is possible to study the proper motions of much fainter and therefore many more knots in the jets and the bow shocks in various emission lines with epoch differences of only a few years. Their strongly increased numbers are necessary for a good sampling of the tangential velocity field.

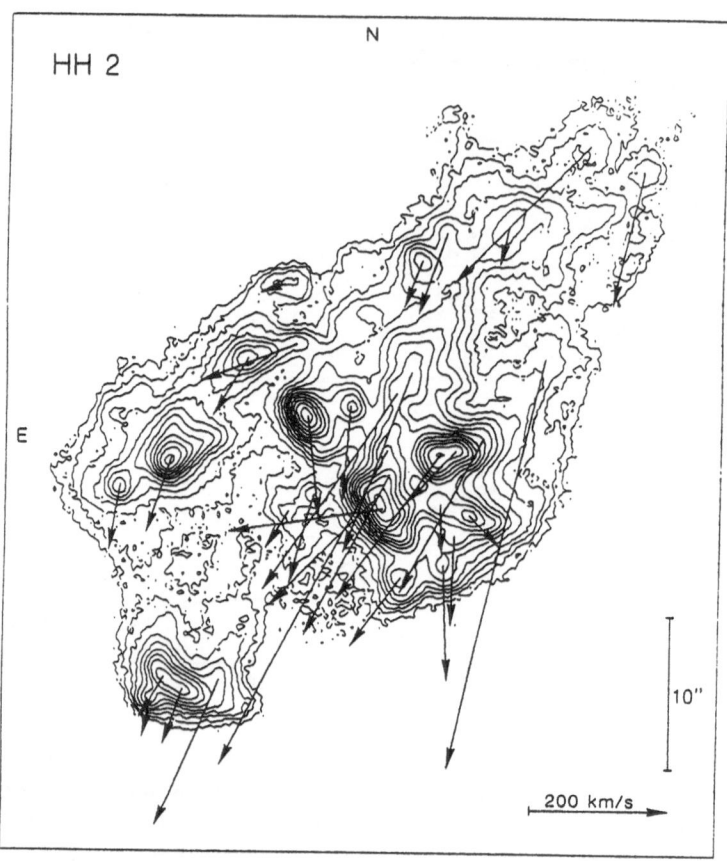

Fig. 3. Proper motions of the condensations in HH 2. Most knots move at velocities of $100 - 300$ km s^{-1} in this object (Eislöffel, Mundt and Böhm 1994). The length of the scale bar corresponds to 0.02 pc for an assumed distance of 450 pc to HH 2.

Recently, such measurements were carried out for a number of objects (e.g. HH 1/2: Eislöffel, Mundt, and Böhm 1994; HH 34: Eislöffel and Mundt 1992; Heathcote and Reipurth 1992; HH 46/47: Eislöffel and Mundt 1994). They showed that all knots move at rather high velocities of $100 - 400$ km s^{-1} (see Fig. 3 for an example). This result was in complete disagreement with models of jet knots, in which the knots were interpreted as oblique stationary crossing shock cells (see, e.g., Falle and Wilson 1985). Such shocks naturally arise in an initially overpressured jet that moves into a medium with a decreasing density gradient. As an alternative scenario it was proposed that the knots are due to temporal variability of the outflow. As the outflow is already highly supersonic, a variation of a few percent in velocity leads to internal shocks in the form of internal working surfaces (e.g., Raga and Kofmann 1992; Hartigan and Raymond 1993). In that model, the knots are thought to be little bow shocks, which develop when faster moving gas in the jet catches up with slower material ahead of it.

The proper motion measurements sometimes also show rather strong variations of the velocity of successive knots within one object. And this leads us to the next point: the pattern motion of the knots.

2.4 Pattern motion of the knots

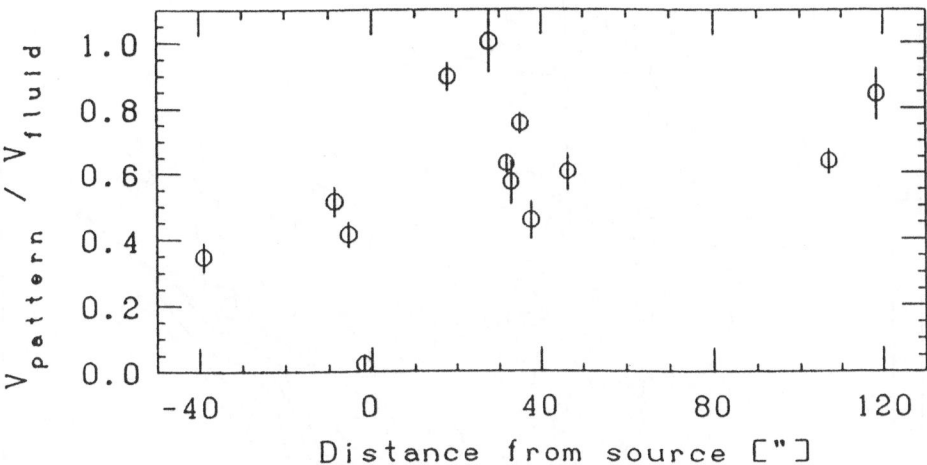

Fig. 4. Pattern motion of the knots in the FS Tau B jet. Sometimes strong variations of a factor of up to two are seen for consecutive knots (Eislöffel and Mundt 1996).

A priori it is not evident if the knots should move at the speed of the material in the flow or if they could move at their own velocity, different from the flow of the gas. As already mentioned, the model of stationary jets even assumed them to be at rest.

How can we measure a possible pattern motion of the knots in the jet flow? If the angle i of the outflow with respect to the plane of the sky is known, then one can directly calculate the spatial velocity of the ions in the flow v_{flow} with that angle and the given radial velocity v_{rad}: $v_{flow} = v_{rad}/sini$. That angle i and the tangential velocity v_{tan} will give us the spatial velocity of the knots: $v_{knot} = v_{tan}/cosi$, and by taking the ratio of the two we get the pattern motion of the knots relative to the ion flow.

The result shows that many of the knots are indeed moving at a pattern speed; in general they are slower than the flow. The values range from 0.2× to 1.0× the velocity of the gas particles in the flow (see Fig. 4). Often they show variations of a factor of 2 – 3 for successsive knots within one object (Eislöffel and Mundt 1992, 1994, 1996). This causes a problem for the above mentioned model of internal working surfaces, because that model predicts that the knots should move close to the flow speed and they should definitely not show such wide variations.

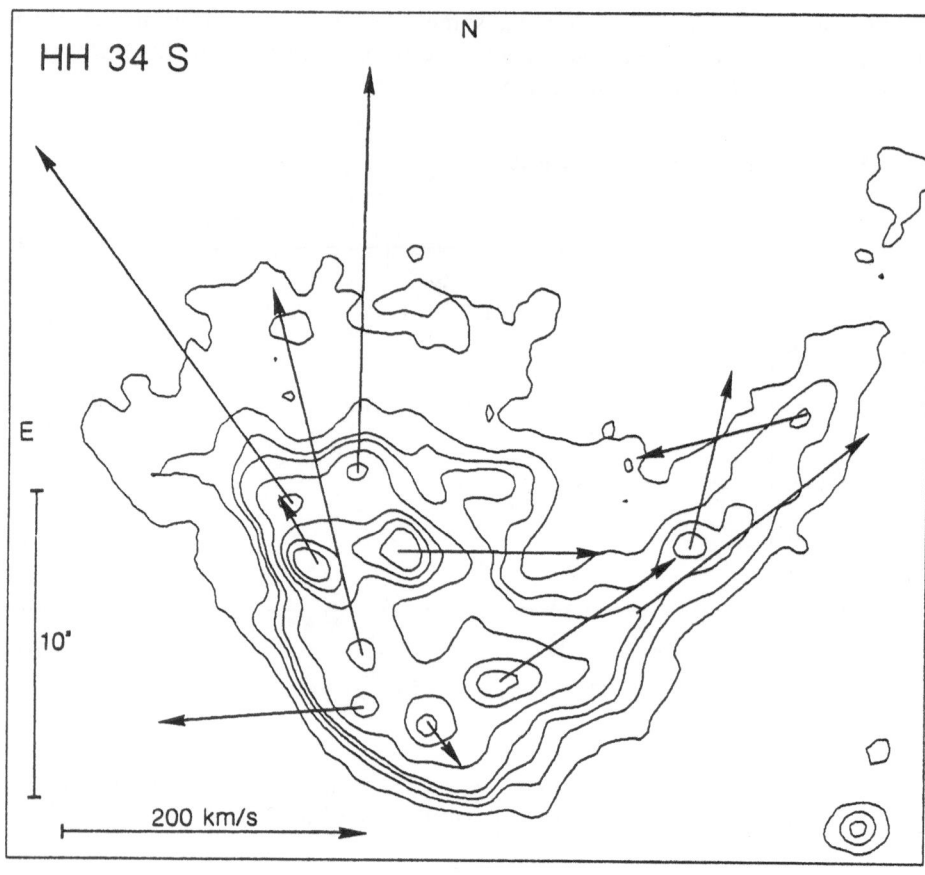

Fig. 5. Pattern motion in the HH 34S bow shock in a reference frame moving with the outflow. The flow is directed backwards and along the bow wings. Some disturbance from this general pattern is seen in the region where the jet hits into the bow (Eislöffel & Mundt 1992).

For the motion of the condensations in the bow shocks, however, a characteristic pattern motion is expected. To see it one takes the measured proper motion vectors and subtracts the vector of the propagation of the apex of the bow from all measurements. That way, the internal motion of the condensations should become apparent. For the bows that so far have been studied using that method, e.g., HH 34S in Fig. 5 (see also Eislöffel and Mundt 1992, 1994; Eislöffel, Mundt, and Böhm 1994) patterns compatible with the predictions from bow models (see, e.g., Hartigan, Raymond and Hartmann 1987; Raga et al. 1988) have been found: the flow is directed backwards and along the wings of the bows.

2.5 Variability

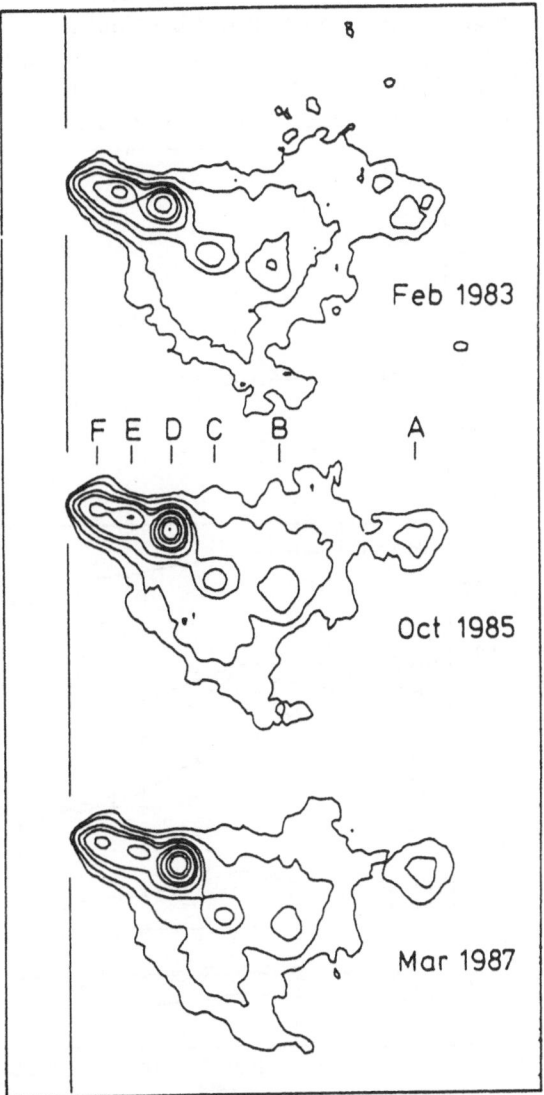

Fig. 6. A series of images of the L1551 IRS5 jet. The new knot F appeared between 1983 and 1985 (Neckel & Staude 1987).

The sources of the outflows, young embedded objects and active T Tauri stars, are found to be variable on all time scales from minutes to centuries and show variability of up to five magnitudes in FU Orionis outbursts. Consequences of this variability may also be seen in the outflows themselves. As already mentioned, episodic variations of the mass loss rate could lead to series of two or more subsequent bow shocks along a flow, which is indeed observed (see, e.g., the two north-eastern bow shocks in HH 46/47 in Fig. 1). Another consequence would be the occasional appearance of new knots close to the outflow sources. So far, however, only a single newly formed optical knot has been found: on a series of images, Neckel & Staude (1987) captured the appearance of this new knot in the jet of the deeply embedded source L1551 IRS5 when it first became visible at a projected distance of about 1".5, or about 225 AU, from the source (Fig. 6).

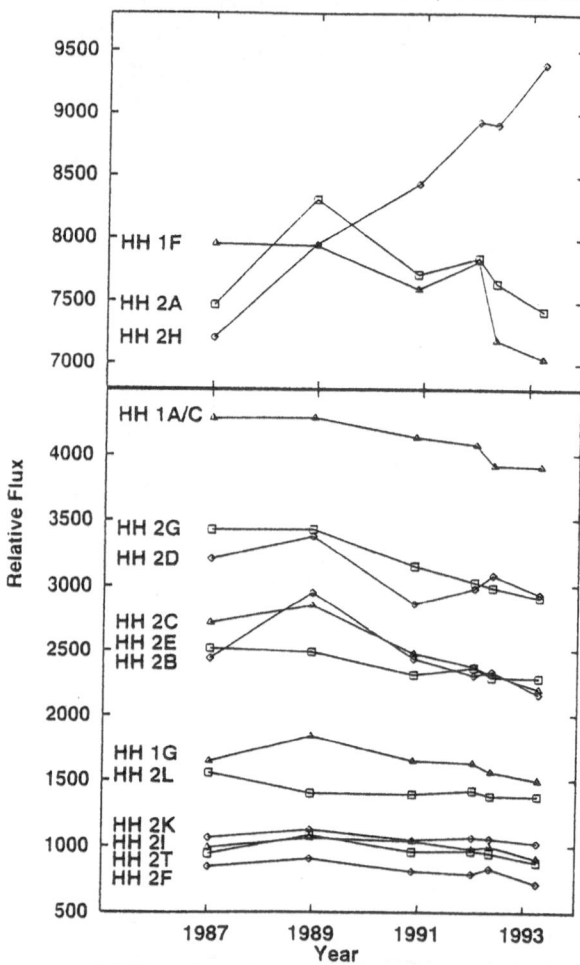

Fig. 7. Photometric variability of condensations in HH 1 and 2 in the [SII]$\lambda\lambda$6716, 6731 lines. Most knots show variability of 10 – 30 % over this six-year-period (Eislöffel, Mundt, and Böhm 1994). HH 2H, one of Herbig's (1969) new knots, now is by far the brightest knot in this outflow and still increasing in brightness, while the other new knots of Herbig (e.g. HH 2G) are fading again.

But variability is also expected for the knots along the length of the jet and for the condensations in the bow shocks. Again, only very few observations are available so far. Gross changes were reported for some condensations in the bow shock HH 2. There, Herbig saw three new knots appear in the 1940's and

1950's (Herbig 1969). These knots must have undergone a brightening by several magnitudes in order to become visible on his plates. A recent study of variability on the basis of CCD images (see Fig. 7) found brightness changes of up to 30% within six years (Eislöffel, Mundt, and Böhm 1994). HH 2H (one of Herbig's new knots) is still brightening and by now is the brightest knot in HH 1/2, while other knots that appeared in the 1940's and 1950's are fading again.

3 Optical spectroscopy

The knots in the jets and the condensations in the bow shocks are seen in the Balmer lines of hydrogen and in forbidden emission lines mainly of sulphur, oxygen, and nitrogen in several ionisation states. Usually Hα and [SII]$\lambda\lambda 6716, 6731$ are the strongest lines in the visible, but in HH 1 Solf, Böhm, and Raga (1988) found a total of 175 emission lines, also including forbidden lines of Fe, Cr, Ni, Ne, Ar, and Cl. Schwartz (1975) first identified HH objects as shock-excited gas. Low-resolution spectroscopy allows us to determine flux ratios between different lines and thereby provides a wealth of information on the excitation of the gas, shock velocities, and electron densities. High-resolution spectroscopy additionally gives us radial velocities, the line profiles and their velocity dispersion, and thereby allows us detailed insights into the kinematics of the flows.

3.1 Radial velocities

Radial velocities of the outflowing gas in the jets and their bow shocks have been studied both by long-slit spectroscopy (see, e.g., Solf, Böhm, and Raga 1986; Mundt, Brugel, and Bührke 1987; Bührke, Mundt, and Ray 1988; Reipurth 1989b; Böhm and Solf 1990) and by Fabry-Perot imaging (see, e.g., Morse et al. 1992, 1993). Often two flow components at different radial velocities are seen: a high-velocity component with radial velocities typically of the order of $100 - 150$ km s^{-1}, but in some objects reaching up to 450 km s^{-1} (in HH 32: Hartigan, Mundt, and Stocke 1986), and a low-velocity component with radial velocities typically of only a few km s^{-1} to a few tens of km s^{-1} (see Fig. 8, and, e.g., Bührke, Mundt, and Ray 1988). The high-velocity component is commonly attributed to the flow in the jet, while the low-velocity component is thought to arise in a turbulent mixing layer around the flow, where ambient material gets accelerated and entrained into the flow.

In many outflows (but not in all) there are no strong radial velocity changes seen along the flow (Mundt 1993). In these objects any velocity variations within their dynamical time scales are relatively small ($\leq 20\%$) and so the energy transport in these flows must be quite efficient. In a few jets, however, a systematic increase or decrease of the radial velocities with distance from the source is observed.

Although there are no strong systematic radial velocity changes seen in many jets, surprisingly often the outflows show different radial velocities in their flow and counterflow: in about 50% of the cases a difference of about a factor of two is found (Hirth et al. 1994). This effect is seen already very close to the source (30 – 50 AU). So this is not an acceleration/deceleration effect that takes place far from the source by interaction with the ambient medium. This effect may be attributed to different opening angles on opposite sides of the flow due to different pressure gradients (Camenzind 1993; Hirth et al. 1994), however, no in-depth modeling has been attempted so far.

3.2 Line profiles and velocity dispersion

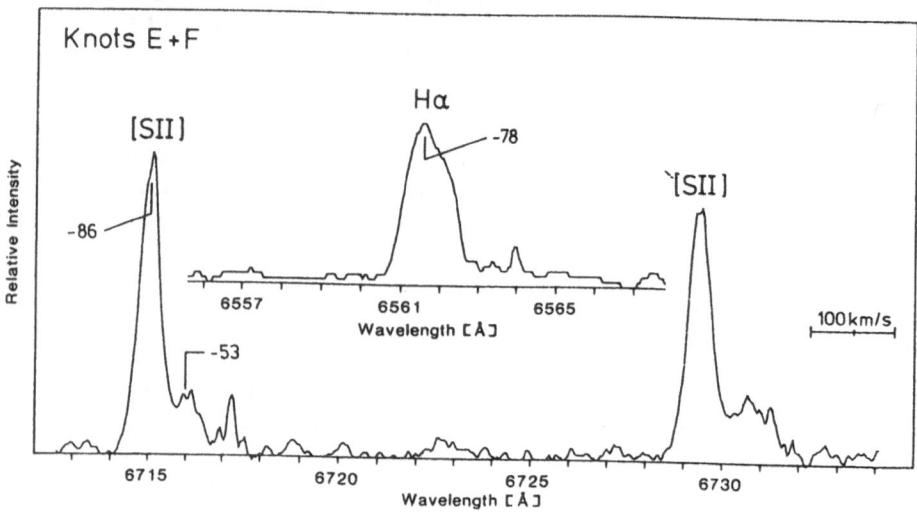

Fig. 8. Hα and [SII]λλ6716, 6731 profiles of two knots in the HH 34 jet. The high-velocity component of the flow and a pedestal at lower velocity are evident in the [SII] lines. The velocity dispersion in the Hα line is much larger than in the high-velocity components of the [SII] lines, which are unresolved at a FWHM $\leq$ 30 km s^{-1} (Bührke, Mundt, and Ray 1988).

As already discussed in the previous section, the line profiles often consist of a high-velocity component and a low-velocity component. It is also interesting to study the width of these components, which may provide us with additional information on the outflowing gas.

In the [SII]λλ6716, 6731 lines a velocity dispersion of typically 10 – 50 km s^{-1} (FWHM) is found (after correction for instrumental broadening), but values of up to 100 km s^{-1} can occur. For an example see Fig. 8. A number of reasons have been suggested to explain this line broadening (Mundt et al. 1990): Thermal broadening (for the light hydrogen atom this can lead to a dispersion of the Hα line of about 20 km s^{-1}); divergence of the flow stream lines; a turbulent boundary layer; scattering of HH emission in surrounding interstellar dust (Noriega-Crespo, Calvet and Böhm 1991). Probably several of these mechanisms, if not all, are at work at the same time in many objects.

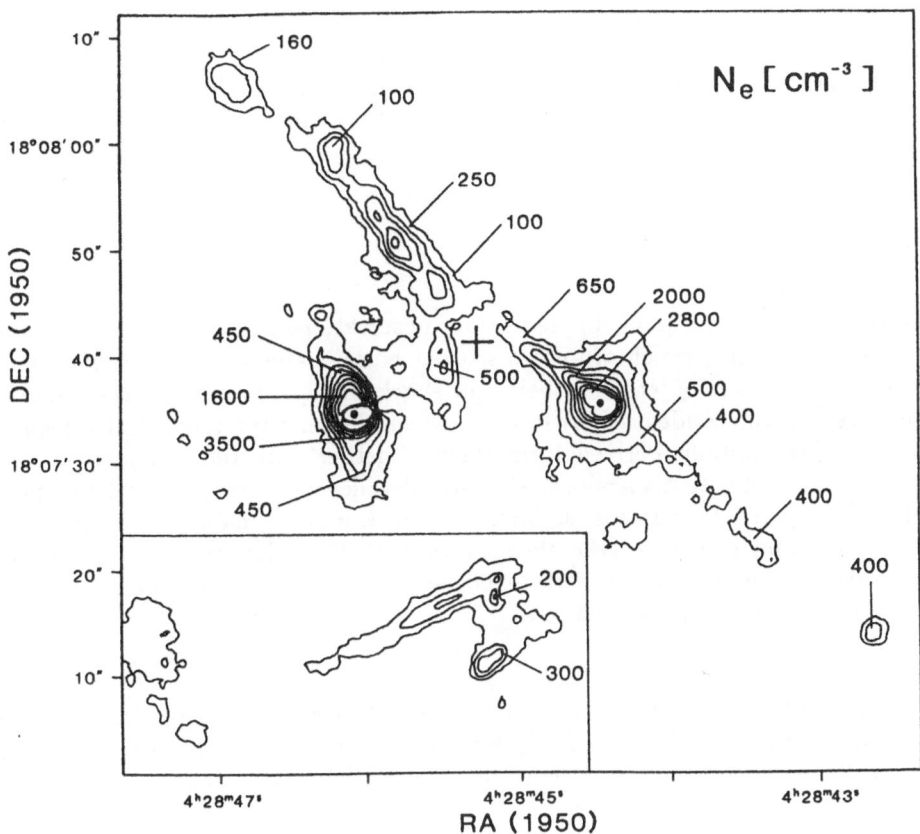

Fig. 9. Electron densities in the outflows from HL Tau (the star to the right), XZ Tau (the star to the left), and the "Hα-jet" (inset), as measured from the [SII]$\lambda\lambda$6716, 6731 line ratio (Mundt et al. 1990). One arcmin corresponds to 0.04 pc for an assumed distance of 150 pc to HL Tau.

3.3 Line ratios: excitation, shock velocities, and electron densities

Line ratios of various lines of ions or atoms with different excitational energies can be used to determine the excitation of the gas. By comparison with shock models, shock velocity and the density of the gas in the outflow can then be worked out (see Hartigan, Raymond, and Hartmann 1987; Raga et al. 1988). Shock velocities of typically 30 – 120 km s^{-1} are found. With the line ratios of molecular hydrogen in the near-infrared also slower shocks can be traced (see Davis and Smith 1995).

Also electron densities have been derived from line ratios, e.g. the ratio between [SII]λ6716 and 6731 (see, e.g., Osterbrock 1989). The electron densities range from a few tens to a few thousands cm^{-3}. See Fig. 9 for examples. They are highest near the source and fall off as one goes away from the source, which is interpreted as an effect of the increasing diameter of the flow (see, e.g., Bührke, Mundt, and Ray 1988; Mundt et al. 1990).

In fact, there is a regrettable lack of good line ratios, because so far good flux calibrated spectra have been taken only for a small number of bow shocks and very few jets.

4 Conclusions

Optical observations (and supplementary observations at UV, near-infrared and radio wavelengths, which are not discussed here) enable us to determine many crucial parameters of jets and their bow shocks. Proper motions of the knots in the jets and the condensations in the bow shocks, and the tangential velocities and pattern motions derived from them, together with radial velocities, line profiles and velocity dispersions, allow for detailed insights into the kinematics of these objects. Line ratios, as input for shock models enable us to determine excitation and density of the shocked gas. All these informations provide us with strong constraints to test the ever more detailed and more sophisticated jet models that are being worked out.

I would like to thank Reinhard Mundt and Tom Ray for many discussions on jets and outflows and for critically reading this manuscript. I am also grateful to Wolfgang Kundt for his patience in awaiting delivery of this manuscript.

References

Bachiller, R. (1996): ARA&A, in press
Bally, J., Devine, D. (1994): ApJ **428**, L65
Böhm, K.H. (1990): in "Evolution in Astrophysics", ESA SP-310, p. 23
Böhm, K.H., Solf, J. (1990): ApJ **348**, 297
Bok, B.J. (1978): PASP **90**, 489
Bührke, T., Mundt, R., Ray, T.P. (1988): A&A **200**, 99
Camenzind, M. (1993): in "Stellar Jets and Bipolar Outflows", ed. L. Errico and A. Vittone (Kluwer, Dordrecht) p. 289
Cudworth, K.M., Herbig, G.H. (1979): AJ **84**, 548
Davis, C.J., Smith, M.D. (1995): A&A, in press
Dopita, M.A., Schwartz, R.D., Evans, I. (1982): ApJ **263**, L73
Eislöffel, J., Mundt, R. (1992): A&A **263**, 292

Eislöffel, J., Mundt, R. (1994): A&A **284**, 530

Eislöffel, J., Mundt, R. (1996): in preparation

Eislöffel, J., Mundt, R., Böhm, K.H. (1994): AJ **108**, 1042

Falle, S.A.E.G., Wilson, M.J. (1985): MNRAS **216**, 79

Hartigan, P., Mundt, R., Stocke, J. (1986): AJ **91**, 1357

Hartigan, P., Raymond, J., Hartmann, L. (1987): ApJ **316**, 323

Hartigan, P., Raymond, J., Meaburn, J. (1990): ApJ **362**, 624

Hartigan, P., Raymond, J. (1993): ApJ **409**,705

Heathcote, S., Reipurth, B. (1992): ApJ **104**, 2193

Herbig, G.H. (1969): in IAU Coll. "Non-Periodic Phenomena in Variable Stars", ed. L. Detre, Reidel, Dordrecht, p.75

Herbig, G.H., Jones, B.F. (1981): AJ **86**, 1232

Herbig, G.H., Jones, B.F. (1983): AJ **88**, 1040

Hirth, G.A., Mundt, R., Solf, J. (1994): A&A **285**, 929

Hirth, G.A., Mundt, R., Solf, J., Ray, T.P. (1994): ApJ **427**, L99

Morse, J.A., Hartigan, P., Cecil, G., Raymond, J.C., Heathcote, S. (1992): ApJ **231**, 198

Morse, J.A., Heathcote, S., Cecil, G., Hartigan, P., Raymond, J.C. (1993): ApJ **410**, 764

Mundt, R., Fried, J.W. (1983): ApJ **274**, L83

Mundt, R., Bührke, T., Fried, J.W., Neckel, T., Sarcander, M., Stocke, J. (1984): A&A **140**, 17

Mundt, R., Brugel, E.W., Bührke, T. (1987): ApJ **319**, 275

Mundt, R. (1988): in "Formation and Evolution of Low Mass Stars", eds. A.K. Dupree, M.T.V.T. Lago (Kluwer, Dordrecht), p. 252

Mundt, R., Ray, T.P., Bührke, T., Raga, A.C., Solf, J. (1990): A&A **232**, 37

Mundt, R., Ray, T.P., Raga, A.C. (1991): A&A **252**, 740

Mundt, R. (1993): in "Stellar Jets and Bipolar Outflows", eds. L. Errico, A.A. Vittone (Kluwer, Dordrecht), p. 91

Noriega-Crespo, A., Calvet, N., Böhm, K.H. (1991): ApJ **379**, 676

Neckel, T., Staude, H.J. (1987): ApJ **322**, L27

Osterbrock, D.E. (1989): "Astrophysics of Gaseous Nebulae and Active Galactic Nuclei" (Mill Valley, California)

Raga, A.C., Mateo, M., Böhm, K.H., Solf, J. (1988): AJ **95**, 1783

Raga, A.C., Kofmann, L. (1992): ApJ **386**, 222

Reipurth, B. (1989a): A&A **220**, 249

Reipurth, B. (1989b): in "Low Mass Star Formation and Pre-Main Sequence Objects", ed. B. Reipurth, (ESO, Garching), p. 247

Reipurth, B., Heathcote, S. (1992): A&A **257**, 693

Schwartz, R.D. (1975): ApJ **195**, 631

Schwartz, R.D. (1977): ApJS **35**, 161

Schwartz, R.D., Jones, B.F., Sirk, M. (1984): AJ **89**, 1735

Solf, J., Böhm, K.H., Raga, A.C. (1986): ApJ **305**, 795

Solf, J., Böhm, K.H., Raga, A.C. (1988): ApJ **334**, 229

Near-infrared imaging in H_2 of molecular (CO) outflows from young stars

Jochen Eislöffel [1], Christopher J. Davis [2]

[1]Laboratoire d'Astrophysique, Observatoire de Grenoble, B.P. 53X,
F-38041 Grenoble Cedex, France
[2]Max-Planck-Institut für Astronomie, Königstuhl 17, D-69117
Heidelberg, Germany

Abstract: We have imaged several known molecular (CO) outflows in H_2 v=1-0 S(1) and wide-band K in order to identify the molecular shocks associated with the acceleration of ambient gas by outflows from young stars. We detected H_2 line emission in all the flows we observed: L 1157, VLA 1623, NGC 6334I, NGC 2264G, L 1641N and Haro 4-255. A comparison of the H_2 data with CO outflow maps strongly suggests that "prompt entrainment" near the head of a collimated jet probably is the dominant mechanism for producing the CO outflows in these low-mass sources. Contrarily, in the outflow from DR 21, one of the most energetic outflows known from a high-mass source, turbulent motions along the interface of the flow with the ambient medium seem to play a far more important role than in its low-mass counterparts.

1 Introduction

Recent near-infrared imaging has revealed extended H_2 emission in a number of Herbig-Haro (HH) outflows and also in a few CO outflows. This opens up a new possibility to study the varying levels of excitation in the shocks throughout each system, because H_2 traces even weaker shocks than the low-excitation optical features seen in Herbig-Haro objects (i.e. those strong in [SII]$\lambda\lambda$6716,6731 or [OI]$\lambda\lambda$6300,6363); for an example see HH46/47 (Eislöffel et al. 1994). Contrary to the optically visible HH outflows, molecular CO outflows are frequently still deeply embedded in the parental cloud. Then H_2 emission is often the only useful tracer of shocks in the flow that can readily be observed from the ground. Very few molecular outflows have been studied in the near-infrared so far and in these the H_2 emission seems to be closely associated with the CO outflows. Therefore it was our goal to investigate this relationship further.

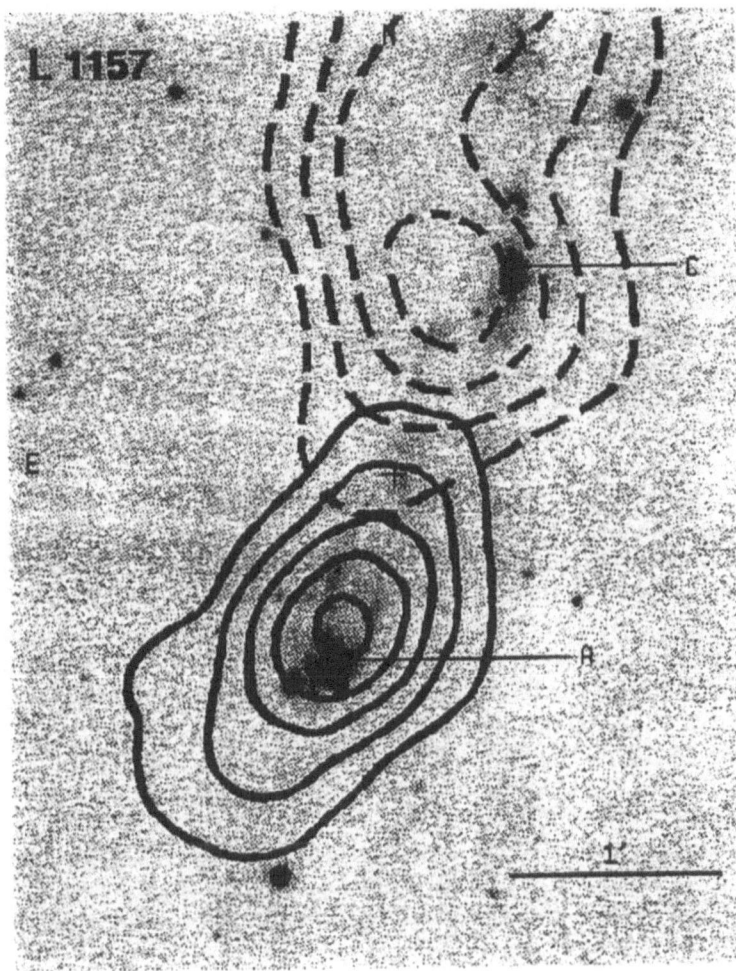

Fig. 1. An H_2 v=1-0 S(1) + continuum image with, superimposed, a CO J = 1–0 map (Umemoto et al. 1992) of the L 1157 outflow region. The position of IRAS 20386+6751 is marked with a cross. The blue-shifted CO contours are drawn in full, the red-shifted ones dashed. The main H_2 line emission features are labelled A and C. The length of the scale bar corresponds to 0.13 pc for an assumed distance of 440 pc.

2 Observations of H_2 emission in CO outflows

We imaged the known molecular CO outflows L 1157, VLA 1623, NGC 6334I, NGC 2264G, L 1641N and Haro 4-255 FIR in the H_2 v=1-0 S(1) line and in wideband K. In all of these flows did we find H_2 line emission, which in general was closely related to the features in the CO outflows. Here we show an H_2 image of the L1157 molecular outflow with, superimposed, a contour plot of the associated CO outflow. Images of the other objects and a more detailed discussion can be found in our paper (Davis and Eislöffel 1995).

In L1157 two bright, bow-shaped H_2 features are seen at a distance of about 1′ north and south of the IRAS 20386+6751 source (Fig. 1). The apices of these "bow shocks" point away from the source, as would be expected for shocks driven into the ambient cloud by a collimated outflow. Apparently these "bow shocks" are situated very close to peaks in the CO outflow maps, indicating a possible relationship between the two.

In the NGC 6334I image much of the spread-out emission in the NGC 6334I cluster is continuum. However, two compact line-emission knots, that are somewhat bow-shaped in appearance, are observed just ahead of the peaks in the blue- and redshifted CO outflow lobes.

In the VLA 1623 outflow the peak in the blue-shifted CO flow lies just behind of (i.e. closer to the source than) a bright, compact, and somewhat "bow-shaped" H_2 knot. This bow is most likely associated with the outflow, as indicated by its head pointing away from the VLA 1623 source. Along the north-western CO flow lobe a number of additional, fainter H_2 knots are seen, which are similarly observed near peaks in the CO outflow.

In the NGC 2264G outflow we found H_2 line-emission in both the blue- and red-shifted CO lobes. While the peak in the blue-shifted CO flow is marked by a compact cluster of H_2 knots, the northern edge of the red-shifted CO flow is outlined by a knotty filament of H_2 emission.

In the L 1641N molecular outflow faint knots of H_2 emission fill much of the space occupied by the blue shifted CO lobe. The overall H_2 morphology suggests a cavity associated with the outflow.

A faint bow-shaped H_2 knot is observed near to the peak in the blue-shifted lobe of the Haro 4-255 FIR outflow. Additional H_2 knots are found closer to the FIR source of the flow; these features clearly delineate the flow direction.

3 Relationship between the H_2 and the CO outflows

In the L 1157, NGC 6334I, VLA 1623, and Haro 4-255 FIR outflows we observed bow-shaped shocks in H_2 and in NGC2264G a compact group of knots just ahead of, i.e. downstream of peaks in the CO flows. In all cases the "bow shocks" are pointing away from the sources as would be expected if they were associated with working surface regions in the flows. It seems likely that the CO peaks in these flows represent clumps of molecular gas that have been swept up by the bow shocks, in a way analogous to a snow-plough. *The CO outflow hence would appear to consist of material entrained by the bow shock at the head of the collimated jet and not of a low collimation flow ejected from the source.*

In NGC 2264G, and perhaps also in L 1641N, H_2 emission outlines also the outer edges of the flows. We interpret these as evidence for a turbulent boundary layer between the flows and the ambient gas in which additional ambient material is entrained into the flow.

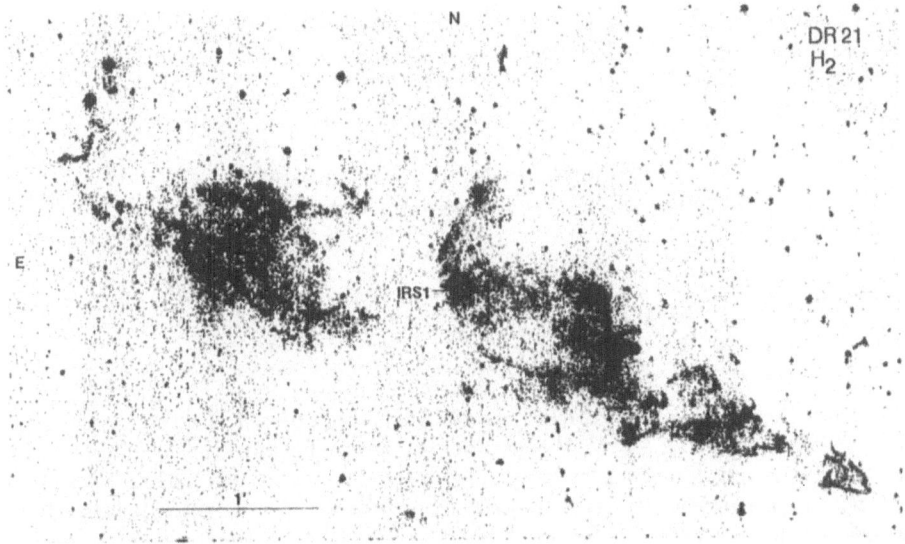

Fig. 2. Narrow-band H_2 + continuum image of the DR 21 outflow. A molecular cloud core (just to the east of IRS 1) separates the two outflow lobes, and harbours a number of ultra-compact HII regions. One of these young, massive stars likely drives this spectacular outflow. The length of the scale bar corresponds to 0.85 pc for an assumed distance of about 3000 pc.

Finally, for comparison with the outflows from low-mass sources discussed above, in Fig. 2 we show a new H_2 image of one of the most massive and most energetic YSO outflows known, DR 21. The complex nature of this outflow is clearly evident. Indeed, turbulent gas motions at the interface between the flow and the ambient medium seem to play a far more important role in this outflow than in its low-mass counterparts. A more detailed discussion of these data, and recently obtained H_2 line profiles, is given in Davis and Smith 1996.

We wish to thank Mike Smith and Reinhard Mundt for much time spent discussing jets and outflows.

References

Davis, C.J., Eislöffel, J. (1995): A&A **300**, 851
Davis, C.J., Smith, M.D. (1996): A&A, submitted
Eislöffel, J. Davis, C.J., Ray, T.P., Mundt, R. (1994): ApJ **422**, L91
Umemoto, T. et al. (1992): ApJ **392**, L83

The Jets in SS 433

René C. Vermeulen

California Institute of Technology,
Astronomy 105–24, Pasadena, CA 91125, USA

Abstract: The unique $0.26\,c$ stellar jets of SS 433 are reviewed, starting at the binary system, and then, using a natural separation by observing wavelength, following them out to successively larger scales over eight orders of magnitude. Recent work is emphasized in this review, and care is taken to provide up-to-date entry points into the extensive older literature on SS 433. This is an update of a similar review presented three years ago (Vermeulen 1993).

1. Introduction

The jets of SS 433 are rather remarkable in that they seem to occupy a part of parameter space which is intermediate between that of the "traditional" stellar jets found in young stellar objects (YSO), and that of the jets in active galactic nuclei (AGN). Matter moving at the bulk speed in the jets emits both thermal radiation, which predominates in YSO, and non-thermal radiation, as in the radio jets of AGN. The jet velocity in SS 433 is much higher than in YSO, but not ultra-relativistic, as in some AGN jets. The jets of SS 433 are very well collimated, a characteristic more commonly found amongst AGN jets than in stellar objects. The jets of SS 433, like those in AGN, originate near a compact object, in contrast to the jets in YSO. One of the main attractions of SS 433, the possibility of studying the dynamics of relativistic jets on a short timescale, is well illustrated by Fig. 1.

Jets in certain types of X-ray binaries may turn out to be a category all by themselves, containing not only objects such as SS 433 and (perhaps) Cyg X-3 (Spencer et al. 1986, Molnar, Reid, & Grindlay 1988, Schalinski, Johnston, & Witzel 1989, Strom, van Paradijs, & van der Klis 1989, van Kerkwijk et al. 1992), but also the fascinating recently discovered Galactic superluminals 1915+105 (Mirabel & Rodrígues 1994) and 1655−40 (Tingay et al. 1995, Hjellming & Rupen 1995). There is no scope here for a discussion of the physical relationships between these objects, but see the review by Mirabel (1996).

Searches for objects similar to SS 433 were started almost as soon as its peculiar nature was recognised. As remarked by Margon (1984), given the relatively

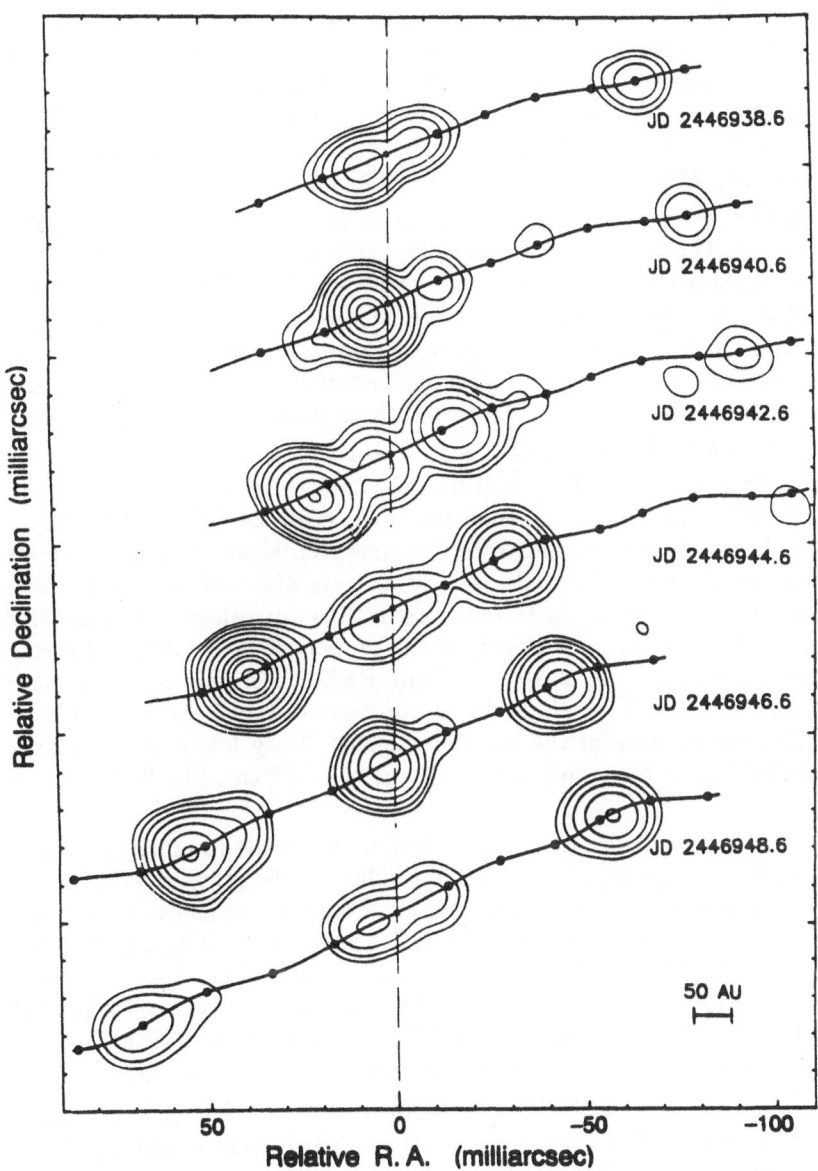

Figure 1. *European VLBI Network images of SS 433 at 5 GHz, observed at two-day intervals in 1987 by Vermeulen et al. (1993a). The same absolute contour levels are shown for all six images: 2, 4, 8, 16, 32, 60, 100, 140, 180, 240 mJy/beam. The same circular restoring beam was used for all images (10 mas or ∼50 AU FWHM, shown in the lower right-hand corner). In keeping with their time spacing, the images are displayed with equal vertical offsets between the adopted centres (shown by the vertical dashed line). The locus of emission predicted by the kinematic model, including nodding motion, is drawn through each image. Markers indicate ejection age intervals of 2 days along that locus.*

large distance of SS 433 (~5 kpc) and, conversely, its relatively large optical and radio luminosity, it is clear that objects like SS 433 are comparatively rare, or else another clear case would already have been discovered closer to us. The one tentative remaining SS 433 candidate from the list proposed by Ryle et al. (1978) is Cir X-1 (e.g. Stewart et al. 1993). Searches for SS 433-like objects in M 33 are currently underway (Fabrika & Sholukhova 1995, Calzetti et al. 1995).

An excellent overview of the first five years of studying SS 433 can be found in Margon (1984). The basic picture of SS 433 sketched there is still valid; Kundt (e.g. 1991) is probably the only one remaining who has a very different view of the system. The present review is far from comprehensive; it is intended to highlight some of the more recent work, and to provide up-to-date entry points into the extensive literature on SS 433 by references to many of the most recent papers. This is an update of a similar review presented three years ago (Vermeulen 1993).

The jets of SS 433 can be studied over at least eight orders of magnitude in linear size. In this review, they will be described roughly from their origin within the binary system to their termination at the radio shell of W 50. Section 2 focuses on the binary system, to illustrate that much remains to be learned about the central engine. A review of recent fascinating X-ray observations which probe the jets on scales $\leq 10^{12}$ cm, closer to their origin than any other observations, is given in Sect. 3. Section 4 is a summary of the properties of the optical moving line emission, which originates at 10^{14}–10^{15} cm. Radio observations, which trace the jets roughly from 10^{14} cm to 10^{17} cm, are described in Sect. 5. Section 6 deals with the continuation of the jets as extended X-ray lobes, and with the supernova radio shell W 50, which has an extent of $\sim 10^{20}$ cm. Finally, Sect. 7 is a short summary.

The division by size scale used in this review also leads roughly to a separation by observing frequency, from X-ray to radio wavelengths, at least for the inner jets. Ultraviolet wavelengths are missing from the sequence, on account of the high interstellar extinction towards SS 433 ($A_V \approx 8$ mag, e.g. Murdin, Clark, & Martin 1980). There were some early reports of moving γ-ray lines (Lamb et al. 1983), but their occurrence, which would have been very hard to understand theoretically (e.g. Brown et al. 1988, and references therein), has not been confirmed (see Geldzahler & Geller 1994 and references therein.).

Discussions of the jets of SS 433 invariably refer to the so-called kinematic model, which assumes ballistic motion of matter in narrow anti-parallel precessing jets, and which was first inferred from the systematic changes in the Doppler shift of some of the optical emission lines (Fabian & Rees 1979, Milgrom 1979). The moving spectral lines result because each new portion of the jets only radiates for a short period (see Sect. 4). In radio emission, which persists for a longer time, the projection onto the plane of the sky leads to a characteristic "corkscrew". However, it is worthwhile to stress that there is no motion along this curve: each part of it moves in a straight line away from the centre (see Sect. 5). Radio imaging (e.g. Hjellming & Johnston 1981) allows a full determination of the kinematics of the jets completely independently from optical spectroscopy, and subject only to the assumption of 2-sided symmetry, for which there is ample morphological, kinematic, and spectral evidence. The fact that an entirely consistent set of parameter values is obtained is, I think, compelling evidence for the validity of this model, and renders more complex hypotheses, such as the orthogonal origin of the radio jets and the optical line emission proposed by Kundt (e.g. 1991) rather unlikely. Indeed, radio imaging, which yields two components of the motion rather than only the radial velocity, allows a resolution of the redundancies left in the spectroscopic kinematic model (e.g. Hjellming & Johnston 1981). Furthermore, by using travel time differences, radio imaging also yields the distance to SS 433, which is $\sim$4.9 kpc (e.g. Vermeulen et al. 1993a). The jet velocity is $\sim$0.26 c, the precession period is $\sim$164 days, and the precession cone has a half-opening angle of $\sim$20° (e.g. Margon & Anderson 1989). The cone axis which points (slightly) towards us, at an inclination of $\sim$80° to the line of sight, is to the east, projected onto the plane of the sky in a position angle of $\sim$100°. The eastern jet precesses clockwise. The jets are collimated to better than $\sim$4° (e.g. Milgrom. Anderson, & Margon 1982). Superimposed on the precession is a $\sim$6 day nodding motion, which results from the interplay of the precession and orbital ($\sim$13 day) periods, but its origin is not well understood and still in dispute; it is discussed, for example, by Katz et al. (1982), Collins & Newsom (1986), Margon & Anderson (1989), and Collins & Garasi (1994). In Vermeulen (1989, p.9) I have extended the radial velocity nodding formalism of Katz et al. (1982) to three dimensions. While, to first approximation, the precession and nodding periods are stable, there is some evidence that the precession period is modulated in a cycle with a length of several years (Baykal, Anderson, & Margon 1993).

2. The Central Engine

SS 433 is an eclipsing binary system with a $\sim$13.1 day period (for ephemeris see e.g. Kemp et al. 1986 and Gladyshev, Goranskij, & Cherepashchuk 1987). In all likelihood it consists of a compact object surrounded by an accretion disk, and an early-type star undergoing copious mass loss (e.g. van den Heuvel 1981), although Fabian et al. (1986) have discussed a triple star system. It has proven difficult to obtain reliable constraints on the system. There are large bodies of broadband photometric data (see Leibowitz 1984, Mazeh et al. 1987, Cherepashchuk & Yarikov 1991, Zwitter, Calvani, & D'Odorico 1991, and references therein), as well as spectrophotometric data (see Anderson, Margon, & Grandi 1983, Wagner 1986, and references therein). However, there is substantial variability, both random and in the form of flares (e.g. Aslanov et al. 1993). Although Chakrabarti & Matsuda (1992) have suggested spiral shocks in the accretion disk as a possible cause for sub-day photometric variability, it is probably at least in part related to a dense outflowing atmosphere enveloping the system. The presence of shrouding matter is also suggested by X-ray data (Zwitter et al. 1991, Brinkmann et al. 1991, Yuan et al. 1995, see Sect. 3), by optical spectroscopy (e.g. Vermeulen et al. 1993b), and by the infrared spectrum (Band 1987, and references therein).

Stewart et al. (1987) first showed that the primary optical eclipse coincides with an X-ray eclipse (see also Sect. 3), demonstrating that the accretion disk is the dominant source of optical continuum radiation. Given the way in which the V-band magnitude varies over the precession period (e.g. Mazeh et al. 1987), the accretion disk is likely to be geometrically thick. Several groups have attempted to use the duration and shape of the eclipses both in visible light and in X-rays to derive the configuration of the binary system (e.g. Zwitter et al. 1991, Antokhina, Seifina, & Cherepashchuk 1992, Sanbuichi & Fukue 1993, and references therein). Quoted values for the mass ratio (q) between the compact object and the mass-losing star have ranged widely; $q \approx 0.25$ (i.e. mass-losing star four times as massive as compact object) seems popular, but contradictions remain, which must be due largely to the difficulty of modeling the evidently complex geometry of the emitting regions.

Some of the optical emission lines (in the so-called stationary spectrum, i.e. not originating in the jets) show a radial velocity curve with the binary orbital period. Several groups have used this phenomenon, using the He II $\lambda 4686$ line in particular, to estimate the mass function (K) of the SS 433 system; see Crampton & Hutchings (1981), Fabrika & Bychkova (1990), and D'Odorico et al. (1991). However, the results have ranged all the way from $K = 190$ kms^{-1} to $K = 112$ kms^{-1}, and seem to depend on how the very complex line profile is analysed. Filippenko et al. (1988) have suggested that double-peaked Paschen line profiles may be the signature of Keplerian rotation in the accretion disk. However, from a comparison of Paschen, Balmer, and He II $\lambda 4686$ profiles, Vermeulen et al. (1993b) suggest that the radial velocity of all these stationary lines could be affected by eclipse phenomena, as well as by emission from outflowing matter. At present, I therefore believe that the claimed values for the mass function should be interpreted with a great deal of caution, especially when used to decide "weighty issues" such as whether SS 433 contains a neutron star or a black hole.

If SS 433 contains a neutron star, it has not (yet ?) revealed itself as a pulsar. Lebedev & Pimonov (1981) have placed stringent limits on the occurrence of periodic optical fluctuations on timescales <1 second. Rand et al. (1988) have found no evidence for coherent radio emission with variability timescales between 4 and 16 seconds, but did not rule out more rapid radio pulsations.

The structure and dynamics of the accretion disk is still poorly understood. In-depth theoretical modeling must await the availability of sufficiently fast three-dimensional codes. It is widely thought that at least the inner part of the accretion disk is perpendicular to the jets at all times, thus providing a symmetry plane which may assist in the collimation of the jets. Hence, the inner disk would partake in the 6-day nodding motion as well as in the 164-day precession. This is most often ascribed to the slaved disk mechanism (e.g. van den Heuvel, Ostriker, & Petterson 1980, Katz et al. 1982, Matese & Whitmire 1982), but for a different view see Collins & Garasi (1994). Such mechanisms only work if the flow of matter through the disk is sufficiently rapid that matter can preserve the angular momentum imparted to it when it left the mass-losing star. Of course, this still leaves the unsolved dynamical problem of making a star precess and nod.

While a binary system at least provides a plausible location in which the release of energy through accretion could lead to the violent bi-polar expulsion of matter, much of the actual physics of the "jet engine" remains the domain of speculation. Observations impose severe constraints on the jets to be produced. They are two-sided and antiparallel. They are subject to 164-day precession and to 6-day nodding motion, and they seem to show other random variations of direction, without loss of collimation (probably better than 4°). Their velocity must be $0.26\,c$, constant to a few percent. However, I will argue in Sect. 4 that the fact that the optical lines are relatively narrow does not absolutely rule out the existence of a broader underlying jet. The radio emission does not have enough compact structure to constrain the jet collimation very well, at least in the inner 10^{15} cm, (see Sect. 5). The best constraints will come from X-ray observations, but the evidence is still ambiguous (see Sect. 3). Arav & Begelman (1993) have developed an interesting model in which the visible, dense portion of the jet is surrounded by a broader, low-density cocoon.

If SS 433 contains a neutron star, it could perhaps more easily harbour a substantial magnetic field than if there is a black hole in this system. While that may be attractive, given that magnetically driven jets currently seem to be in vogue, radiation pressure may provide better jet engine scenarios in SS 433, since it is quite probable that in this object the accretion rate exceeds the Eddington rate, perhaps by a factor of ten ($\dot{M} = 10^{-5}$–10^{-4} $M_\odot$ yr^{-1}, van den Heuvel et al. 1980). Studies concerning radiative acceleration in SS 433 have been done for example by Bodo et al. (1985), Eggum, Coroniti, & Katz (1985), and Peter & Eichler (1995), and see also the work of Icke (1989). Of course, there is a well-known problem: in the suggested thick disk environment, efficient acceleration and efficient collimation seem to be mutually exclusive. The acceleration efficiency can increase if the matter is highly clumped (see also Katz 1986, 1987). Indeed, Begelman & Rees (1984) suggested a scenario involving the violent mixing of incoming and outgoing matter and radiation in a "cauldron".

At the jet velocity of $0.26\,c$, the Lyman continuum edge in the spectrum is Doppler shifted almost exactly to the rest wavelength of Lyα (e.g. Milgrom 1979). However, it is unlikely that line-locking can be the dominant source of acceleration for the jets of SS 433, unless they are highly clumped as well as highly enriched with iron, such that line-locking could work on hydrogenic iron ions (e.g. Pekarevich, Piran, & Shaham 1984, Shapiro, Milgrom, & Rees 1986). Thus, while Lyα line-locking may play a role in fine-tuning the terminal jet velocity, it is probably not the dominant acceleration mechanism.

3. X-ray Observations of the Inner Jets

X-ray observations have contributed significantly to our knowledge of the inner jets of SS 433. All of the more recent work, starting with Exosat (Watson et al. 1986), has supported the idea that the X-rays are at least predominantly due to thermal bremsstrahlung from the inner jets themselves, and this now seems firmly established.

The most impressive spectroscopic observations to date are those recently obtained with ASCA by Kotani et al. (1994). They have identified at least 18 different lines, 9 from each jet, all at a Doppler shift corresponding closely to the predictions of the optical kinematic model. Every indication is still that the jets are fully accelerated and collimated by the time they are visible in X-rays ($\lesssim 10^{12}$ cm). Kotani et al. (1994) show that the previously observed "iron line" actually has contributions from two different ionization states, Fe xxv Kα and Fe xxvi Kα, and that additional contributions from Ni xxvii Kα and possibly Fe xxv Kβ may also have been part of the blend in older data. These ASCA observations were taken when the jets of SS 433 were fairly closely pointed into the plane of the sky. At that precession phase there is no evidence for the broad line component fitted to earlier Ginga data (Brinkmann et al. 1991). Also, both jets can be detected, in contrast to the earlier view that only the jet on the approaching side was visible at any time. An analysis by Yuan et al. (1995) of multiple Ginga observations, which had poorer spectral resolution, shows that at the same precession phase as the ASCA data a fit with multiple narrow lines to the older data is indeed possible. Curiously, however, Yuan et al. (1995) maintain that at larger inclinations from the plane of the sky, only the approaching jet will be prominently visible, and that a broad line component may yet show up.

Several (partial) X-ray eclipses have been observed, first by Stewart et al. (1987). The spectrum softens in eclipse, indicative of a temperature gradient outward along the jets. Attempts have been made to constrain the geometry of the binary system by using the length and depth of the eclipses. However, the analyses of Ginga data by Brinkmann et al. (1991) and Yuan et al. (1995) have shown that the situation is rather more complicated than was first assumed. There is a suggestion of opaque matter in the binary system which obscures part of the X-ray jet(s) at orbital phases other than the primary minimum. However, a long track covering a substantial portion of an orbital cycle including both ingress and egress from all claimed eclipses would be highly desirable before firm conclusions are drawn about the geometry of the binary system.

There are no pronounced X-ray intensity variations on timescales of hours (e.g. Yuan et al. 1995), which suggests that the jets have a more or less steady outflow, and do not show X-ray events corresponding to the ejection of "bullets", such as the entities seen in optical emission lines (see Sect. 4). On the other hand, modeling of the emissivity and temperature profile by Brinkmann et al. (1988) suggests that the X-ray emitting portions of the jets may consist of a large number of very small clumps, for example with a radius of $\sim 10^8$ cm, and a particle density of $\sim 10^{14}$ cm^{-3}. Such clumps would be produced at a rate of

$\sim 10^5$ per second. It is interesting to note that highly clumped jets may also enhance the acceleration process of the jets (see Sect. 2).

It is clear that further X-ray observations of SS 433 will be of the utmost importance in studying the geometry of the inner jets and the mass-transferring binary, and to obtain the best possible constraints on the distance at which the jets are fully accelerated and collimated.

4. The Optical Line-Emitting Region

The moving emission lines originate at a distance of 10^{14}–10^{15} cm from the binary system. The literature on this subject is much too extensive to review here. Instead, as a summary of the observational data, I will present the results of our 1987 intensive monitoring campaign, which has confirmed and extended much of the earlier work. The evolution of the moving emission lines was followed for several weeks on timescales of hours; 28 blueshifted and 26 redshifted Hα line components were found. The resultant paper by Vermeulen et al. (1993b), complemented by the review of Margon (1984), may be consulted for extensive further references to papers on the moving lines (as well as on the other optical spectral components).

The (changing) Doppler shift of the emission lines allows the SS 433 jet kinematics to be derived with an accuracy unparalleled in any other object (e.g. Margon & Anderson 1989). However, the analysis is significantly complicated by the fact that, in individual measurements, the Doppler shifts may "jitter" by several thousand $km s^{-1}$ around the value predicted by the kinematic model; a recent analysis is given by Baykal et al. (1993), but note that Collins & Garasi (1994) have a dissenting view. The jitter phenomenon is likely to be related at least in part to the fact that the lines do not shift smoothly. Instead, there seem to be separate "bullets" (e.g. Grandi 1981), parcels of matter moving ballistically at $0.26\,c$, each of which gives rise to a discrete line component which brightens and fades at a particular wavelength corresponding to its projected radial velocity. The occurrence of bullets is not correlated on a one-to-one basis between the approaching and the receding jet, but, statistically, the bullet properties are indistinguishable between the two jets, after taking into account the effects of orientation. The lines are sometimes temporarily absent for several days; these so-called "switch-offs" do occur in both jets at the same time (Vermeulen et al. 1993b; see also Iijima 1993).

The individual line profiles have reasonably well-defined parameters: a line width of $\sim 1700\ km s^{-1}$ (FWHM), and a maximum equivalent width of ~ 25 Å at 6797 Å (Hα at the transverse Doppler shift). Doppler boosting and de-boosting clearly affect both parameters; the average ratios found between the two jets are in good agreement with those expected for optically thin bullets. Substantial individual variations in bullet lifetimes have thus far masked the expected light travel time difference between the approaching and receding jet. The lines typically take ~ 10 hours to reach their maximum brightness. The total radiative lifetime is typically ~ 2 days, but some bullets persist for up to 6 days.

Hence, since the bullet generation interval is typically <1 day, more than one line component is usually visible in the spectrum at any given time. Occasional rather longer intervals between bullets lead to temporary switch-offs. Some of the brightest lines occur at the nodding cusps in radial velocity, which suggests that projection effects play a role, i.e. that not all "bullets" are single physical entities which have a fixed size. However, it is likely that bullets do form the "basic units" to emit the moving lines, since their properties, as summarised above from the work of Vermeulen et al. (1993b), seem to be reasonably well-defined, and since there is real intermittence in the optical lines, with more than one line often being present at the same time.

The intermittence of the optical moving lines contrasts with the continuity of the jets both in X-rays (Sect. 3) and in radio emission (Sect. 5). It is possible that the moving lines are due to another, differently distributed, particle population than the matter which gives rise to the X-ray and radio emission. The jets could have a two-phase composition, with cool condensations (10^4 K) embedded in a hotter, more continuous medium. It should be pointed out that, if only part of the cross-section or volume of the jets partakes in the optical line emission, then the relatively small line width may not necessarily be strong evidence for a very tight collimation of the jets. In fact, the jitter in ejection angle shown by individual bullets could be interpreted to indicate that they occupy only a part of the cross-section of rather broader jets, as suggested in Vermeulen et al. (1993b).

The Balmer series predominates in the moving line spectrum, as it does in the stationary lines. Shifted He I lines (particularly $\lambda 5876$) are also often seen, as well as Paschen and Brackett lines in the infrared. However, shifted forbidden lines, and significantly, He II $\lambda 4686$, have not been observed (e.g. Vermeulen et al. 1993b). Hence, the line emitting matter in the jets has a density $n_e > 10^6$ cm^{-3}, and a temperature of $\sim 10^4$ K. In order to allow the observed Hα luminosity ($\sim 10^{35}$ erg s^{-1}) to be emitted without requiring a kinetic luminosity far in excess of 10^{40} erg s^{-1}, it is often thought that the line emitting matter must be clumped (e.g. Begelman et al. 1980, Davidson & McCray 1980, Bodo et al. 1985), since the radiation efficiency per unit mass increases with density.

One could hypothesize that these clumps are identical to the clumps of radius $\sim 10^8$ cm discussed in the X-ray analysis of Brinkmann et al. (1988). In that case the optical bullets could be due to modulations, on a scale of $\sim 10^{14}$ cm, of some property of these clumps; for example their number density in the jets, or their mean internal density or temperature. Such variability could plausibly be related to slight changes in the initial conditions during the ejection from the central source, thus leading to a partial correlation between bullets in the two jets, as is required by the observations (Vermeulen et al. 1993b). These ideas also underly the suggestion by Brown et al. (1995) that the bullets are related to a radiative instability phenomenon. Considering the balance between radiative cooling and collisional heating due to interactions of the jet with the stellar wind from SS 433, they show that there is a regime in which a slight change in the jet filling factor causes a transition from jets which cool efficiently to jets which stay at X-ray temperatures throughout the inner 10^{15} cm. Indeed, given the rarity, perhaps even uniqueness, of the SS 433 jet phenomenon, it seems reasonable to appeal to finely tuned conditions, on the border between cold and hot jets. The model leads to a predicted anti-correlation between the X-ray and optical line intensitites, which should be testable (Brown et al. 1995).

It was realized early on (e.g. Davidson & McCray 1980, Begelman et al. 1980) that the interaction of the jets with the stellar wind from SS 433 is likely to assist in exciting the lines. Brown & Fletcher (1992) have suggested that collisional excitation would produce polarization of the moving lines perpendicular to the jet axis; this should soon be testable. Wind-related collisional excitation seems more plausible than excitation by internal shocks, since the comparatively small line widths do not point to substantial internal velocity gradients. Excitation by radiation beamed along the jets has also been proposed (e.g. Panferov & Fabrika 1993). However, it is difficult to see how the geometric effects related to the swinging of both the physical jets and the light cone would explain the frequent simultaneous presence of multiple bullets, the lack of detailed correlation between the two jets, the variable bullet generation rate, and the shape of the light curves.

For plausible densities, the cooling timescale in the optical line emitting matter must be of order seconds to minutes. Hence, this is not the factor which directly sets the evolutionary timescales of the individual bullets. Rather, from the turn-on timescale of ~ 10 hours, it would seem that this might be the timescale over which the jet filling factor (or clump density, etc.) varies significantly. The excitation mechanism must then continue to operate for several days while the bullets slowly fade. As mentioned earlier in Vermeulen et al. (1993b), the final turn-off could be due to the fact that at several 10^{15} cm from the centre, the jets enter the area where the passage of jet matter in a previous precession cycle might have already evacuated a channel. There have also been suggestions (e.g. Bodo et al. 1985) that the bullets may in fact stop radiating after a few days because the clumps in the jets have evaporated.

5. The Radio Jets

For a binary stellar system, SS 433 is an unusually bright radio source. It has a typical quiescent flux density of 0.5 Jy at 5 GHz; flux densities well above 1 Jy occur during flares. At high frequency (>1 GHz), the spectral index is usually $\alpha \approx -0.6$ ($S \propto \nu^\alpha$), typical for optically thin synchrotron emission. Extensive monitoring datasets exist at multiple frequencies (Bonsignori-Facondi et al. 1986, Fiedler et al. 1987, Waltman et al. 1991). This work has shown that, while flares do occur in clusters, separated by quiescent intervals, there is no clearly defined periodicity; to date, there is no evidence in radio flux data for the precession, orbital, or nodding periods. The radio jets of SS 433 can be traced out to $\sim 2 \times 10^{17}$ cm (several arcseconds). They have been intensively observed with VLBI networks, with MERLIN, and with the VLA; the results will now be summarised from the smallest to the largest scales.

Numerous extensive VLBI studies of SS 433 have been made, culminating in two sequences of six 5 GHz European VLBI Network images obtained at two-day intervals (Vermeulen et al. 1987, Vermeulen et al. 1993a; see those papers for references to the earlier work). The later of those sequences is shown in Fig. 1. There is little compact structure on scales $\leq 10^{14}$ cm (e.g. Walker et al. 1981, Vermeulen et al. 1993a). Thus, the central radio feature (the core) is much larger than the binary stellar system. Furthermore, much of the radio emission originates on scales which are several times more extended than the typical size of the optical "bullets" (see Sect. 4). Indeed, no detailed correspondence has been found between specific features in the VLBI maps of Vermeulen et al. (1993a), and specific optical moving line components tracked by Vermeulen et al. (1993b).

The core region has wing-like extensions, each spanning ~ 100 AU. It is likely that these wings, while variable in flux density, are permanently present, though they may sometimes be hidden by moving knots. While the extent does not seem to vary much, so that there is no kinematic evidence for motion, the observed flux density ratio between the two sides does suggest Doppler favouritism, and Vermeulen et al. (1993a) have concluded that a substantial fraction of the radio emission from the complex core region probably originates in quasi-continuously ejected matter moving at $0.26 c$. It should be noted that the somewhat extended VLBI morphology in the inner 10^{15} cm does not place tight constraints on the jet collimation.

Occasionally the jets of SS 433 show side-to-side differences (e.g. Romney et al. 1987, Vermeulen et al. 1987), which may be intrinsic, or could result from transient differences in the local environment. Mostly, however, the extended radio jets closely follow the bulk flow described by the kinematic model, both in Doppler asymmetry and in projected velocity, even including the six-day nodding motion (Vermeulen et al. 1993a). There is no sign of deceleration from $0.26 c$. All the kinematic model parameters can be verified (or determined independently) by tracking the proper motion of individual approaching and receding features in the radio images (Vermeulen et al. 1993a). The distance to SS 433, 4.85 ± 0.2 kpc, is another parameter which follows independently from these observations.

The permanent existence of a brightening zone at a distance of $\sim 4 \times 10^{15}$ cm now seems firmly established; the evidence has been discussed by Vermeulen et al. (1987), Fejes, Schilizzi, & Vermeulen 1988 , and Vermeulen et al. (1993a). When distinct radio knots are present, these become brighter while they travel out to that distance, and they fade rapidly thereafter. They also tend to develop tail-like features, which, however, do not point back in a straight line to the core, and are therefore probably due to new portions of the jets which pass through the brightening zone. This type of phase effect is probably responsible for the fact that, in single images, there is very often a bright patch close to the brightening zone. Evidently, while the amount of flux can vary a lot between successive portions of the jets, the brightening and fading pattern stays the same.

The features seen with VLBI fade rapidly beyond $\sim 5 \times 10^{15}$ cm, presumably due to adiabatic expansion. However, this expansion cannot continue unchecked for long, since MERLIN images, with a resolution of $\sim 10^{16}$ cm, typically show substantial compact jet emission along the curved kinematic model locus (e.g. Spencer 1984, Spencer, Vermeulen, & Schilizzi 1992). Again, the jets do not have a continuous appearance, but contain brighter and fainter patches, which move as discrete entities at $0.26\,c$. Bright MERLIN knots can be linked to flaring episodes during the time when they originated at the binary system. They fade as they move out, and while the data have not yet allowed an unambiguous discrimination between power law and exponential decay, it is plausible that the knots undergo sub-adiabatic expansion (Spencer 1984). However, it is currently not clear how these knots, with scale sizes of $\sim 10^{16}$ cm, are related to the VLBI features discussed above, which are perhaps 10 times smaller. Presumably, several VLBI knots, formed over an outburst period, can expand and merge to form MERLIN-scale features (Spencer et al. 1992).

VLA observations of SS 433 (e.g. Hjellming & Johnston 1986) show twisted lobes, which extend out to $\sim 2 \times 10^{17}$ cm (2–3 arcseconds) on both sides of the centre. This corresponds to ~ 300 days of travel at $0.26\,c$, which is less than two full precession cycles. No radio emission has been detected from the jets at greater distances, as they propagate to the shell of W 50 (but see Sect. 6). Early VLA images provided a very important independent verification and extension of the kinematic model parameters (e.g. Hjellming & Johnston 1981, see also Sect. 1). The lobes have a spectral index $\alpha = -0.6$ ($S \propto \nu^{\alpha}$), characteristic of optically thin synchrotron emission. They usually appear to be continuous. Knots on these scales, such as those discussed by Hjellming & Johnston (1985), result at least in part from projection effects. Hjellming & Johnston (1988) have studied the expected emission profile of precessing jets undergoing lateral expansion in a conical geometry. Interestingly, in order to match the observed morphology, they had to postulate that the radio emitting material undergoes slowed adiabatic expansion out to $\sim 7 \times 10^{16}$ cm, followed by free expansion at greater distances. The cause of this transition is unknown. Also, it should be pointed out that, since the VLA lobes only cover about two precession loops, it is not clear whether models in which the jet material fills up the mantle of a cone are applicable, particularly in view of the rather discrete radio morphology on smaller scales.

Attempts to model the generation of radio emission from SS 433 are hampered by the evident morphological complexity on different size scales. Also, as demonstrated by the observations of Vermeulen et al. (1993a) and Vermeulen et al. (1993c), radio flares, found in total flux density measurements, will often have contributions from several simultaneously evolving features in the jets. Flares can originate both in the radio core, and in the brightening zone (Vermeulen et al. 1993a, Vermeulen et al. 1993c). There is a sustained generation of relativistic electrons over several days. As the flares evolve, it is probable that the energy spectrum of the generated particles softens, since the flares evolve to show an excess flux density below 1 GHz. These properties again emerged in recent observations of one of the largest flaring sequences ever observed in SS 433 (Bursov & Trushkin 1995). There is also a substantial quiescent radio flux density, which presumably has a contribution both from the core region, with features on scales of $\geq 10^{14}$ cm, and from the lobes, on a scale of $\geq 10^{16}$ cm.

It should be noted that, under any plausible physical conditions, the radiative lifetime of electrons in the jets of SS 433 is at least 1000 years, perhaps much longer. This is of the same order as the travel time at $0.26\,c$ between SS 433 and the shell of W 50 (see Sect. 6). Therefore, once electrons have been accelerated at some place along the jets, they are able to produce synchrotron emission for the rest of their journey to W 50, given appropriate conditions (compression, enhanced magnetic field). However, the radio emission from the shell itself is probably due to a different population of relativistic electrons (see Sect. 6).

Hjellming & Johnston (e.g. 1986) have suggested that acceleration of relativistic electrons takes place in a sheath around the jets, due to the fact that they are laterally expanding into the external medium. However, as noted before, the restriction to lateral expansion only holds if the jet material has already expanded radially to fill up the mantle of the precession cone. Heavens, Ballard, & Kirk (1990) have attempted to take this into account, by suggesting that electrons are accelerated in a shock, set up as each section of the jet propagates into the surrounding medium. They think that particle acceleration could eventually terminate because of the changing shock geometry: the angle of incidence of any section of the jets gradually becomes less and less oblique as it moves out. However, such mechanisms seem inappropriate for generating the discrete knots of radio emission on scales of 10^{15}–10^{16} cm. On the other hand, even single flares are certainly not formed in instantaneous injection events, and flaring episodes may last for weeks. Possibly, the brightening of the knots in the inner $\sim 4 \times 10^{15}$ cm is partly due to a falling optical depth. On the other hand, Fejes et al. (1988) have suggested that the rapid fading of knots just beyond the brightening zone could be due to adiabatic expansion resulting from the fact that, at this distance, the jets must have emerged from the freshly deposited stellar wind material into a "channel" evacuated by the jets one precession cycle earlier. Thus, the earlier brightening could be related to shocks set up by jet-wind interaction. Further combined VLBI imaging and multi-wavelength total flux density monitoring would be very helpful in studying this complex issue.

6. The Jets on Parsec Scales; W 50

Evidence for the continued existence of the jets at a distance of up to 100 pc has come from X-ray imaging (Watson et al. 1983, Yamauchi, Kawai, & Aoki 1994). There are elongated, fan-shaped lobes in a position angle of $\sim100°$, exactly overlapping the extension of the precession cones observed in radio maps on (sub)parsec scales. Furthermore, the whole system is embedded in a plerion, an elliptically shaped shell-like radio nebula, which extends over $2° \times 1°$. Its structure has been mapped at several frequencies and resolutions (see Downes, Pauls, & Salter 1986, Elston & Baum 1987, and references therein). It is quite likely to be a supernova remnant, related to the event in which the compact object in SS 433 was formed; SS 433 is located almost exactly in the centre of W 50. However, it has also been suggested that the radio shell is instead due to a blast wave generated by the stellar wind emanating from SS 433 (e.g. Königl 1983). The radio shell of W 50 is interrupted in the east and west over ~30°, centred around position angle 100°, by regions of radio emission somewhat further out, which have been called "ansae", "wings", or "ears". The alignment of these wings, on scales of many arcminutes, with the orientation of the precession cone axis on a scale of (sub)arcseconds, is further evidence that the jets emanating from SS 433 extend over ~100 pc, where they then interact with the shell of W 50 to form the wings.

X-ray studies have concentrated on the eastern lobe. Its spectrum softens with distance from SS 433, indicating that the energy input occurs (predominantly ?) at the base. As already suggested by Watson et al. (1983), the spectrum may also soften away from the precession cone axis. Yamauchi et al. (1994) believe that the X-ray emission is due the synchrotron radiation, because no emission lines could be detected in their ASCA spectra. However, the continuum spectrum would also allow a thermal origin (see also Watson et al. 1983).

The X-ray lobes are centre-brightened, which would not be expected for jets which propagate along a hollow (precession) cone. This is an indication that the jets are deflected towards the cone axis (focused). That is corroborated by the morphology of the radio wings of W 50, which, as found by Icke (1988), could not have been formed by hydrodynamic jets propagating along the surface of a (precession) cone. Whether the focusing mechanism is hydrodynamic (e.g. Peter & Eichler 1993) or due to a toroidal magnetic field (e.g. Kochanek 1991) remains unclear.

Attempts have been made to constrain the physical conditions in and around the outer jets or lobes by studying a number of optical filaments (see Mazeh et al. 1983, and references therein) and infrared knots (see Wang et al. 1990, and references therein) which lie close to the projected path of the jets. Again, however, many uncertain assumptions are involved. From their detailed radio images, Elston & Baum (1987) derive an estimate of the minimum total energy in relativistic (radio synchrotron) particles and magnetic fields in W 50: 4×10^{48} erg. They calculate that the pressure is $\sim 10^{-10}$ dyne cm^{-2} in the interior of W 50 and about 3 times higher in the wings, and that $\sim 10^{53}$ erg of mechanical energy was needed to form the wings. If the X-ray lobes have a thermal origin their total (radiative) energy content would be close to 10^{51} erg (Watson et al. 1983). This would indicate, perhaps not surprisingly, that the jets deposit far more mechanical energy than radiative energy. An energy flux in the jets of $\sim 3 \times 10^{40}$ erg s^{-1} is deduced if W 50 is $\sim 10^5$ years old.

7. Summary

The central engine of the SS 433 jets is not at all well understood. Given the probably supercritical accretion rate, radiative acceleration models seem possible. The dynamics of the SS 433 binary system are complicated by outflows, a geometrically thick accretion disk, and the occurrence of precession and nodding motion. It is still unclear whether SS 433 contains a neutron star or a black hole.

X-ray observations probe the inner 10^{11}–10^{12} cm, and are therefore very important in studying the jets as close as possible to their origin. Recent ASCA data show a host of spectral lines, appearing in both jets at the Doppler shift predicted by the optical kinematic model. Further observations are sure to provide a wealth of information on the physical conditions in the inner jets as well as limits on their acceleration and collimation. Prolonged X-ray photometry is likely to yield important constraints on the configuration of the binary system.

Intensive monitoring of the moving optical lines has revealed many of their properties in detail. They originate at a distance of 10^{14}–10^{15} cm, in "bullets", individual portions of the jet, moving ballistically, which radiate for several days, probably as a result of collisional excitation due to interactions with ambient matter. More than one bullet is often simultaneously visible in each jet, but their occurrence is not correlated on a one-to-one basis between the jets. Bullets may delineate portions of the jets in which conditions, such as clumping, permit the occurrence of matter at 10^4 K, capable of producing line emission.

Radio images show evolving structures in the jets on scales of $\geq 10^{14}$ cm to $\sim 10^{17}$ cm. They are a direct continuation of the inner jets, with matter moving ballistically at $0.26\,c$. They allow the distance to SS 433 to be determined accurately (~ 4.9 kpc). Complex multiple brightening and fading phases occur. Flares can occur in the radio core region, as well as in individual knots. The mechanism through which radio emission is produced is not well understood. While there are large brightness differences between individual patches, there is also evidence for the continuous ejection of radio emitting material.

Elongated X-ray lobes, which connect SS 433 to the shell of W 50, show the continued existence of the jets for a total distance of ~ 100 pc. Where the jets impact the radio shell, they have blown out wings. The morphology of these wings, as well as the centre-brightened profile of the X-ray lobes, indicate that, on their way to W 50, the jets have been focused from the precession cone mantle onto the cone axis, by a mechanism which is not yet clear.

Acknowledgements. I am grateful to Wolfgang Kundt for inviting me to his pleasant meeting in Bad Honnef. Travel support was provided by the Stiftung Volkswagenwerk. The writing of this review was supported in part by the NSF of the USA under grant AST-9420018.

References

Anderson, S. F., Margon, B. H., & Grandi, S. A. 1983, ApJ, 273, 697
Antokhina, É. A., Seifina, E. V., & Cherepashchuk, A. M.. 1992, SvA, 36, 143
Arav, N., & Begelman, M. C. 1993, ApJ, 413, 700
Aslanov, A. A., Cherepashchuk, A. M., Goranskij, V. P., Rakhimov, V. Yu., & Vermeulen, R. C. 1993, A&A, 270, 200
Band, D. L. 1987, PASP, 99, 1269
Baykal, A., Anderson, S. F., & Margon, B. 1993, AJ, 106, 2359
Begelman, M. C., & Rees, M. J. 1984, MNRAS, 206, 209

Begelman, M. C., Sarazin, C. L., Hatchett, S. P., McKee, C. F., & Arons, J. 1980, ApJ, 238, 722

Bodo, G., Ferrari, A., Massaglia, S., & Tsinganos K. 1985, A&A, 149, 246

Bonsignori-Facondi, S. R., Padrielli, L., Montebugnoli, S., & Barbieri, R. 1986, A&A, 166, 157

Brinkmann, W., Fink, H. H., Massaglia, S., Bodo, G., & Ferrari, A. 1988, A&A, 196, 313

Brinkmann, W., Kawai, N., Matsuoka, M., & Fink, H. H. 1991, A&A, 241, 112

Brown, J. C., & Fletcher, L. 1992, A&A, 259, L43

Brown, J. C., Carlaw, V. A., Cawthorne, T. V., & Icke, V. 1988, Ap Sp Sci, 143, 153

Brown, J. C., Mundell, C. G., Petkaki, P., & Jenkins, G. 1995, A&A, 296, L45

Bursov, N. N., & Trushkin, S. A. 1995, Astron Lett, 21, 145

Calzetti, D., Kinney, A. L., Ford, H., Doggett, J., & Long, K. S. 1995, AJ, in the press

Chakrabarti, S. K., & Matsuda, T. 1992, ApJ, 390, 639

Cherepashchuk, A. M., & Yarikov, S. F. 1991, SvA Lett, 17, 258

Collins II, G. W., & Garasi, C. J. 1994, ApJ, 431, 836

Collins II, G. W., & Newsom, G. H. 1986, ApJ, 308, 144

Crampton D., & Hutchings, J. B. 1981, ApJ, 251, 604

Davidson K., & McCray, R. 1980, ApJ, 242, 306

D'Odorico, S., Oosterloo, T., Zwitter, T., & Calvani, M. 1991, Nat, 353, 329

Downes, A. J. B., Pauls, T., & Salter, C. J. 1986, MNRAS, 218, 393

Eggum, G. E., Coroniti, F. V., & Katz, J. I. 1985, ApJ, 298, L41

Elston R., & Baum S. 1987, AJ, 94, 1633

Fabian, A., & Rees, M. J. 1979, MNRAS, 187, 13P

Fabian, A. C., Eggleton, P. P., Hut, P., & Pringle, J. E. 1986, ApJ, 305, 333

Fabrika S. N., & Bychkova, L. V. 1990, A&A, 240, L5

Fabrika S. N., & Sholukhova, O. 1995, Ap Sp Sci, 226, 229

Fejes, I., Schilizzi, R. T., & Vermeulen, R.C. 1988, A&A, 189, 124

Fiedler, R. L., Johnston, K. J., Spencer, J. H., Waltman, E. B., Florkowski, D. R., Matsakis, D. N., Josties, F. J., Angerhofer, P. E., et al. 1987, AJ, 94, 1244

Filippenko, A. V., Romani, R. W., Sargent, W. L. W., & Blandford, R. D. 1988, AJ, 96, 242

Geldzahler, B. J., & Geller, H. A. 1994, ApJ, 420, 655

Gladyshev, S. A., Goranskij, V. P., & Cherepashchuk, A. M. 1987, SvA, 31, 541

Grandi, S. A. 1981, Vistas Astron, 25, 7

Heavens, A. F., Ballard, K. R., & Kirk, J. G. 1990, MNRAS, 244, 474

Hjellming, R. M., & Johnston, K.J. 1981, ApJ, 246, L141

Hjellming, R. M., & Johnston, K. J. 1985, in *Radio Stars*, eds. R.M. Hjellming & D.M. Gibson (Reidel: Dordrecht), p. 309

Hjellming, R. M., & Johnston, K. J. 1986, in *The Physics of Accretion onto Compact Objects*, eds. K.O. Mason, M.G. Watson, & N.E. White (Springer: Berlin), p. 287

Hjellming, R. M., & Johnston, K. J. 1988, ApJ, 328, 600

Hjellming, R. M., & Rupen, M. P. 1995, Nat, 375, 464

Icke, V. 1988, A&A, 202, 177

Icke, V. 1989, A&A, 216, 294

Iijima, T. 1993, ApJ, 410, 295

Katz, J. I. 1986, Comm Astrophys, 11, 201

Katz, J. I. 1987, ApJ, 317, 264

Katz, J. I., Anderson, S. F., Margon, B., & Grandi, S. 1982, ApJ, 260, 780

138

Kemp, J. C., Henson, G. D., Kraus, D. J., Carroll, L. C., Beardsley, I. S., Takagishi, K., Jugaku, J., Matsuoka, M., et al. 1986, ApJ, 305, 808

Kochanek, C. S. 1991, ApJ, 371, 289

Königl, A. 1983, MNRAS, 205, 471

Kotani, T., Kawai, N., Aoki, T., Doty, J., Matsuoka, M., Mitsuda, K., Nagase, F., Ricker, G., & White, N. E. 1994, PASJ, 46, L147

Kundt, W. 1991, Comm Astrophys, 15, 255

Lamb, R. C., Ling, J. C., Mahoney, W. A., Riegler, G. R., Wheaton, W. A., & Jacobson, A. S. 1983, Nat, 305, 37

Lebedev, V. S., & Pimonov, A. A. 1981, SvA Lett, 7, 333

Leibowitz, E. M. 1984, MNRAS, 210, 279

Margon, B. 1984, ARA&A, 22, 507

Margon, B., & Anderson, S. F. 1989, ApJ, 347, 448

Matese, J. J., & Whitmire, D. P. 1982, A&A, 106, L9

Mazeh, T., Aguilar, L. A., Treffers, R. R., Königl, A., & Sparke, L. S. 1983, ApJ, 265, 235

Mazeh, T., Kemp, J. C., Leibowitz, E. M., Meningher, H., & Mendelson H. 1987, ApJ, 317, 824

Milgrom, M. 1979, A&A, 76, L3

Milgrom, M., Anderson, S. F., & Margon, B. 1982, ApJ, 256, 222

Mirabel, I. F. 1996, in *Jets from Stars and Galactic Nuclei* (These Proceedings), ed. W. Kundt (Springer: Berlin), in the press

Mirabel, I. F., & Rodríguez, L. F. 1994, Nat, 371, 46358 215

Molnar, L. A., Reid, M. J., & Grindlay, J. E. 1988, ApJ, 331, 494

Murdin, P. G., Clark, D. H., & Martin, P. G. 1980, MNRAS, 193, 135

Panferov, A. A., & Fabrika, S. N. 1993, Astron Lett, 19, 41

Pekarevich, M., Piran, T., & Shaham, J. 1984, ApJ, 283, 295

Peter, W., & Eichler, D. 1993, ApJ, 417, 170

Peter, W., & Eichler, D. 1995, ApJ, 438, 244

Rand, R. J., Kulkarni, S. R., Backer, D. C., & Clifton, T.R. 1988, A&A, 196, 185

Romney, J. D., Schilizzi, R. T., Fejes, I., & Spencer, R. E. 1987, ApJ, 321, 822

Ryle, M., Caswell, J. L., Hine, G., & Shakeshaft, J. 1978, Nat, 276, 571

Sanbuichi, K., & Fukue, J. 1993, PASJ, 45,, 727

Schalinski, C. J., Johnston, K. J., & Witzel, A. 1989, in *Parsec-scale Radio Jets*, eds. J. A. Zensus & T. J. Pearson (Cambridge University Press: Cambridge), p. 141

Shapiro, P. R., Milgrom, M., & Rees, M. 1986, ApJS, 60, 393

Spencer, R. E. 1984, MNRAS, 209, 869

Spencer, R. E., Swinney, R. W., Johnston, K. J., & Hjellming, R. M. 1986, ApJ, 309, 604

Spencer, R. E., Vermeulen, R. C., & Schilizzi, R. T. 1993, in *Stellar Jets and Bipolar Outflows*, eds. L. Errico & A.A. Vittone (Kluwer: Dordrecht), p. 203

Stewart, G. C., Watson, M. G., Matsuoka, M., Brinkmann, W., Jugaku, J., Takagishi, K., Omodaka, T., Kemp, J. C., et al. 1987, MNRAS, 228, 293

Stewart, R. T., Caswell, J. L., Haynes, R. F., & Nelson, G. J. 1993, MNRAS, 261, 593

Strom, R. G., van Paradijs, J., & van der Klis, M. 1989, Nat, 337, 234

Tingay, S. J., Jauncey, D. L., Preston, R. A., Reynolds, J. E., Meier, D. L., Murphy, D. W., Tzioumis, A. K., Mckay, D. J., et al. 1995, Nat, 374, 141

van den Heuvel, E. P. J. 1981, Vistas Astron, 25, 95

van den Heuvel, E. P. J., Ostriker, J. P., & Petterson J.A., 1980, A&A, 81, L7 van Kerkwijk, M. H., Charles, P. A., Geballe, T. R., Kings, D. L., Miley, G. K., Molnar, L. A., van den Heuvel, E. P. J., van der Klis, M., & van Paradijs, J. 1992, Nat, 355, 703

Vermeulen, R. C. 1989, *Multi-Wavelength Studies of SS 433*, Ph.D. Thesis, University of Leiden, The Netherlands.

Vermeulen, R. C. 1993, in *Astrophysical Jets*, eds. D. Burgarella, M. Livio, & C. O'Dea (Cambridge University Press: Cambridge), p. 241

Vermeulen, R. C., Schilizzi, R. T., Icke, V., Fejes, I., & Spencer, R. E. 1987, Nat, 328, 309

Vermeulen, R. C., Schilizzi, R. T., Spencer, R. E., Romney, J. D., & Fejes, I. 1993a, A&A, 270, 177

Vermeulen, R. C., Murdin, P. G., van den Heuvel, E. P. J., Fabrika, S. N., Wagner, R. M., Margon, B., Hutchings, J. B., Schilizzi, R. T., et al. 1993b, A&A, 270, 204

Vermeulen, R. C., McAdam, W. B., Trushkin, S. A., Facondi, S. R., Fiedler, R. L., Hjellming, R. M., Johnston, K. J., & Corbin, J. 1993c, A&A, 270, 189

Wagner, R. M. 1986, ApJ, 308, 152

Walker, R. C., Readhead, A. C. S., Seielstad, G. A., Preston, R. A., Niell, A. E., Resch, G. M., Crane, P. C., Shaffer, et al. 1981, ApJ, 243, 589

Waltman, E. B., Fiedler, R. L., Johnston, K. J., Spencer, J. H., Florkowski, D. R., Josties, F. J., McCarthy, D. D., & Matsakis, D. N. 1991, ApJS, 77, 379

Wang, Z.-R., McCray, R., Chen, Y., & Qu, Q.-Y. 1990, A&A, 240, 98

Watson, M. G., Willingale, R., Grindlay, J. E., & Seward, F. D. 1983, ApJ, 273, 688

Watson, M. G., Stewart, G. C., Brinkmann, W., & King, A. R. 1986, MNRAS, 222, 261

Yamauchi, S., Kawai, N., & Aoki, T. 1994, PASJ, 46, L109

Yuan, W., Kawai, N., Brinkmann, W., & Matsuoka, M. 1995, A&A, 297, 451

Zwitter, T., Calvani, M., & D'Odorico, S. 1991, A&A, 251, 92

The SS 433 System

Wolfgang Kundt

Institut für Astrophysik und Extraterrestrische Forschung, Auf dem Hügel 71, 53121 Bonn, Germany

Abstract. The Standard Model for SS 433 - reviewed by René Vermeulen (in the preceding contribution) - is confronted with the Model Preferred by (at least) me.

1 Motivation

SS 433 ranks among the best-studied astrophysical objects. It is a compact stellar binary system, probably composed of a $10^4\,yr$-old neutron star with a B-star companion, located at the center of the SNR W 50 (whose formation gave birth to the neutron star), and emitting a pair of jets observed at radio and X-ray frequencies. SS 433 is a key object for understanding jet sources, because of its multiple (multi-wavelength) periodicities: orbit, precession, nodding plus further beat frequencies, and because of its correlated variabilities, down to the timescale of minutes. My understanding of this fascinating astrophysical object is best summarized in (Kundt, 1991).

In this book, René Vermeulen has given an up-to-date review of "the jets in SS 433" under which title he includes the emission-line structure - which I prefer to be divorced from the jets. The assumption that the lines are emitted by the jets causes all sorts of problems: it raises the energetics of the system by some 10^3, doubles its distance, reverses its orientation and sense of precession, forms jets composed of 'bullets' rather than (relativistic) pair plasma, leaves the minute-scale temporal correlations unexplained, and bereaves the multi-wavelength campaigns of their capability of finding coincidences. Yet this assumption has been made almost unanimously in the literature. As it is the distinct goal of this book to find a uniform understanding of all the jet sources - if feasible - and as the (daily) 'bullets' of SS 433 would lead the wrong way, I appreciate René's asking me, at Bad Honnef, to confront his review with mine. In section 2, I shall repeat the main differences between the two interpretations, and in section 3, I shall highlight what I consider 'problems' of the standard model.

2 The different building blocks of the two models

In tables I and II of (Kundt, 1991), I have summarized and documented what I will review (in parts) in this section; see also figure 1 for clarification.

The *distance* of the SS 433 system has been found to be $(3 \pm 0.5)Kpc$ from (i) 21 cm absorption, (ii) interstellar (optical) absorption, (iii) soft X-ray cutoff, (iv) CO mapping, (v) size of W 50 (as an upper limit to all well-mapped Galactic SNRs), (vi) distance from the Galactic plane, and (vii) brightness of the system (controlled by the Eddington limit). It conflicts with René's $(4.85 \pm 0.2)Kpc$ (obtained on the VLBI scale), and with a yet larger distance reported on the VLA scale - both based on the assumption that the radio knots coincide with bullets moving radially at $v/c = 0.260$.

The overall *power* of the system would largely exceed that of W 50 (including its ansae) if the jets stored the $10^{42} C^{-1/2} erg/s$ required by the standard model: why is their dumping site invisible ? (Pair-plasma jets involve less than $10^{36} erg/s$).

The VLA and VLBI *radio jets* involve highly relativistic electrons (and positrons) which - in the standard model - require in-situ acceleration, (a likely offence against the Second Law). Figure 1 is a snapshot which shows sidedness, with west approaching (in agreement with the optical filaments in W 50), whereas the standard model places the center such that the jets appear symmetrical (Kundt, 1987). (Sidedness - in the preferred model - is due to relativistic beaming). There have been reports on blob motion, and linear polarization along the 'corkscrew', inconsistent with a straight-line motion of bullets.

Knotty, focussed *X-ray jets* are seen on scales between $10^{19.5}$ and $10^{20} cm (d/3 Kpc)$, and *radio lobes* on scales between 10^{20} and $10^{20.3} cm (d/3 Kpc)$. Their morphology differs from Milgrom's (hollow) precession cones but likens that of (the radio galaxy) Cyg A . The inner (and dominant) X-ray morphology is unresolved.

The emission sites of the *'moving' lines* - both optical and X-ray - are unknown. Their correlated variability with the 'stationary' lines and with the continuum, on timescales down to five minutes, argues in favour of an origin inside the central binary system. For reasons of their (marginally relativistic) velocity, power, and width (showing no recoil), an origin near the inner edge of the disk, at some $10^7 cm$, is suggested. For the optical lines, this implies coherent emission - the 'exotic' building block of the preferred model. On the other hand, coherent emission offers a way out for an understanding of their various deviations from the expected line ratios of an incoherent emitter. These properties contrast with the separations of $\leq 10^{15} cm$ and $\leq 10^{12} cm$ adopted respectively for their emission sites in the standard model.

What is the origin of the *optical continuum*? Both its strength and its strong - and partially stochastic - variability, on the timescale of hours and at (almost) constant colours, argue against an origin in the (B-star) companion or the accretion disk. Involved is a source comparable in size to the binary orbit, most likely the illuminated windzone. This identification influences conclusions on the masses in the system, its orientation, precession, and nodding. The literature often talks of a 'slaved, thick disk' which has dynamical problems: only the inner edge of the (thin) disk may perform the nodding.

A wealth of data on the *lightcurves* at various frequencies, and on various *radial velocities* (in emission and absorption) multiply overdetermines the system. As far as I know, there is no attempt at their simultaneous interpretation other than mine in 1991. In particular, no strict 'eclipses' have ever been found, only 'dippings'.

142

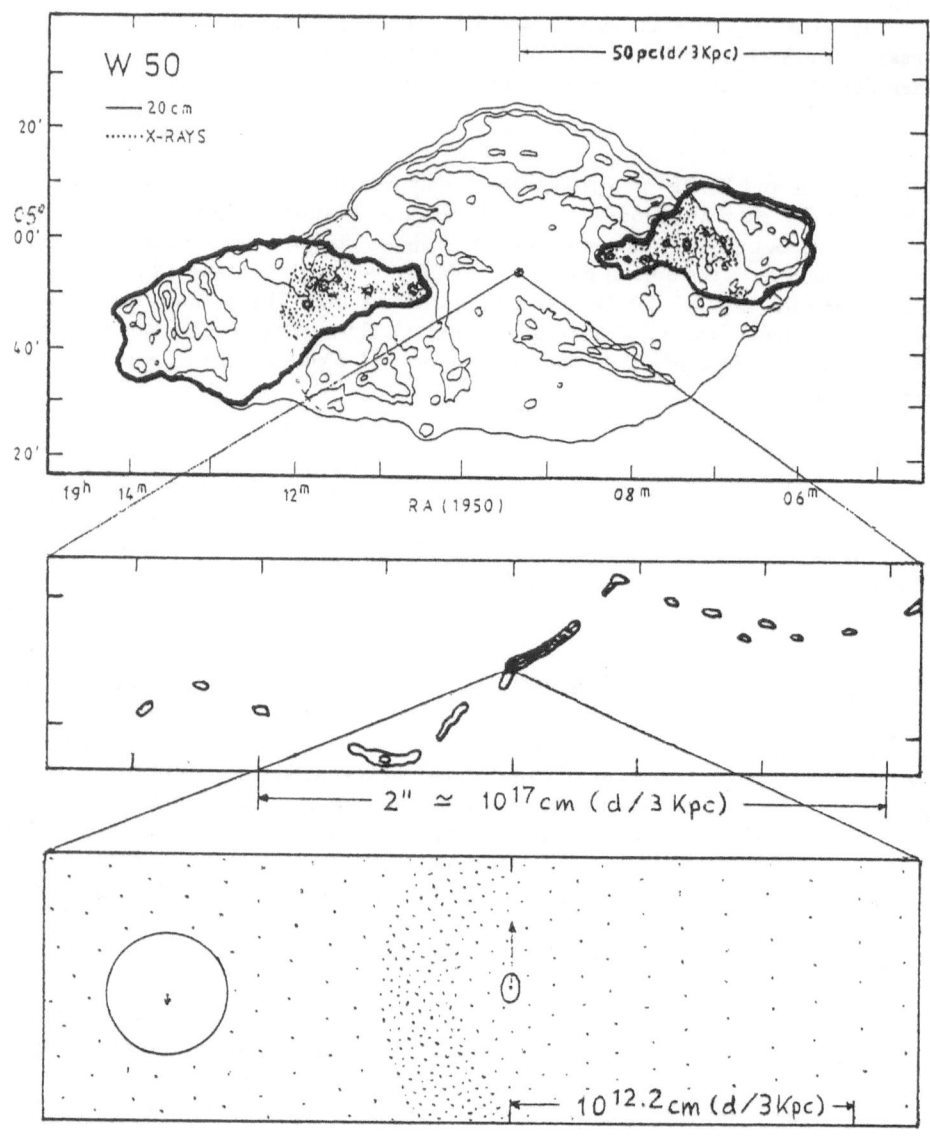

Fig. 1. Sketch of the SS 433 system on three different length scales, produced from (Kundt, 1987 and 1991), and aligned in the preferred way. Fat lines emphasize the jet-lobe morphology.

3   The Cinderella phenomenon

Whenever one tries to model a set of data with an incorrect ansatz, one sooner or later runs into difficulties; in analogy to the problem of Cinderella's step sisters who did not fit into her shoe, I like to talk of the 'Cinderella phenomenon'. A lucid example is the SS 433 system. René has reviewed many of its properties in a clear and careful way, and yet I cannot identify myself with his interpretation. In order to make myself understood, let me give one example per each of his 15 pages.

• Page 1 (of his review) starts with the "unique" jets of SS 433. Certainly, SS 433 is a rare object in the sky - as are $10^4 yr$-young binary Galactic neutron stars. But we know some 10 more neutron-star binaries with radio jets - both subluminal and superluminal - among them Cyg X-3 (as the most similar-looking one) and a few BH-candidates. What makes SS 433 'unique' is its relativistically moving lines which do not involve much power, cannot easily be detected, and may be difficult to produce. Whether or not a radio jet appears superluminal may solely depend on its viewing angle. SS 433 may not be as unique as you think: once we understand it, we may expect it on statistical grounds.

• Page 2 shows VLBI maps in which successive radio snapshots are aligned vertically in a symmetrical fashion (not specified in either the caption, or the text), and criticized in (Kundt, 1987). Subsequent conclusions depend strongly on the symmetry assumption.

• Page 3 "stresses" that "there is no motion along ... the 'corkscrew' ", against earlier reports of blob displacements, and linear polarization along it.

• Page 4 admits "(at least in part) ... a dense outflowing atmosphere enveloping the system", on top of a "geometrically thick accretion disk".

• Page 5 admits that "contradictions remain" in modelling the emitting regions, and also that "the structure and dynamics of the accretion disk is still poorly understood".

• Page 6 mentions the "well-known problem" that "in the suggested thick disk environment, efficient acceleration and efficient collimation seem to be mutually exclusive".

• Page 7 mentions a suggestion of "opaque matter in the binary system which obscures part of the X-ray jet(s)", at separations of $\leq 10^{12} cm$ (!).

• Page 8 reports on "switchoffs" occurring "in both jets at the same time"(!) - even though the 'red' jet should react with a light-travel delay of order one hour.

• Page 9 admits that "the intermittence of the optical moving lines" (of order day) "contrasts with the continuity of the jets both in X-ray and radio emission".

• Page 10 appeals to "finely tuned conditions" of the jet's clumpiness, "on the border between cold and hot jets".

• Page 11 worries (rightly) that "no detailed correspondence has been found between ... the VLBI maps ... and specific optical moving-line components".

• Page 12 worries that "it is currently not clear how these (MERLIN) knots, with scale sizes of $\approx 10^{16} cm$, are related to the VLBI features ..., which are perhaps 10 times smaller".

• Page 13 suggests that "electrons are accelerated in a shock, set up as each section of the jet propagates into the surrounding medium" - a Münchhausen effect ?!

• Page 14 speaks of "an indication that the jets are deflected towards the cone axis", i.e. of deflected "ballistically moving bullets" - another Münchhausen effect ?!

• The Summary (pages 14/15) combines "supercritical accretion" with "radiative acceleration, a geometrically thick accretion disk, and a possible black hole", admitting, though, that the mechanisms of radio-emission production and of jet focussing are "not well understood" and "not yet clear", respectively.

4 Summary

The jets in SS 433 have received two very different explanations: Whereas the standard model talks of ballistically moving hydrogen-helium bullets, the preferred model describes them by extremely relativistic pair plasma traversing distorted channels.

References

Kundt, W., 1987: Astrophys. Space Sci. **134**, 407.
Kundt, W., 1991: Comments on Astrophysics **15**, 255.

Southern Hemisphere VLBI Observations of GRO J1655-40

S.J. Tingay[1], D.L. Jauncey[2], R.A. Preston[3], J.E. Reynolds[2], D.L. Meier[3],
D.W. Murphy[3], A.K. Tzioumis[2], D.J. McKay[2], M.J. Kesteven[2], J.E.J. Lovell[4],
D. Campbell-Wilson[5], S.P. Ellingsen[4], R.G. Gough[2], R.W. Hunstead[5],
D.L. Jones[3], P.M. McCulloch[4], V. Migenes[2], J. Quick[6], M.W. Sinclair[2], and
D. Smits[6]

[1] Mount Stromlo and Siding Spring Observatories, Canberra, ACT 2611, Australia
[2] Australia Telescope National Facility, Epping, NSW 2121, Australia
[3] Jet Propulsion Laboratory, Pasadena, CA 91109, USA
[4] University of Tasmania, Hobart, Tasmania 7001, Australia
[5] University of Sydney, Sydney, NSW 2006, Australia
[6] Hartebeesthoek Radio Astronomy Observatory, Krugersdorp 1740, South Africa

Abstract. Here we present very long baseline interferometry observations of the Galactic radio and X-ray source GRO J1655-40. These observations show that the radio source which appeared approximately two weeks after the initial X-ray outburst consisted of two prominent components which separated with an apparent speed of $1.5 \pm 0.4c$. When the various possibilities for the geometry of the radio source are taken into account the apparent speed implies an intrinsic speed between 0.5c and 0.9c.

Our results and those of other investigators imply a strong link between the accretion of material onto a highly compact object and the ejection of relativistic components of radio emission.

1 Introduction

GRO J1655-40 was discovered as a new and bright X-ray source on July 27, 1994 by the Burst and Transient Source Experiment on board the Compton Gamma-ray Observatory (Harmon et al. 1995). Immediately after the announcement of discovery we began searching for a possible radio counterpart to the X-ray source. On August 6 a new radio source appeared coincident with the gamma-ray position (Campbell-Wilson et al. 1994). Around August 12 the radio counterpart began to flare to approximately 7 Jy. Following the radio flare a series of observations involving the Australia Telescope Compact Array (ATCA) and the Southern Hemisphere VLBI Experiment (SHEVE) array were instigated.

2 Observations and results

The radio counterpart to GRO J1655-40 was observed with the SHEVE array (Preston et al., 1993; Jauncey et al., 1994) over four consecutive days (August 21 - 24) 9 days after the radio outburst, and approximately 25 days after the initial X-ray outburst. We used radio telescopes at Tidbinbilla and Goldstone (DSN),

Parkes, Culgoora, Mopra (ATNF), Hobart (University of Tasmania), Mauna Kea (NRAO) and Hartebeesthoek (HRAO) and observed at a frequency of 2.3 GHz. The radio source was detected on the short baselines of the internal Australian array, but not on any of the intercontinental baselines.

The VLBI data were recorded with the MK II system and correlated at the Caltech/JPL Block II processor in Pasadena, California. The data were subsequently fringe-fit in AIPS and averaged, calibrated and edited in the Caltech VLBI package. The images from the four epochs were made using the DIFMAP software and are shown in figure 1. The source consisted of two prominent components, each of which have substructure. We have registered the four images so that the brightest component is aligned vertically. This is not an astrometric registration since in the process of fringe-fitting VLBI data absolute positional information is lost. However it is clear that the two components changed their relative positions over the four day period. The components separated linearly along a position angle which is the same as the position angle which defines the individual components, at a rate of 65 ± 5 mas/day.

This angular expansion corresponds to an apparent speed of motion of $1.5 \pm 0.4c$ at the 3-5 kpc distance we have estimated for the radio source (Tingay et al., 1995). Extrapolating the motion of the components back in time we find that zero separation coincided well with the begining of the radio outburst, approximately 16 days after the outburst in X-ray emission.

3 Discussion

GRO J1655-40 was the second apparently superluminal radio source to be discovered in our Galaxy, after GRS 1915+105 (Mirabel and Rodriguez, 1994). For GRS 1915+105 an accurate estimate of the intrinsic speeds of the components could be made since the stationary core could be astrometrically identified and components were seen in motion on opposite sides of the core. Since we have no absolute positions for the two components in GRO J1655-40 we cannot tell which, if either, of the components were stationary. This leads us to consider two possibilities to explain the relative motion in this radio source. Firstly, one of the components may be a stationary core. Alternatively, both of the components may be in motion away from an unseen common origin. A range in component speed of 0.5c - 0.9c can account for both scenarios.

For this source we have shown that a link can be made between the accretion of material onto a compact object (evidenced by the X-ray activity) and the ejection of relativistic components during a subsequent radio outburst. From more extensive VLBA observations (Harmon et al., 1995; Hjellming and Rupen, 1995) of GRO J1655-40 this scenario has been confirmed. In addition, observations of the optical counterpart to GRO J1655-40 during a quiescent state (Bailyn et al., 1995) reveal that it is part of a binary sytem with a mass function which suggests that the primary is a black hole.

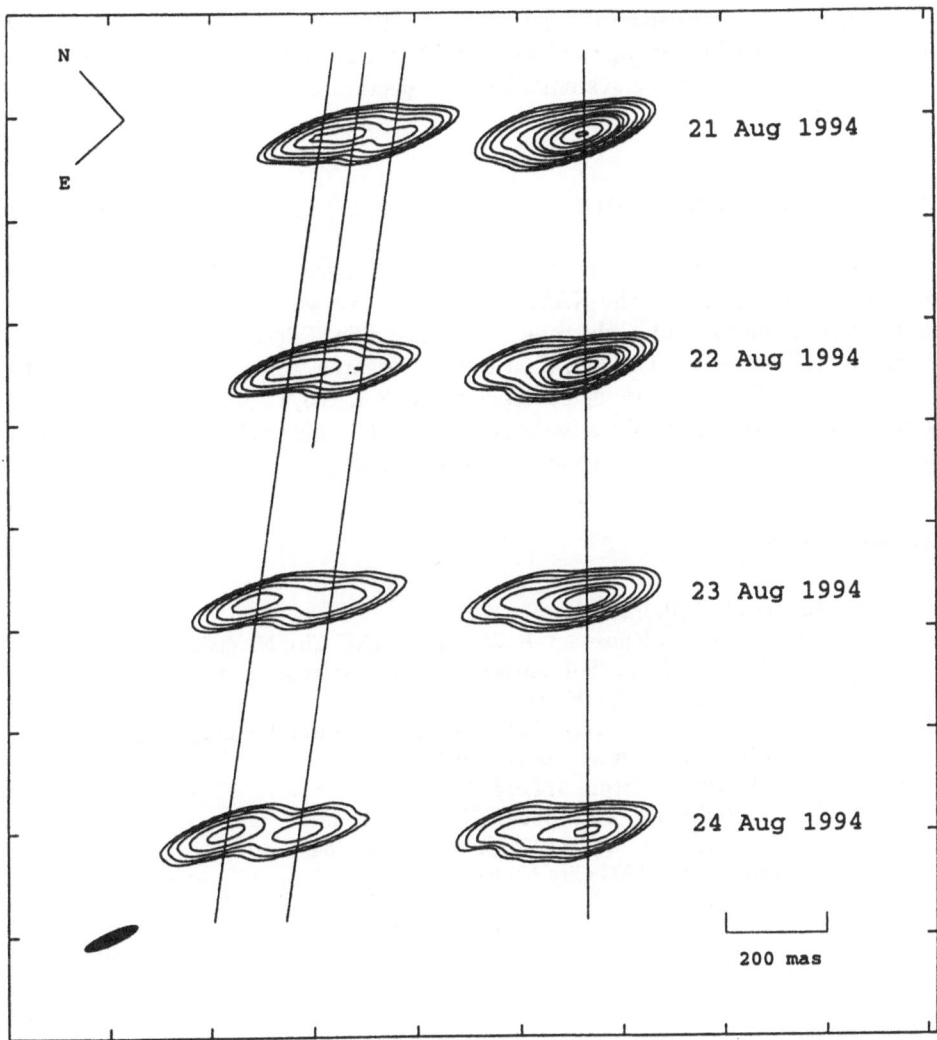

Fig. 1. The four VLBI images at 2.3 GHz. The contour levels are 1,2,4,8,16,24,32,64,80 and 95 percent of the peak brightness in the montage of 0.59 Jy per beam. Each image has been rotated anti-clockwise by 42° and aligned to register the south-west component vertically. The restoring beam is 123 × 26 mas with major axis position angle of 72°.

Thus, multi-wavelength observations of GRO J1655-40 and high resolution imaging has provided strong evidence linking accretion onto a highly compact object in a stellar binary system with the ejection of relativistic radio components.

4   Acknowledgements

We thank the observatory directors for allowing these observations to be scheduled at short notice, and the NASA Deep Space Network for making time available at Tidbinbilla and Goldstone. The Australia Telescope is operated as a national facility by the CSIRO. Part of this research was carried out at the Jet Propulsion Laboratory, under contract to NASA. S.J.T. and J.E.J.L. are supported by Australian Postgraduate Awards. S.J.T. also acknowledges generous support from the organisers to attend this workshop.

References

Harmon, B.A. et al. (1995): Nature 374, 703
Campbell-Wilson, D. and Hunstead, R.W. (1994): IAU Circ No 6052
Preston, R.A. et al. (1993): in 'Sub-Arcsecond Radio Astronomy '(eds Davis, R.J. and Booth, R.J.) 428 (Cambridge University Press)
Jauncey et al. et al (1994): in 'Very High Angular Resolution Imaging '(eds Robertson, J. and Tango, W.) 131 (Kluwer, Dordrecht)
Tingay, S.J. et al. (1995): Nature 374, 141
Mirabel, I.F. and Rodriguez, L.F. (1994): Nature 371, 46
Hjellming, R.M and Rupen, M.P. (1995): Nature 375, 464
Bailyn, C.D. et al. (1995): IAU Circ No 6173

Jets in Planetary Nebulae

Garrelt Mellema [1,2]

[1]Stockholm Observatory, S-13336 Saltsjöbaden, Sweden
[2]Dept. of Physics, UMIST, Manchester M60 1QD, UK

Abstract: The occurrence of highly collimated outflow phenomena in Planetary Nebulae is reviewed. The observations are diverse. Sometimes simple pairs of fast knots are seen, sometimes more extended and complicated structures. Closer analysis of the simpler cases shows that they are anomalous not only in their outflow velocity, but also in their (N) abundance. The presence of a hot star further complicates their structure. Several models for the formation of these structures have been proposed but none go into much detail. These models are reviewed as well.

1 Introduction

Since the beginning of the 1980s there has been a renewed interest in Planetary Nebulae (PNe). The advance of CCD technology has made the study of this phase in the later evolution of low to intermediate mass stars (the main sequence mass of PN progenitors is thought to be around 6 $M_\odot$ and lower) much easier, both from the narrow band imaging as from the spectroscopic point of view. With modern CCD techniques it is possible to study the fainter parts of the objects, and often there is much to be found there (see e.g. the narrow band catalogue of southern hemisphere PNe by Schwarz et al. 1992). The availability of these catalogues of narrow band images (see Balick 1987 for northern hemisphere nebulae) has proved to be very useful both for the study of individual objects (see e.g. Corradi & Schwarz 1993) and for statistical studies (see e.g. Corradi & Schwarz 1995).

At the same time the opening up of other wavelength ranges has helped to further the understanding of the formation and evolution of these objects. The IRAS results have helped to substantially increase our knowledge about the Asymptotic Giant Branch (AGB) phase, which precedes the PN phase. The IUE spectra have made it possible to study the fast winds that come off the central stars of PNe (Perinotto 1993), ROSAT has shown the hot gas contained within the nebula (Kreysing et al. 1992), and millimeter range observations have shown the presence of large quantities of molecular gas (Huggins 1993).

In this paper I address one aspect of PNe that has come up from all this observational work, namely that a fraction of these objects show signs of well collimated outflows. In Sect. 2 I describe the observations of the different types of 'jets' seen in PNe. In Sect. 3 deals with the internal structure of one particular type of PN 'jets' and Sect. 4 contains an overview of the different explanations that have been put forth. I end with the conclusions in Sect. 5.

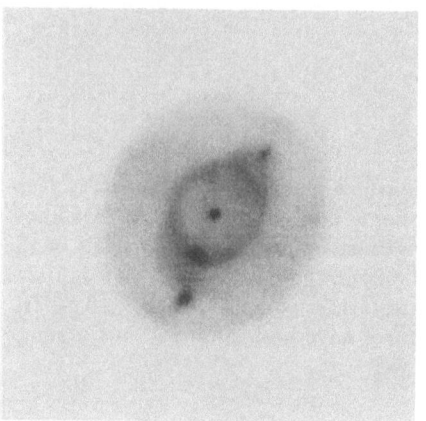

Fig. 1. Image of NGC 3242 in [NII]6583, in negative, non-linear grey scales, from Balick (1987). Notice the two very bright FLIERs. They are located at 17″ from the central star which corresponds to a separation of 0.066(d/0.8kpc) pc. Distances to individual PNe are not well known, the values used in this paper are taken from Maciel (1984).

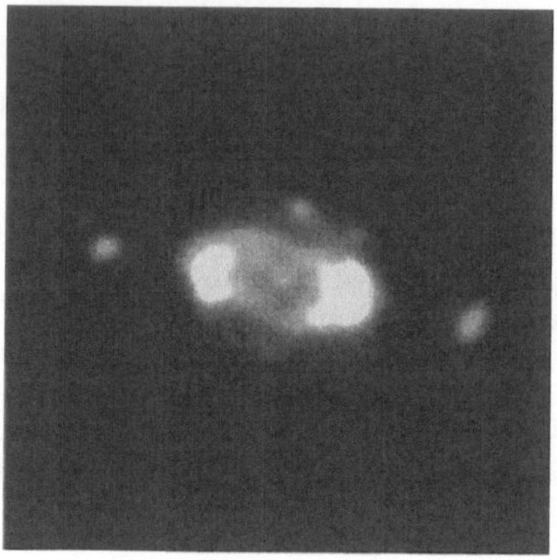

Fig. 2. Image of NGC 7009 in [NII]6583, in positive, non-linear grey scales, from Balick (1987). Notice the outer FLIERs and the inner 'caps'. The outer FLIERs are located at 25″ from the central star which corresponds to 0.11(d/0.9kpc) pc.

2 Description of PN jets

The phenomenon of collimated outflows in PNe is quite diverse. In general one can say that no PN looks like another, though sometimes the similarities can be striking and there are sufficient similarities among virtually all PNe to justify simple morphological classification schemes. For the bright core nebula the most popular morphological classification at this moment is the one presented in Stanghellini et al. (1993), which is a variation on the scheme used by Balick (1987). It distinguishes between elliptical (E), which also includes round ones, bipolar (B), defined as objects with a clear waist, point-symmetric (P), for objects that show more spiral shapes, and irregular (I) shapes. Added to these letters can be an M for multiple shell, if the nebula has a shell around it, and S if the nebula shows a lot of inner structure (knots, filaments). In the sample Stanghellini et al. use, 65% are E, 14% B, 6% P, and 15% I. These fractions should be used with caution, but still seem to be fairly typical.

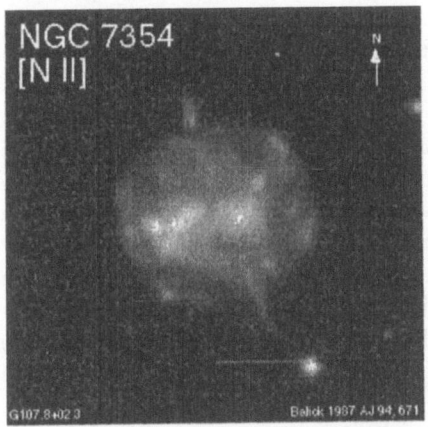

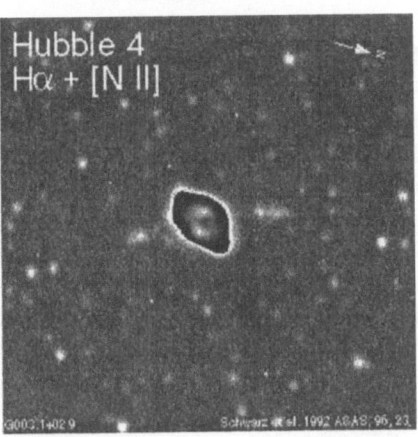

Fig. 3. Image of NGC 7354 in [NII]6583, from Balick (1987). The grey scales have been non-linearly adjusted to bring out most features. Note the two faint tails sticking out to the north and the south. Their tips are 23″ from the central star which corresponds to a separation of 0.11(d/1.0kpc) pc. The bright spot near the lower right hand corner is a star

Fig. 4. Image of Hb4 in Hα+[NII] from Schwarz et al. (1993). Two grey scale ranges are used in order to bring out the two faint tails. They run more or less horizontally in the figure, on either side of the short axis of the central elliptical nebula. The tips are 15″ away from the central star which corresponds to a separation of 0.16(d/2.2kpc) pc

There is no fixed morphological scheme for the 'jets' in PNe. I should stress that the word 'jet' should not be taken too literally, and in fact it could be argued that it should not be used at all. The word 'jet' implies a continuous stream of fast flowing gas and it is far from clear that this is the case in most PNe. In fact it is not clear that this is the explanation for a lot of astrophysical phenomena which are called 'jets'. I will use the term 'collimated outflow phenomena' (or COP) here, which is sufficiently broad to cover all that is needed.

Below I will distinguish between three types of COP in PNe:

1. Ansae and FLIERs

2. Tails

3. Extended COP and BRETs

The distinction between these types is sometimes rather hazy. Often the only clear distinction is how well the phenomenon has been studied. This shows our limited understanding of what these COP really are. The field is clearly just beginning to be explored, and different people will use different terms to describe the same phenomenon (or the same term to describe different phenomena). This should always be kept in mind when going through the literature.

2.1 Ansae and FLIERs

It has been known for a long time that some PNe show pairs of knotty condensations, commonly called 'ansae'. This term was used already by Curtis (1918). It actually means 'handles', but it tends to be used for any obvious knot in PNe. In the seventies it became clear that some showed higher velocities than the core nebula they belong to. In order to distinguish between these and other types of ansae, Balick et al. (1993) introduced the acronym FLIER (Fast Low Ionization Emission line Region) to describe those ansae which distinguish themselves by their low ionization character (they are mostly very bright in NII lines), the fact that they occur in pairs along the major axis of the core nebula, and their high velocity. Typical velocities are 50–200 $km s^{-1}$. The observed velocities are of course velocities along the line of sight, so some low velocity ansae might be FLIERs in disguise.

Clear cases of ansae in E-type PNe in the catalogue of Balick (1987) are NGC 3242, NGC 6826, NGC 6905, and NGC 7009. The first three show one pair, NGC 7009 is a bit more complicated with two ansae on either side of the core nebula. The inner pair is slightly larger and they are therefore sometimes called 'caps'. All these have been shown to be real FLIERs in the sense that they have large velocities (Balick et al. 1993; Balick et al. 1994; Cuesta et al. 1993). The outer pair in NGC 7009 does not show a very high velocity, but this may be a projection effect. Note also that the classification of NGC 7009 as an E-type is somewhat questionable, one might argue it is really a P-type.

A more complicated example is offered by the E-type PN NGC 7662, which is surrounded by a dozen or so ansae which are very bright in the [NII] lines. But a kinematic study by Balick et al. (1994) shows that only the two pairs along the major axis of the core nebula show high velocities and are therefore proper FLIERs.

Another special case is NGC 6543, which I describe in Sect. 2.4.

In the image catalogue of Schwarz et al. (1992) one can find some more examples of ansae, but unfortunately the representation of the images in the paper is such that one does not easily discern ansae. See e.g. their image of NGC 3242, on the basis of which one would *not* conclude that this nebula has ansae. From the point of view of COP, it is a pity that the Schwarz et al. survey contains $H\alpha$+[NII] images instead of separate $H\alpha$ and [NII]. Objects that do seem to have ansae are: CTS 1, J 320, He 2-84, He 2-104, He 2-141, and IC 4634. Of these He 2-84 and He 2-104 are B-type (bipolar) and the ansae appear to be the brighter tips of the lobes. A kinematic study by Corradi & Schwarz (1993a, 1993b) shows that they have high velocities, up to 200 $km s^{-1}$ and are bright in [NII], so they would classify as FLIERs. On the other hand the central star of

K 1-2   [O III]

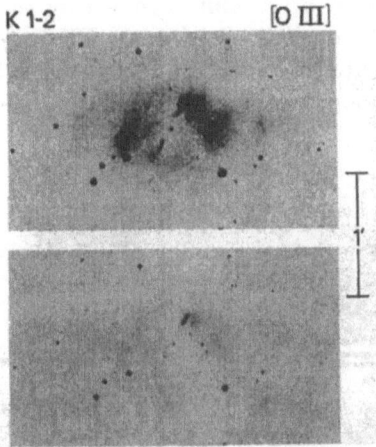

Fig. 5. Images of K 1-2 in [OIII]5007, from Bond & Livio (1990). The lower image shows a bright tail pointing to the upper righthand corner. The upper image shows the fainter parts of the nebula, including the two knots which might constitute a counter-tail. The tip of the tail is 24″ from the central star at the centre of the image. This corresponds to a separation of 0.39(d/3.3kpc) pc. Copyright 1990 American Astronomical Society, reproduced with permission

He 2-104 has been classified as a symbiotic system, so one could debate whether this object belongs in the PN category.

For the other two P-type nebulae there is no kinematic data available (as far as I know). Also, the ansae clearly show up in [OIII], so in that sense they are dissimilar from the FLIERs discussed above. Some other objects in the catalogue might be worth further study: NGC 3918, NGC 5882, He 2-123, H 1-7.

2.2 Tails

In a few cases the COP do not look like (unresolved) knots, but are clearly elongated. This type of COP could be called 'tails' (Balick 1987). These tails are still quite small compared to the core nebula. Examples can be found in NGC 7354 (Balick 1987, see Fig. 3) and Hb 4 (Schwarz et al. 1992, see Fig. 4), both ellipticals. These tails still occur in pairs and are at least in the case of NGC 7354, bright in [NII] (we can not tell in the case of Hb 4). So in these respects they resemble FLIERs. Neither of these two COP has been investigated kinematically, so we do not have velocity information.

Bond & Livio (1990) presented an image of the elliptical nebula K 1-2 which shows a clear tail on one side and two bright knots on the other side, presumably the counter-tail (see Fig. 5). Again no kinematic follow-up was done.

That one does have to be careful is shown by the case of NGC 40, which shows a single-sided tail on deep Hα images (Balick 1987, see also Fig. 6). A kinematic study by Meaburn et al. (1996) shows this feature to have no radial velocity, so unless its velocity direction is exactly in the plane of the sky, this is not a real COP.

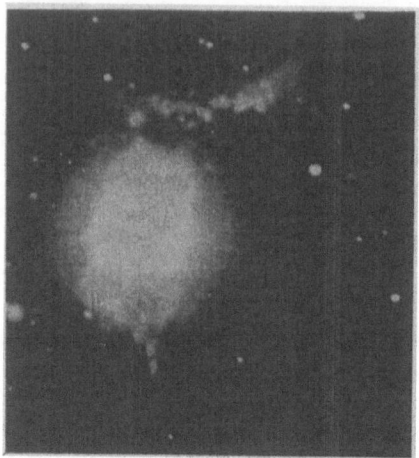 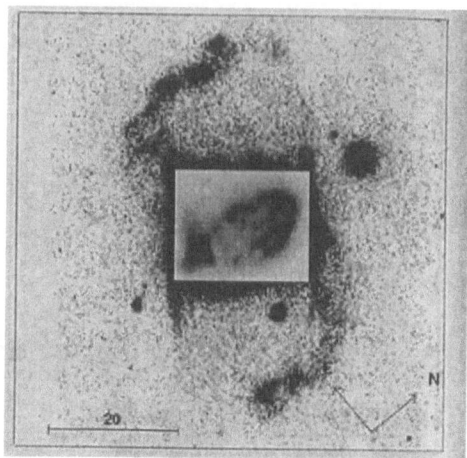

Fig. 6. Image of NGC 40 in Hα from Meaburn et al. (1996), taken by J.A. López. The grey scales have been modified to bring out the fainter structures. Note the one sided faint 'jet' to the south, looking like two drops falling from the nebula. The separation from the central star is 17″, which corresponds to 0.065(d/0.8kpc) pc. The material north of the nebula is also associated with it, see Meaburn et al. (1996)

Fig. 7. Image of NGC 6309 in [OIII] from Schwarz et al. (1993). An example of a spiral shaped COP, suggesting a precessing source

2.3 Extended COP and BRETs

A third (and more spectacular) class of COP in PNe was discovered recently by López and collaborators. In these the COP has a size much larger than the PN it belongs to. To date only two objects clearly belong to this class, Fg 1 (López et al 1993; López et al 1994; Palmer et al. 1996, see also Fig. 8) and KjPn 8 (López et al. 1995, see also Fig. 9). There are some objects whose images looks similar to Fg 1, but have a smaller size, such as NGC 6309 (Schwarz et al. 1992, see Fig. 7), which might or might not belong to this class. As it is difficult to detect these faint extended structures, for which both a large field of view and high sensitivity are needed, I suspect that a systematic search will turn up more examples.

Fg 1 shows a spiral shaped string of knots emanating from both sides of the core PN. The whole structure is extremely symmetric, also in velocity space and can be explained with a simple ballistic model with precession in which knots are ejected with a constant speed of 85 $km\,s^{-1}$, while their source precesses with 7″ per year (López et al 1994; Palmer et al. 1996). This behaviour has let to the introduction of the acronym BRET (Bipolar Rotating Episodic jeT) to describe this class of COP. The full size of the BRET in Fg 1 is estimated at 2.8 pc.

The BRET associated with KjPn 8 is even more spectacular, it's estimated size is 4.1 by 2.1 pc. This BRET is different from Fg 1 in that it looks more tubular shaped. But as Fg 1 it shows signs of a precessing source at the centre. Analysis of line ratios shows that the BRET material is predominantly shock excited (López et al. 1995). To date no kinematic results have been published.

BRETs are probably related to the other COP, but how is not entirely clear. In Fg 1, the BRET is very symmetric, especially bright in [NII], and expanding fast, properties shared with the other COP discussed above.

Two objects that also have COP that are much larger than the the core nebula are M 1-16 and M 2-9 (images can be found in Schwarz et al. 1993). Both show high outflow velocities ($\sim$ 200 $km\,s^{-1}$, see Corradi & Schwarz 1993b; Schwarz et al. 1995). M 2-9 is peculiar in that both end of the two-sided jets are red-shifted. According to Schwarz et al. (1995) this is because we are actually seeing reflected light from much further in. This is backed up by polarization measurements. Based on the the amount of dust and the molecular material observed in both these objects (Huggins 1993), they are presumed to be young PN, or maybe even proto-PN, related to the objects mentioned in Sect. 2.5.

2.4 NGC 6543

The case of NGC 6543 (nicknamed the Cat's Eye Nebula) is special in that this PN has been imaged using the Hubble Space Telescope (HST). Before what from ground based images appeared to be ansae, had been classified as FLIERs by Balick et al. (1994). The HST pictures show that these simple structures are actually much more complicated (Harrington 1995), see Fig. 10. Although the deprojected structure of this object is still a matter of debate, it is obvious that its dynamics is quite complex, and one could speak of a COP system.

More or less orientated along the major axis of the whole system there are two pairs of systems. The inner pair shows an S-shape and was shown to have a line of sight velocity of about 25 $km\,s^{-1}$ (Miranda & Solf 1992).

Further out there is a pair of elongated structures ending in a brighter knot. The structures look very much like proper jets and resemble the tails discussed in Sect. 2.2, but they also seem to be split. This pair is also seen in the ground-based observations and was classified as FLIERs by Balick et al. (1993). The derived intrinsic velocity is between 100 and 200 $km\,s^{-1}$.

The HST images also show a series of very weak radial streaks all pointing towards the central star. For these no kinematic data exists. They might be related to similar radial streaks seen in the Helix nebula, although those appear to be located inside the main nebula.

It is important to note that also in this source the directions of the COP are not constant (but all point back to the centre of the nebula). In this sense the COP in NGC 6543 are like BRETs (but at a small scale). At the same time they resemble the tails from Sect. 2.2, but before the HST images, they were (sometimes) classified as FLIERs. This shows how much the taxonomy still depends on the quality of the observations.

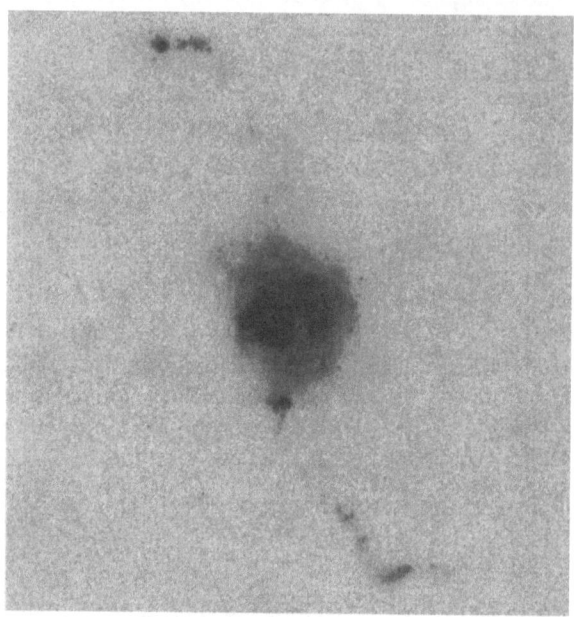

Fig. 8. Image of Fg 1 in Hα+[NII], taken by J.A. López, from Palmer et al. (1996). The brightest stars have been removed in order to bring out the 'jet' somewhat better. The end of the jet is 2′ from the central star which corresponds to a separation of 1.4(d/2.4kpc) pc

2.5 COP in proto-planetary nebulae

In a number of objects suspected to be on the way to becoming a PN, fast outflows have been detected in CO. The best known example is CRL 618 (Neri et al. 1992). Others are CRL 2688, and OH231.8+4.2 (Morris et al. 1987). IRAS 1634-3814 shows high velocity water and hydroxyl masers (te Lintel Hekkert et al. 1992). Some show ansae-like features (IRAS 09371+1212, Morris & Reipurth 1990), and already have some ionized material in which high velocities have been detected (IRAS 17423-1755, Riera et al. 1995)

The connection between these outflows and the ones observed in proper PNe is not clear. It is tempting to suggest that these outflows will for instance produce the later FLIERs (Morris 1990), but there is no hard evidence for that.

3 Analysis of FLIERs

Balick et al. (1993) and Balick et al. (1994) made a closer study of some well known FLIERs. Using images, kinematic data, and line ratios an attempt was made to find a more coherent picture of this enigmatic phenomenon. The authors find that closer analysis produces more puzzles.

Studying the FLIERs in NGC 3242, 6543, 6826, 7009, and 7662 they derive some basic properties. FLIERs are small at least in one dimension (~ 0.01 pc), and have masses of about 10^{-4}–10^{-5} $M_\odot$. One can define an age by dividing the distance from the centre of the PN by the (deprojected) velocity; this is called the kinematic age. The kinematic age of some FLIERs is less than that of the rest of the PN. They also have an apparent enhancement of N abundance (relative to H). The derived abundance is 2 to 5 times higher than in the rest of the nebula. Only N shows such 'anomalous' behaviour.

On the other hand the line ratio analysis shows that temperatures and densities are not substantially different from their surroundings. This is a most curious result, which puts severe restraints on possible models. One should of course realise that the densities measured are the densities in the zone where the ion under consideration is most abundant. The fact that one integrates line emissivities along the line of sight may also slightly influence the outcome. But despite this it is obvious that the FLIERs do not have conspicuously different densities and temperatures from the rest of the nebula.

Another curious fact is that the ionization stratification is such that the higher ionization stages are seen nearer to the star; the peak of the [NII] distribution is slightly displaced from the [OI] peak (see Balick et al.1994, Fig. 4).

The case of NGC 6543 has shown that high resolution observations reveal a wealth of details in the COP. A necessary step in the further study of these objects is therefore imaging with the HST. Observations of this kind have been scheduled and maybe a year from now we will have some more answers.

3.1 What are FLIERs?

Balick et al. (1994) tried to find an analogy between FLIERs and other astronomical phenomena. The answer remains inconclusive. In many ways FLIERs are like HH objects, but unlike HH objects, they are also influenced by ionizing stellar photons. This complicates the analysis since one needs a combination of bow shock and photo-ionization models to properly describe them, and these types of models have not been worked out in much detail.

But even if one sees them as related to HH objects, it is still unclear whether they are discrete clumps of gas, or the knots and heads of jets. If FLIERs and BRETs are indeed closely related, then this would seem to point to a more continuous process.

4 Explanations

In this section I describe some of the models that have been suggested for the origin and formation of the COP. In coming up with explanation most people have mostly restricted themselves to the simpler case of ansae and FLIERs, partly because they have been known for the longest time, and also because they are the most common. Before I discuss the models for the COP, I first give a short overview of what is nowadays the standard model for the formation of PNe.

4.1 Standard PN formation model

The generally accepted model to explain PNe and their shapes is the so called interacting winds model. In this model the nebula forms from the interaction of a fast stellar wind (2000 km s^{-1}) with surrounding slower moving material. This surrounding material (or slow wind) was deposited there when the star was still an AGB red giant. The fast wind blows a bubble into the surrounding material and the high density edge of the bubble is the PN. If the material was deposited in a non-uniform way, for instance because of the presence of a companion star, the interaction will result in an aspherical nebula. See e.g. Balick (1987) for the general idea of the interacting winds model, and Livio (1995) for a review on the origin of the asphericity.

This interacting stellar winds scenario has been extensively studied, initially by analytic methods (Kahn & West 1985; Icke 1988), later using numerical methods (Livio & Soker 1989; Mellema et al. 1991; Frank et al. 1993; Frank & Mellema 1994; Mellema & Frank 1995a, see also the review by Mellema & Frank 1995b). This work has shown that this scenario does explain many features of PNe. However, the aspherical density distribution of the slow wind is assumed in all of these models since its exact nature is not well known.

The different shapes of PNe are explained by different slow wind density distributions. If the bulk of the slow wind is concentrated in a thin disk, a bipolar or butterfly type nebula forms, if it is more spread out, an elliptical nebula forms.

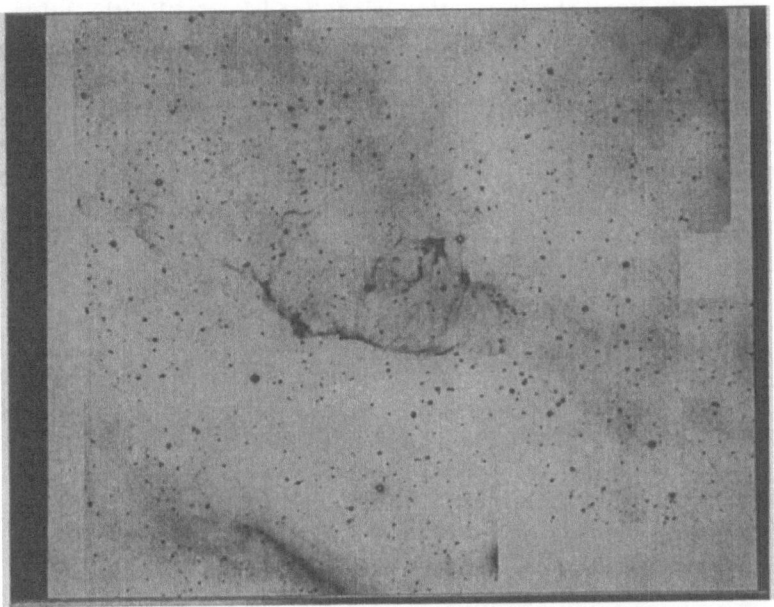

Fig. 9. Mosaic of KjPn 8 in Hα, from López et al. 1995. The whole structure is 14′ by 4′ which at an estimated distance of 1 kpc corresponds to 4.1 by 1.2 pc. Copyright 1995 American Astronomical Society, reproduced with permission

4.2 Explanation within the framework of the standard model

At some point it seemed that simple COP (like FLIERs) could be explained by the interacting stellar winds model, see Balick et al. (1987). The idea was that the flow would be focused to the top of the bubble and form a dense clump with high velocity. However, this scenario has never been seen to work in any of the numerical simulations. There are some cases in which structures do form on the symmetry axis, but these are not convincing FLIERs. Part of the problem is that these structures normally form just inside the bubble whereas most COP are clearly located outside the bubble. The more recent models are quite sophisticated in that they take heating and cooling of the gas into account and this does not make a lot of difference as far as the formation of COP is concerned.

More generally speaking it is not so difficult to form jets in the interacting winds scenario (Icke et al. 1992). If one interpreted the observed COP as parts of jets, there is still a possibility to form them using interacting winds, this does however require a rather specific density distribution. It seems more likely that this mechanism is applicable to the formation of jets from YSOs (Frank & Mellema 1996).

Recently while studying the interacting winds model in a slightly different situation with fast winds of 100–200 kms^{-1}, I found that a fast, dense knot can form, and appear to be located outside the bubble. The difference between this situation and the one studied previously is that at these lower velocities the bubble is momentum driven, which leads to stronger focusing. This is essentially the mechanism described by Cantó (1980). It might be possible that the COP are formed in the post-AGB phase when the stellar wind still has modest velocities and cooling plays an important role, though it is unclear why they would differ in abundance from the rest of the nebula. More study is needed here.

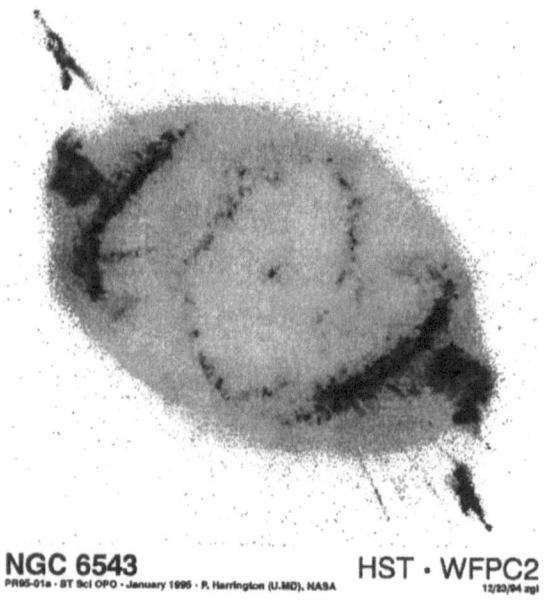

NGC 6543   **HST · WFPC2**

PR95-01a · ST ScI OPO · January 1995 · P. Harrington (U.MD), NASA   12/23/94 zgi

Fig. 10. The HST image of NGC 6543 in Hα. The knots furthest away from the central star are 18″ away from it. This corresponds to a separation of $0.06(d/0.7\text{kpc})$ pc. Image courtesy of J.P. Harrington and K.J. Borkowski – University of Maryland and NASA

4.3 Jet formation in common envelope binaries

Three models exist which use a common envelope situation to explain COP. This is an attractive approach since a companion object (star, brown dwarf or massive planet) when orbiting through the extended envelope of the AGB star can cause aspherical mass loss through a variety of mechanisms (Livio 1995; Soker 1995a). As was explained above, aspherical mass loss is needed to explain the elliptical shapes of the PNe in most COP are seen. So, if one can explain both the shape of the nebula and the formation of the COP within the same model, then this would be very satisfactory.

In the first model, presented by Soker (1993), the COP form because of the strong deformation of the stellar envelope in the later stages of the common envelope phase. Because of the spin up by the companion object the envelope acquires a donut-like shape. Material flowing away from the polar holes collides at the symmetry axis and is supposed to form a jet this way. The observed COP are the heads of these jets.

In the second model (Soker 1995b) the companion object evaporates while orbiting within the common envelope. The companion material is then assumed to form a disk around the stellar core, from which a jet emerges. The head of this jet is the observed COP.

A third model is described in Soker & Livio (1994). In this again an accretion disk forms around the primary, but this time after the ejection of the common envelope. The idea is that after that event the secondary will expand and since the binary can have become quite close during the common envelope phase, mass transfer can occur and a disk may form. This disk is then proposed to lead to jets.

In all three models the jet velocities are shown to be consistent with observed COP velocities and it is likely that processed material from the stellar core is mixed in, which would account for the higher N abundance. Also, the dynamical ages of the COP formed are about right.

But the models are very sketchy when it comes to the details of the actual formation of the jet and hence the COP. Especially the second model first needs to assume the formation of a disk and then assume the formation of the jet from this disk, and both assumptions are not very well motivated in the paper. Also in the first model it is not obvious that a jet will form. A similar line of reasoning (flow focused towards the axis) used in Balick et al. (1987) to explain the formation of FLIERs, was shown not to work by the more detailed numerical models. So there is reason to be cautious.

Then again no established model exists for any kind of astrophysical jet formation, so one should not judge these models too harshly. A slightly different approach was followed by Cliffe et al. (1995) who did three dimensional numerical modelling of a precessing jet in order to explain the point-symmetry seen in some PNe. Here the jet was assumed to be there and its influence on the environment studied.

4.4 Jets from disks

Another possible explanation for the formation of jets is the one proposed by Morris (1987, 1990), see also Soker & Livio (1994). In this scenario the AGB star has a companion star which is too far away to result in a common envelope situation. However, the secondary does accrete material from the AGB star's slow wind and thus will be surrounded by an accretion disk. Then claiming analogy with models for accretion disks around young stars, it is supposed that jets will form from this disk around the secondary. The jets then lead to the formation of observable COP at their heads. A weak point of this model is that it does not explain the anomalous N abundance.

5 Conclusions

PNe show a variety of collimated, fast outflow phenomena (COP). 'Fast' means faster than the normal PN expansion velocity, so faster than ~ 50 km s^{-1}. Most common are the pairs of small knots (ansae and FLIERs), but in some cases we observe small elongated tails or very extended structures (BRETs). Most COP come in pairs which are (point-)symmetric with respect to the central star and lie along the major axis of the core nebula. For the FLIERs the line ratios seem to indicate an over-abundance of N.

The study of these phenomena is severely hampered by resolution problems as was dramatically shown by the HST picture of NGC 6543. Also in a lot of cases there is no good kinematic data making it impossible to say whether it is or is not a fast outflow. A few cases have been studied well but this has not led to a satisfactory description of the phenomenon, so more observations are clearly needed. Hopefully the HST will be of assistance.

Given the problems in interpreting the observations, it is not surprising that no definite model for their origin is available. Since the phenomena occur uniquely along the core nebula's symmetry axis, the origin of the COP should be related to the origin of the asphericity. Accepting the standard model for PN formation this leaves two options: either the COP are a result of the interacting winds, or they are another effect created by whatever causes aspherical mass loss on the AGB, presumably the presence of a companion object. If the N overabundance is real, it seems to point to an origin close to the star, i.e. the latter of the two options mentioned. However, given the complexities of a common envelope system, it may be very hard to come up with a quantitative explanation for COP within this context.

6 Future

Given all the uncertainties about the COP in PNe, there is a lot of work to be done. I already mentioned further high resolution observations with the HST. Further spectroscopic investigation of some of the COP candidates mentioned above is clearly needed. It would also be interesting to search for molecular gas.

From the theoretical point of view, there is a need for combined shock and photo ionization models. If we cannot understand the structure of the COP then it will be even harder to come up with models for their origin. Clearly the origin of the COP and the origin of the aspherical shape are related, the problem is that there is not a definite model for either one of the two. But they do put severe constraints on each other, since any model for either should be able to explain the other.

In this way a study of COP in PNe is hoped to lead to a better understanding of PN formation and late stages of stellar evolution in general. Then these peculiar features may turn out to be not annoying details, but a useful way of studying late stages of stellar evolution.

Acknowledgements

I would like to thank the organiser of the workshop for inviting me and the opportunity to write this review. It is a pleasure to thank Bruce Balick, Adam Frank, Alberto López, and John Maeburn for commenting on this paper, and Bruce and Alberto letting me use some of their images. The usage of other images was kindly granted by Hugo Schwarz, Patrick Harrington, Kazik Borkowski, Howard Bond, and Mario Livio.

References

Balick, B., 1987, AJ 94, 671

Balick, B., Preston, H.L., Icke, V., 1987, AJ 94, 1641

Balick, B., Rugers, M., Terzian, Y., Chengalur, J.N., 1993, ApJ 411, 778

Balick, B., Perinotto, M., Maccioni, A., Terzian, Y., Hajian, A., 1994, ApJ 424, 800

Bond H.E., Livio, M. 1990 ApJ 355, 568

Cantó, J. 1980 A&A 86, 327

Cliffe, J.A., Frank, A., Livio, M., Jones, T.W. 1995, ApJ 447, L49

Corradi, R.L.M., Schwarz H.E. 1993a, A&A 269, 462

Corradi, R.L.M., Schwarz H.E. 1993b, A&A 278, 247

Corradi, R.L.M., Schwarz H.E. 1995, A&A 293, 871

Cuesta, L., Phillips, J.P., Mampaso, A., 1993, A&A 267, 199

Curtis, H.D. 1918, Publ. Lick Obs. XIII, p. 57

Frank, A., Balick, B., Icke, V., Mellema, G., 1993, ApJ 404, L25

Frank, A., Mellema, G. 1994, ApJ 430, 800

Frank, A., Mellema, G. 1996, submitted to ApJ

Harrington, P. 1995, STScI Newsletter 12 (1), 3

Huggins, P.J. 1993, in: IAU symposium 155: Planetary Nebulae, R. Weinberger, A. Acker (eds.). Kluwer, Dordrecht, p 147

Icke, V. 1988, A&A 202, 177

Icke, V., Mellema, G., Balick B., Eulderink, F., Frank, A. 1992, Nat 355, 524

Kahn, F.D., West, K.A., 1985, MNRAS 212, 837

Kreysing, H.C., Diesch, C., Zweigle, J., Staubert, R., Grewing, M., Hasinger, G., 1992, A&A 264, 623

Livio, M. 1995, Annals of the Israel Phys. Soc. 11, 51

López, J.A., Roth, M., Tapia, M., 1993, A&A 267, 194

López, J.A., Meaburn, J., Palmer, J. 1994, ApJ 415, L135

López, J.A., Vázquez, R., Rodríguez, L.F. 1995, ApJ in press

Maciel, W.J. 1984, A&AS 55, 253

Meaburn, J., López, J.A.,Bryce, M., Mellema, G. 1996, A&A in press

Mellema, G., Eulderink, F., Icke, V. 1991 A&A 252, 718

Mellema, G., Frank, A. 1995a, MNRAS 273, 401

Mellema, G., Frank, A. 1995b, Annals of the Israel Phys. Soc. 11, 229

Miranda L.F., Solf J., 1992, A&A 260, 397

Morris, M., 1987, PASP 99, 1115

Morris, M. 1990, in: From Miras to Planetary Nebulae, M.O. Mennessier, A. Omont (eds.). Editions Frontières, Gif-sur-Yvette, p. 520

Morris, M., Guilloteau, S., Lucas, R., Omont, A. 1987 ApJ 321, 888

Morris, M., Reipurth, B. 1990 PASP 102, 446

Neri, R., Garcia-Burillo, S., Guelin, M., Cernicharo, J., Guilloteau, S., Lucas, R. 1992, A&A 262, 544

Palmer, J.W., López, J.A., Meaburn, J., Lloyd, H.M. 1996, A&A in press

Perinotto, M., 1993, in: IAU symposium 155: Planetary Nebulae, R. Weinberger, A. Acker (eds.). Kluwer, Dordrecht, p. 57

Riera, A., Garcia-Lario, P., Manchado, A., Pottasch, S.R., Raga, A.C. 1995, A&A 302, 137

Schwarz, H.E., Corradi, R.L.M., Melnick, J. 1992, A&A Suppl. 96, 23

Soker, N., 1992, ApJ 389, 628

Soker, N. 1995, preprint: Destruction of brown dwarfs and jet formation in planetary nebulae

Soker, N. 1995, preprint: Comments on the formation of elliptical planetary nebulae

Soker, N., Livio, M. 1989, ApJ 339, 268

Soker, N., Livio, M. 1994, ApJ 421, 219

Stanghellini, L., Corradi, R.L.M., Schwarz, H.E., 1993, A&A 279, 521 te Lintel Hekkert et al. 1992

Supermassive Binaries and Extragalactic Jets

C. Martin GASKELL

Physics & Astronomy Dept., University of Nebraska, Lincoln, NE 68504-0111, USA

Abstract. Some quasars show Doppler shifted broad emission line peaks. I give new statistics of the occurrence of these peaks and show that, while the most spectacular cases are in quasars with strong jets inclined to the line of sight, they are also almost as common in radio-quiet quasars. Theories of the origin of the peaks are reviewed and it is argued that the displaced peaks are most likely produced by supermassive binaries. 3C 390.3 shows precisely the change in radial velocity predicted by the supermassive binary model. The separations of the peaks in 3C 390.3-type objects are consistent with orientation-dependent "unified models" of quasar activity. If the supermassive binary model is correct, all members of "the jet set" (astrophysical objects showing jets) could be binaries.

1. Introduction

Many quasars show the strong jets considered in detail elsewhere in this volume. However, at a conference like this it is all too easy to forget that *many quasars do not have strong jets at all*. These quasars are the radio-quiet quasars (I will use the word "quasar" to refer to all active galaxies – see Gaskell 1987 for an introduction to quasar taxonomy). To understand how and why extragalactic jets are produced and sustained I believe it is important to understand the differences between when jets are produced and when they are not. The optical and UV spectra of quasars show strong broad emission lines coming from dense rapidly moving gas (the "broad line region"; BLR). From light-echo studies we now have a good idea of the location of the BLR (see Gaskell 1994 for a review). It extends from within a few light days of the central engine (CE) to a few light-weeks or light-months away. The size of the region depends on the luminosity (Koratkar & Gaskell 1991b). This range of radius is *precisely* the range in which jets form. It is reasonable to ask therefore whether we see differences in the BLR between radio-loud quasars (*i.e.*, quasars with strong jets) and radio-quiet ones.

There are a number of differences between radio-loud and radio-quiet quasars – line profiles, Balmer decrements, strength of the extended emission, optical Fe II emission, broad absorption lines, *etc*. Not all of these are necessarily fundamental; some differences might be consequences of differing viewing angles (see Antonucci 1993, 1996 and Wills 1996). The difference I want to focus on here is the line profile difference: quasars with strong jets, especially those not believed to be viewed face-on, have broad emission line profiles that are broader and more complex than those of radio-quiet objects (Osterbrock, Koski & Phillips 1975, 1976; Osterbrock 1977; Miley & Miller 1979). I want to explore here possible explanations of these line profiles. The explanation(s) will clearly be relevant to the question of why some quasars show strong jets while others do not.

The majority of quasars have a smooth symmetric profile that can be represented by the classic logarithmic profile (see Blumenthal & Mathews 1975, and Mathews & Capriotti 1985), except in the very core. The apparent symmetry is often illusory because the lines can be shifted relative to the rest-frame of the galaxy. This is almost always the case for the high ionization lines (Gaskell 1982) which show a 600 km s^{-1} blueshift on average. Sulentic (1989) found that only ~ 15 % of a sample of quasars had Hβ profiles that were truly symmetric. In addition to these subtle asymmetries, I drew attention (Gaskell 1983b) to a class of quasars, which I will call "3C 390.3 quasars", in which the broad line peaks were displaced by *substantial* amounts relative to the rest-frame of the galaxy and in which two displaced peaks (one blueshifted and one redshifted) were sometimes present. I proposed that the displaced peaks were caused by two separate BLRs, each associated with a separate CE forming a supermassive binary (SMB) in the center of the host galaxy. I further suggested that the presence of such a binary may well be essential for the formation of collimated large-scale radio jets. In this paper I review and update the evidence for the SMB model and discuss these ideas further.

2. Properties of 3C 390.3 Objects

The classic example of a displaced broad-line peak object is 3C 390.3. It was the first example discovered (Sandage 1966; Lynds 1968), is by far the best studied, and shows all the features of the class. Following astronomical tradition I will call members of the class of quasars with shifted emission line peaks "3C 390.3 objects". Before discussing models of the 3C 390.3

phenomenon I first summarize the properties of the broad lines in 3C 390.3 objects, including some properties that are quite obvious.

1. *The relative velocities are large.* Relative velocities of thousands of kilometers per second are relatively common. 3C 390.3 itself has a displaced blue Balmer peak at ~ -4000 km s^{-1} and a displaced red peak at $\sim +2000$ km s^{-1}. Blending of profiles makes the determination of the distribution of relative velocities difficult and small velocity differences will be missed[1].

2. *All quasars show an undisplaced peak.* Without exception, every 3C 390.3 object shows a central (undisplaced) peak. The profile of this peak is uncertain since it depends on the subtraction of the displaced components and of the narrow line region (NLR). In some cases at least, it seems to be broader than the NLR lines.

3. *A substantial part of the line flux comes from the displaced peaks.* When the peaks are well separated it is obvious that *the majority of the flux* in the Balmer lines comes from the displaced peaks (see, for example, figures 2 & 5 of Zheng, Veilleux & Grandi 1991 or figures 4*a-j* of Eracleous & Halpern 1994).

4. *Quasars with displaced peaks have broader lines than ordinary quasars.* This has been known for some time (Lynds 1968; Osterbrock, Koski & Phillips 1975, 1976; Osterbrock 1977), but I give a new analysis of this in section 4 below.

5. *3C 390.3 objects have steeper Balmer decrements.* It is well known that lobe-dominated quasars have steeper Balmer decrements than radio-quiet quasars (Grandi & Osterbrock 1978). It is therefore not surprising that radio-loud 3C 390.3 objects have steep Balmer decrements, but I show below (see section 6.7) that *radio-quiet* 3C 390.3 objects also have slightly steeper Balmer decrements.

6. *The peaks vary independently in brightness.* Like probably all quasars, 3C 390.3 is variable and the lines vary as well as the continuum. Veilleux & Zheng (1991) give a useful display of many Lick Observatory spectra. At some epochs (such as the summer of 1975) the blue displaced peak is prominent; at others (such as June 1980), the red peak is more prominent.

7. *There could always be two displaced peaks.* Not all displaced BLR peak objects show two peaks − some only show one − but I believe that if we look carefully enough or wait long enough most of the *single* displaced peak objects will show two peaks. 3C 390.3 itself is important because it shows all the characteristics of the class at various times as far as relative strengths of peaks go.

---

[1] The situation is very similar to the problem of estimating the distribution of apparent separations of visual double stars.

8. *There is no unambiguous case of more than one peak with the same sign of the displacement.* I believe that no spectra taken *at one epoch* show more than two displaced peaks or the two peaks being displaced in the same direction. I emphasize "at one epoch" because it is easy to see all sorts of peaks in difference spectra (the difference between spectra taken at different epochs). For example, if a symmetric line simply becomes broader the difference spectrum will show at least one peak in each wing. Noise bumps are also magnified in difference spectra. Peaks in difference spectra therefore need to be interpreted with great caution. The presence or absence of more than two displaced peaks is a very important issue since some models only predict two peaks and any claim of more than two peaks needs critical examination[2].

9. *The central peak varies less than the displaced peak.* If spectra of different epochs are subtracted, the central peak often subtracts out perfectly (see Wamsteker et al. 1985, figure 3, for example).

10. *The central peak has different line ratios from the displaced peaks.* This is particularly noticeable for Ly α which is much stronger in the central peak in 3C 390.3 (Ferland et al. 1979; Zheng, Veilleux & Grandi 1991). This is also true for non-3C 390.3 quasars (Zheng 1992).

11. *All broad lines show the displaced peak effect.* The C IV λ1549, C III] λ1909 and Mg II λ2798 lines of 3C 390.3 have profiles which are consistent with the Hβ profiles of the same epoch (Perez et al 1988). The displaced blue peak is clearly visible in the C IV line, albeit shifted by $\sim$ 900 km s^{-1} (see figure 3 of Perez et al. 1988). The redward peak is harder to separate, but Perez et al. think it is also blueshifted relative to the one in Hβ. The 900 km s^{-1} shift is the normal blueshift of high ionization lines with respect to low ionization lines (Gaskell 1982). The cause of this shift is not fully understood, but it is interesting that, at least in 3C 390.3, the separate peaks show it. It will be interesting to see whether this remains generally true as *HST* data become available for more 3C 390.3 objects. Perez et al. (1988) find other broad emission line profiles in 3C 390.3 to be basically similar to those of C IV and Hβ. Unfortunately all lines have blending problems that hinder detection of displaced peaks. The best line for detecting displaced peaks is probably Hβ, but even it is not perfect. Hα has the [N II] lines on either side of its peak. Hγ is weak and had [O III] λ4363 on its red wing. Ly α is blended

---

[2] Veilleux & Zheng (1991) claim that 3C 390.3 shows a highly redshifted peak at +4600 km s^{-1} in the 1974-75 Lick spectra (the main redshifted peak is at about +2000 km s^{-1}). However, this is almost certainly an instrumental effect caused by a spurious dip in the spectrum. This is demonstrated by the alleged +4600 km s^{-1} feature being strongest in two spectra taken on the (UT) nights of June 4 and June 5, but vanishing in a longer observation with a slightly different setup the very next night. For further discussion of this, see Gaskell (1995).

with N V λ1240 on the red side and tends to have absorption on the blue side. The He II lines are badly blended (He II λ1640 with C IV and [O III] λ1663; He II λ4686 with Fe II). C IV λ1549 has He II on the red wing. The higher ionization lines in quasars in general tend to be wider than the low ionization lines (Shuder 1982; Mathews & Wampler 1985) and Ly α is often con-siderably broader than the Balmer lines (Zheng 1992). This also makes detection of displaced peaks harder. The strength of the central component of Ly α noted in the previous section is a further problem for detecting displaced peaks.

In addition to these properties of the broad emission lines in 3C 390.3 quasars, there are some other differences between 3C 390.3 quasars and non-3C 390.3 quasars in their other properties:

1. *3C 390.3 objects have a larger contribution of starlight to the optical continuum.* The median percentage of starlight is about three-times greater on average in the most extreme radio-loud 3C 390.3 objects compared with other radio-loud quasars. In a number of extreme 3C 390.3 quasars the fraction is close to 100%. Eracleous & Halpern (1994) interpret this to be indicative of a weaker "blue-bump" implying a different accretion-disk structure in 3C 390.3 objects. I believe, however, that a simpler explanation, since 3C 390.3 objects also have steeper Balmer decrements, is that there is simply more reddening. This has a natural orientation explanation in unified schemes (see section 6.7 below).

2. *The equivalent widths of the forbidden lines are greater in 3C 390.3 quasars.* The median equivalent widths of the low-ionization [O I] λ6300 and [S II] λλ6716, 6731 lines of the most extreme 3C 390.3 quasars are about double those of other radio-loud quasars. In contrast, for the higher ionization [O III] λλ4959, 5007 lines, Eracleous & Halpern (1994) find no such difference in the equivalent width distributions. They conclude that the ionization parameter is systematically smaller in the extreme 3C 390.3 quasars. I believe, however, that the difference in equivalent widths can also be readily explained as an orientation effect. The low-ionization lines come from a very extended, often conical, region that will not be easily obscured from any angle. Their equivalent width is expected to increase when they are viewed edge-on. The [O III] lines, on the other hand, come from much closer in. They will suffer almost as much extinction as the continuum when the quasar is viewed close to edge-on. Their equivalent width will be much less orientation dependent. In this picture, the ionization parameter is the same on average, but the closer-in higher-ionization lines suffer greater extinction. This is supported by De Zotti & Gaskell (1984) finding broad-line

Balmer decrements to be orientation dependent while narrow-line decrements are not.

One additional alleged difference needs addressing: Halpern & Eracleous (1994) claim that the far-IR properties of 3C 390.3 quasars are significantly different from the far-IR properties of "normal" radio-galaxies. They point out that the three quasars with the flattest spectra between 25 and 60 μm "among the sample of 131 radio galaxies observed by IRAS" (Golombek, Miley & Neugebauer 1988), are all extreme 3C 390.3 quasars[3]. If this is really a significant effect, there are a number of possible explanations, but I believe there is not a significant effect. At first sight it looks as though the significance level is 1 in 9 x 10^4, but there is a serious problem in how the comparison is made. Naturally, Eracleous & Halpern (1994) observed quasars which they initially thought would show broad lines. The IRAS sample was not so chosen. It includes "narrow-line radio galaxies", LINERs and many quasars with only stellar absorption line spectra. Broad-line quasars are in the minority. Inspection of the Golombek et al. (1988) data shows that broad-line quasars have a mean 25 to 60 μm spectral index of -0.22, while the narrow-line quasars have a mean index of -0.99, and the LINERs and non-emission-line galaxies have a mean index of -1.86. The differences between these are significant at about the 90 – 95 % level. The Eracleous & Halpern (1994) objects in common with the IRAS sample are, of course, only from the broad-line subset and it is only with these that a comparison should be made. *The 25 to 60 μm spectral indices of the 3C 390.3 quasars in the IRAS sample are not significantly different from the other broad-line quasars in the IRAS sample.* In other words, I believe the 3C 390.3 objects have flat far-IR spectra *because they are broad line objects*, not because they are 3C 390.3 objects. A preliminary check of some of the 25 to 60 μm spectral indices of radio-*quiet* 3C 390.3 objects shows that they are similar to the radio-loud ones and that, again, there is no statistically significant difference between these objects and the "non-3C 390.3 type" ones.

Why does the far-IR emission depend on whether we can readily see the BLR? Since far-IR emission is essentially completely isotropic the explanation cannot be in orientation effects. Miley et al. (1985) and Golombek et al. (1988) argue that a combination of two dust components is the likeliest interpretation of the radio galaxy data. The main body of the galaxy produces the coldest emission while the nuclear region has warmer dust that raises the 25 μm flux. They do not point out the difference between broad-line and narrow-line quasars, but I believe it can be explained as follows: The main variable is the amount of dust in the host galaxy. This

---

[3] Actually, "131 radio galaxies" is misleading. Although Golombek et al. (1988) observed this number, they only *detected* 58 of them.

dust produces the 60 μm or longer wavelength emission; the more cool dust there is the steeper the far-IR spectrum. A lot of dust also makes it less likely that we will be able to see BLR; even more dust will make it impossible even to see the NLR.

3. Frequency of Occurrence of 3C 390.3 Objects

Gaskell (1983b) noted that the best cases of displaced broad lines were in extended radio sources (= "lobe-dominated" quasars). This still remains true, but the availability of high-quality profiles has revealed that many radio-*quiet* quasars also have displaced peaks. Osterbrock & Shuder (1982) and Stirpe (1990) give good samples of profiles. These are particularly valuable because of the careful subtraction of narrow lines, and in case of Stirpe (1990), of Fe II emission and the stellar continuum as well. I have visually classified the profiles in the Stirpe atlas into "no displaced peak", "single displaced peak" and "double displaced peak". If a displaced peak has a small displacement it will appear only as an inflection on the side of the central peak unless it is very strong. If it is very strong, the line peak will be shifted from the systemic redshift of the host galaxy. I therefore identified such inflections as displaced peaks. I summarize the results in the table 1 (the Osterbrock & Shuder 1982 spectra are consistent with these results).

Table 1. Statistics of displaced emission lines in radio-quiet quasars.

Profile Type	Percentage
No obvious displaced peaks	40%
Single displaced peak	25%
Two displaced peaks	35%
Blue peak strongest	29%
Red peak strongest	25%
Peaks approximately equal	6%

The most striking thing in this table is the high percentage of displaced peaks. This is particularly so since the percentage of displaced peaks must have been underestimated, possibly by a lot. Displaced peaks vary. For example, Mrk 6 (not a member of the Stirpe sample) shows no displaced peak when it is in a low state, but when it is bright it has a clear single blueward-displaced peak (see Rosenblatt et al. 1994, figure 14). The percentages of displaced peaks would also certainly be slightly higher still if my search had been conducted by subtracting off an appropriate central peak or subtracting spectra taken at different epochs. Likewise, such analyses would reveal more second peaks in the objects that are currently classified as only having single displaced peaks. The other thing to notice from table 1 is that the percentages of blue and red peaks are about the same. The excess of blue peaks is not statistically significant, but models invoking organized bulk motions to explain the displaced peaks predict that the blue peaks will be slightly stronger because of relativistic effects (see Mathews 1982).

Eracleous & Halpern (1994) have undertaken a useful large survey of radio-loud quasars. Their motive was to search for quasars with emission lines that might have arisen from a disk, but their sample can be used to look at the statistics of displaced emission line peaks. Unfortunately, unlike Stirpe (1992), they do not report Hβ as well, nor do they subtract the narrow emission lines (the [N II] lines are particularly bothersome), and they only show starlight subtracted profiles for quasars which they considered "disk-like". This makes their survey a little less sensitive to displaced peaks than the Stirpe atlas. I classified the Eracleous & Halpern spectra in the same way as the Stirpe spectra. I excluded a few spectra from the sample because they were not classifiable. This was either because of the strength of [N II] emission, the narrowness of Hα, or a poor signal to noise ratio. The remaining spectra give the following statistics:

Table 2. Statistics of displaced emission lines in radio-loud quasars.

Profile Type	Percentage
No obvious displaced peaks	30%
Single displaced peak	32%
Two displaced peaks	38%
Blue peak strongest	26%
Red peak strongest	28%
Peaks approximately equal	16%

The first thing to notice from this table is the higher percentage of displaced peaks. This is particularly significant in light of the caveats just given. What is not clear from table 2 is that the displaced peaks are also more obvious and spectacular. Although *most* of the spectacular cases of displaced broad-line peak objects are strong extended radio sources, *not all are* (this contrary to what Eracleous & Halpern 1994 state). IC 4329A is a clear counter-example (Disney 1973). IC 4329A has a strong red displaced peak at a Δz of +2000 km s^{-1}, but IC 4329A is not a strong radio source[4]. The second thing to notice from table 2 is that apart from the higher overall percentages, the ratio of occurrence of blue to red peaks is again approximately equal[5]. As was noted for the radio-quiet objects, the percentages of displaced peaks given in table 2 *must be underestimates* because of variability. Pictor A is a good illustration of this. If it were classified using the 1983 spectrum of Filippenko (1985), it would be classified as having no obvious displaced peaks, but the 1994 spectrum taken by Halpern and Eracleous (1994) shows clear double displaced peaks. It is important to remember, when considering profile classification, that essentially all quasars vary and the prototypical object, 3C 390.3 itself, could be placed in *any* of the categories in table 2 depending on when the observation was made!

3C 390.3 objects cover a range in optical luminosity running all the way up to high-luminosity objects. OX 169 with $M_{abs} = -24.7$ (for $H_0 = 50$ km s^{-1} Mpc^{-1}) is a high-luminosity example (Gaskell 1981). A couple of selection effects are expected to be working against detecting high-luminosity 3C 390.3 objects. Because they are rare, high-luminosity quasars are not seen nearby. They therefore have high redshifts and only the rest-frame UV lines are readily seen. The difficulties in detecting displaced peaks in the UV lines (see 2.11) make the detection of high-redshift 3C 390.3 objects difficult, but they do exist. PKS 0119-046 (z = 1.953, Mabs = −28.5) has a narrow component to Ly α. Ly α and C IV have broad peaks blueshifted by −3000 km s^{-1} relative to the narrow peaks (Gaskell 1983a). This *could* be an extreme example of the high ionization blueshifting effect (Gaskell 1982), but the relative velocity is more consistent with PKS 0119–046 being a 3C 390.3 object.

Although 3C 390.3 objects tend to be lobe-dominated quasars, there are exceptions. OX 169, for example, is core dominated. As Eracleous & Halpern (1994) point out, 3C 390.3 objects fall into both Fanaroff & Riley classes and they "have no special characteristics in the radio-band that would distinguish them from typical broad-line radio galaxies."

---

[4] IC 4329A does however have a compact core with a luminosity that places it at the upper end of the Seyfert galaxy radio luminosity function (Giuricin et al. 1990a).
[5] Eracleous & Halpern (1994) themselves identify a lower number of displaced peaks, because their motivation is fitting them with their disk model. Eracleous & Halpern mention an excess of dominant blue peaks but this is only the case when only the double displaced peak objects are considered. If we consider the single displaced peak objects as well, the numbers of red and blue dominant peaks are the same.

4. Widths of Emission Lines

3C 390.3 objects have long been noted for their very broad lines (e.g., Lynds 1968; Osterbrock, Koski & Phillips 1975, 1976). It is also well known that lobe-dominated radio-loud quasars have broader lines than core-dominated or radio quiet ones (Miley & Miller 1979). There is good evidence that this is an orientation effect (Wills & Browne 1986). However, I want to show here that the width of lines in 3C 390.3 objects is not just a result of their predilection for radio-loud quasars. This can be seen if we look at the median line widths of Hα in the two radio-loud and radio-quiet quasar samples considered above (see table 3). From table 3 it is obvious that *regardless of whether there are strong radio jets or not*, the displaced-peak quasars have wider lines.

Note the progression from no displaced peak to single displaced peak to double peaks. It is important to note that most of the differences in line width can be accounted for *by the velocity shifts of the displaced peaks themselves*. For the radio-quiet objects a typical shift is about 1250 km s^{-1}. Thus the FWHM goes from 2150 km s^{-1} when no peak is present to 2150 + 1250 = 3700 km s^{-1} when one displaced peak is seen and to 2150 + 1250 + 1250 = 4650 km s^{-1} when two are seen. The separations are larger for the broader line objects. Average increases of 2650 km s^{-1} will similarly explain increases in both the FWHM and FWZI in the radio-loud sample.

Table 3. Median FWHM and FWZI (km s^{-1}) of Hα for different profile types.

	Radio-quiet FWHM	Radio-loud FWHM	Radio-loud FWZI
No obvious displaced peak	2150	4300	17800
Single displaced peak	3400	5000	19800
Double displaced peaks	4700	9800	12900

5. Models

In this section I review the main models proposed to explain 3C 390.3 objects. But first, one consensus should be noted: everybody agrees that the central peak is due to a separate component of gas that is probably somewhat further out than the gas producing the displaced peaks. This is because of the differing line ratios and lack of variability. I believe that this gas is the "intermediate line region" of Wills et al. (1993) and Brotherton et al. (1994). I discuss this further in section 6.8.

5.1 Obscuration Effects

Capriotti et al. (1979) and Ferland, Netzer & Shields (1979) suggested very asymmetric profiles were the result of obscuration and radial motion. Although good fits can be obtained to some lines (see Capriotti et al. 1979) the problem with this explanation is that blue and red asymmetries are equally prevalent (see section 3). The model is in the unsatisfactory situation of requiring BLR inflow in some objects but outflow in others (Gaskell 1983b).

5.2 Self-Absorption

A number of quasars show intrinsic absorption line systems with redshifts close to the emission-line redshift (e.g., NGC 4151; Anderson 1974). A number of people suggested that the dips in the profiles of 3C 390.3 objects are due to self-absorption (e.g., Boksenberg & Netzer 1977; Smith 1980). UV studies have shown that when there *is* such absorption, it is quite narrow and does not produce the broad gentle dips seen in 3C 390.3 objects. Also, self-absorption is usually on the blue side of an emission line so it is an unlikely explanation of the very redshifted displaced peaks. Nonetheless, the role of self-absorption should not be forgotten since it does occur to some degree (see, for example, the Hβ and C IV profiles in NGC 3516; Wanders et al. 1993)

5.3 Light Echo Effects

Capriotti et al. (1982) suggested that displaced BLR peaks were the result of pulses of the photoionizing continuum in an ordered BLR velocity field. While such a model can fit some lines, the light travel times across BLRs are now known to be much too small by more than an order of magnitude (Gaskell & Sparke 1986) for this idea to be tenable. The wavelengths of displaced peaks do not vary on a timescale as short as the light-crossing timescale (Gaskell 1983b).

5.4 Ejection of Clumps of BLR Material from the Central Engine

This was the first explanation of displaced peaks offered (Burbidge & Burbidge 1972; Osterbrock, Koski & Phillips 1975). The idea is attractive because radio-emitting plasma is clearly ejected in the jets and we see broad blueshifted UV absorption lines in some radio-quiet quasars. Bi-conical ejection was suggested for 3C 390.3 itself by Oke (1987) and explored in detail by Zheng, Binette & Sulentic (1990) and Zheng, Veilleux & Grandi (1991). The latter were able to obtain good fits to a couple of epochs[6]. The main problem with the jet model (Oke 1987, Gaskell 1988b) is that it is the *opposite* of what is expected in unified models. Core-dominated sources (where the jet is coming towards us) would be expected always to show strong displaced peaks while lobe-dominated sources where the jet is at an angle to the line of sight would show smaller shifts of the peaks. As we have noted above, some of the most spectacular cases are in lobe-dominated quasars. Unified models therefore imply that the bulk motions producing displaced peaks are in the plane *perpendicular* to the jet axis.

Unfortunately, a large number of variability studies of quasars show that the net motion of BLR gas is *not* a simple outflow (Gaskell 1988a, Koratkar & Gaskell 1989, 1991a, Crenshaw & Blackwell 1990, Maoz et al. 1991, Korista et al. 1995)[7]. These studies all show that in the majority of cases, if anything, there is a slight net *inflow*, because the redshifted C IV wing tends to lead the blue wing slightly (see also Done & Krolik 1996).

Bi-conical outflow models also have difficulty explaining why there are so many *single* displaced peak objects (especially when the peak is on the red side).

5.5 Anisotropic Illumination of the BLR Gas

Zheng, Veilleux & Grandi (1991) also mention the possibility that double peaks could be produced by anisotropic illumination of the BLR (a "searchlight"). This also requires net inflow or outflow of the BLR. There is now some evidence that the ionizing radiation is confined to a wide conical beam as it illuminates the BLR in at least one object (NGC 5548; Wanders et al. 1995). Goad et al (1996) show that double-peaked profiles arise when the observer lies outside this cone. The further the observer is outside the cone the more double peaked the lines appear and the weaker the continuum becomes. It is not clear yet whether a satisfactory fit to the profiles of the

---

[6] However, everybody always gets good fits to the line profiles! The ability to reproduce the profiles is unfortunately not a way to distinguish between models.

[7] For a possible way of reconciling these observations with outflow see Kundt (1988).

extreme 3C 390.3 objects can be obtained with parameters that are consistent with the observed slight inflows (since the variability studies imply that the *dominant* motion is not pure inflow). The model also has to explain the smooth shift in wavelength of the blue peak in 3C 390.3 (see figure 1 below).

5.6 Accretion Disk Models

Since quasars often show a clear axis of symmetry (the line the jets emerge along) something in the inner region is rotating, probably the central engine itself. Not surprisingly, as Mathews (1982) puts it, "Rotational energy production and accretion disks as explanations of broad emission lines have been regarded as possibilities since the earliest quasar literature (Woltjer 1959; Lynden-Bell 1969)." He gives an already quite long list of papers up to 1981 that give models employing rotational broadening and mentions that there are "numerous other studies" in which favorable allusions to the possibility appear. Now, 14 years later, the list would be many times longer and I will not attempt to give one. The interested reader is referred to Collin-Souffrin et al. (1988), Collin & Dumont (1989), Dumont & Joly 1992 and Rokaki (1994) for theoretical arguments in favor of line emission from disks and further references. Mathews (1982) calculates theoretical profiles from disks and concludes for a variety of reasons that "rotation is an extremely unlikely means of producing broad emission lines". One of Mathews's biggest arguments is that quasar emission line profiles do not resemble those expected from disks. The realization about five years later that quite a number of quasars actually *do* show double-peaked profiles, or at least could be made to show double profiles by suitable differencing of spectra led many authors to propose that the displaced peaks were produced in disks (Oke 1987, Alloin, Boisson & Pelat 1988, Stirpe, de Bruyn & van Groningen 1988, Halpern & Filippenko 1988 and Perez et al. 1988). There have subsequently been a number of attempts to fit disk models to observations. The largest attempt has been by Eracleous & Halpern (1994).

Gaskell (1988b) identified three problems with disk models:

1. Although fits can be obtained in *some* cases[8], the ratio of peak intensities is often not right (this point is also made by Sulentic et al. 1990). Usually only one displaced peak is prominent, and in many cases the other cannot even be detected (see tables 1 and 2 above). Also, about 50% of the time it is the *red* peak which is strongest. As Eracleous & Halpern (1994) concede, these objects cannot be explained by the simple disk model. These are a large fraction of the displaced peak quasars. Eracleous & Halpern

---

[8] See footnote 6.

suggest that the SMB model might be appropriate for some of these.

2. There is no natural explanation of the "normal" ("classic" or "logarithmic") line profile in the disk model.

3. Line variability rules out disks. This is the strongest argument against disks. The continuum of quasars is seen to vary and the lines vary in response to the continuum changes (this is the whole basis of the quasar reverberation mapping cottage industry!). If the lines arise in a disk, and are ionized by a central source, then *both peaks must always vary up and down together*. In general they do not (Oke 1987 - see his figure 1; Peterson, Korista & Cota 1987; Miller & Peterson 1990; Veilleux & Zheng 1991 - see their figure 2e)[9]. A disk model which fits at one epoch will not fit at another epoch! Even if a disk model can be made to fit at more than one epoch, an unphysical change of parameters might be needed (e.g., in the inclination of the disk).

In order to circumvent the first of these problems Zheng, Veilleux & Grandi (1991) modified the simple disk model to include a "hot spot". There is precedence from cataclysmic variable stars for including such hot spots (see Zheng et al. 1991). They obtain satisfactory fits to two epochs of 3C 390.3 in this manner[10]. In a similar vein, Eracleous et al. (1995) introduced highly elliptical disks to get round problem number one. They are able to fit some of their profiles that could not be fit with simple circular disks[11], but I think that both this model, and the Zheng et al (1991) one, are probably already excluded by existing variability observations.

There are a couple of additional problems for disk models:

1. All disk models need to introduce an arbitrary broadening by some additional mechanism.

2. Although the fits to the two humps can be acceptable, the disk models do not fit the high velocity wings, as Eracleous & Halpern (1994) point out (this is despite having 7 free parameters in the circular disk model and 9 in the elliptical disk model).

3. The disk model makes very specific predictions of the percentage polarization of the broad lines, the polarization position angle within the line, and the shape of the polarized flux (Chen & Halpern 1990). Unfortunately observations do not confirm these specific predictions (Antonucci, Hurt & Agol 1995). The polarization observations do, however, show that the wings are

---

[9] Some authors *have* claimed that some profile variations are consistent with a disk origin however (Alloin, Boisson & Pellat 1988; Stirpe, de Bruyn & van Groningen 1988, but see also Marziani, Calvani & Sulentic 1992).

[10] See footnote 6

[11] Again, see footnote 6!

polarized and probably do arise in a region separate from the line core (but this is the one point that almost all models agree upon!)

4. It is difficult to get a disk to emit lines. Most workers on this model favor an external source of ionization shining or reflecting back onto the disk.

5.7 Supermassive Binary (SMB) Model

Gaskell (1983b) suggested that the displaced peaks in 3C 390.3 objects were due to two separate BLRs each with its associated central engine. The associated central engines are necessary because the BLRs will not survive without them. The high displacement velocities of several thousand km s^{-1} are the projected velocities as the engines orbit each other. The idea that quasars consist of some sort of binary had been originally proposed by Komberg (1968). In the late 1970's the discovery of jets from the galactic stellar-mass binary system SS 433 followed by radio observations suggesting that the radio axes of some quasars wobble or precess (Ekers et al. 1978) led to further consideration of the idea (Collins 1980; Begelman, Blandford, & Rees 1980; Whitmire & Matese 1981).

There are several ways a supermassive binary in the center of a galaxy could form, but the most sure-fire one is through mergers (Blandford, Begelman & Rees 1980; Roos 1981). Galaxy mergers are very common and still going on today. A large fraction of galaxies, perhaps 50%, show at least low-level quasar activity, (see Keel 1985 for a review), so some sort of central engine must be present. Also, even when there is no quasar activity, we see signs of inactive supermassive objects (e.g., in M 31 and M 32 in the local group). Therefore many mergers *must* form supermassive binaries. As two galaxies merge, the two nuclei will spiral together. This is beautifully shown in the video of Barnes (1993). It is important to realize that this is something common that is happening all the time. There can be no doubt that supermassive binaries form. Indeed, if 3C 390.3 objects eventually prove *not* to be a consequence of the formation of supermassive binaries, it is reasonable to ask, "what *are* the consequences?"

The SMB model correctly predicts several things:

1. *The BLR profiles*. Because of its non-uniqueness, profile fitting does not "prove" the correctness of a model, but displaced peaks in difference-spectra, where one hopes that the central stationary component has cancelled out, can be fit well with just a pair of Gaussians (six parameters). Good examples can be seen in Akn 120 (Alloin, Boisson & Pelat 1988) and Pictor A (Halpern & Filippenko 1994). In both of these cases, a disk model is a poor-fit, even for $i = 90°$. This is because the disk models do not reproduce the lack of emission between the peaks. For

Pictor A, the two Gaussian-like components in the difference spectrum are well separated. Roughly equal peaks, as in the case of Pictor A are not the rule. The red peak in the Alloin, Boisson & Pelat (1988) difference spectrum of Akn 120 is 50% stronger than the blue peak. The Wamsteker et al. (1985) difference spectrum of Fairall 9 in the early 1980's is flat apart from one approximately Gaussian, well-defined, red displaced peak. The NGC 5548 difference spectra from the mid-1980's are flat apart from a similarly well-defined peak to the blue (Peterson, Korista & Cota 1987).

2. *The velocities seen*. An SMB is most likely only to be seen in a limited velocity range because there are natural lower and upper limits to the orbital velocity (Gaskell 1983b). The initial merger takes place on the dynamical time-scale and the orbit decays fairly rapidly because of tidal friction (Begelman, Blandford & Rees 1980). Again, this is well illustrated in the Barnes (1993) video. Although we do see a few cases of mergers of quasars in progress (such as Abell 400, Mrk 78 and Mrk 266), we expect them to be rare, simply because they happen so quickly. As the orbital velocity of the binary goes up, the Rutherford gravitational cross-section goes down and the rate of evolution decreases (Begelman, Blandford & Rees 1980). The timescale will increase when the orbital velocity exceeds the stellar velocity dispersion (a few hundred km s^{-1}), so this is a *natural lower limit* to the orbital velocity of observed SMBs (Gaskell 1983b). At high velocities, gravitational radiation causes rapid decay of the orbit when the speed starts to become relativistic (Begelman, Blandford & Rees 1980). This causes a *natural upper limit* to the orbital velocity of a few percent of the speed of light (Gaskell 1983b).

3. *There are only two displaced peaks at most* (Gaskell 1983b). Multiple mergers are possible, but three-body systems are unstable except in certain particular cases. One or more components will be ejected from the galaxy via the "gravitational slingshot" mechanism (Saslaw, Valtonen & Aarseth 1974; Makino & Ebisuzaki 1994). Any stable hierarchical systems (supermassive analogs of α Geminorum *etc.*) which do form will have relatively short lifetimes because of gravitational radiation.

4. *The peaks can be of very different brightness*. Since the luminosity of the BLR correlates well with the luminosity of the central engine (Yee 1980; Shuder 1981), and the luminosity of the central engine correlates with its mass (Koratkar & Gaskell 1991b and references therein), the model predicts that when the average luminosities of the two components are unequal, the brightest component will have the smallest velocity shift.

5. *The two peaks will vary independently in brightness* because they have independent continuum sources. This seems to be the case in every 3C 390.3 object seen to vary so far (Wamsteker et al. 1985; Peterson, Korista & Cota 1987; Miller & Peterson 1990; Veilleux & Zheng 1991; Marziani

et al. 1993). Each BLR will vary with its continuum. We will see the combined effects. The most variable component, in the absolute sense (which is probably also the brightest one) will dominate the observed continuum variability. Only one of the two peaks will be observed to follow the continuum. This is obviously the case for 3C 390.3 (Veilleux & Zheng 1991)[12] and has also been claimed for NGC 5548 (Peterson, Korista & Cota 1987).

6. *There will be smooth radial velocity changes.* Gaskell (1983) estimated that orbital periods would be of the order of centuries and therefore that radial velocity changes could be detectable in a few decades. Although Halpern & Filippenko (1988) claimed that the lack of a radial velocity change in Arp 102B ruled out the SMB model, Veilleux & Zheng (1991) reported what seemed to be a shift in the wavelength of the blue peak in 3C 390.3 in Lick Observatory spectra taken by D.E. Osterbrock, J.S. Miller and their collaborators. Marziani et al. (1993) report a variation of the displacement of the peak of Hβ in OQ 208 *with luminosity* (rather than with time). I believe this is not a real effect but the result of the blending of two components. My analysis of over two decades of 3C 390.3 data (see figure 1) shows that there is indeed a very clear and convincing radial velocity curve (Gaskell 1995). This is compatible with the period of a few centuries predicted by Gaskell (1983b). The weakness of the Filippenko & Halpern (1988) argument is the assumed total mass.

6. Discussion

6.1 System Parameters for 3C 390.3

The radial-velocity curve for 3C 390.3 (Gaskell 1995 – see figure 1) allows the derivation of a number of parameters. The minimum orbital period that provides an acceptable fit to the curve is 210 yr, with 300 yr being more likely ($\chi^2 = 0.8$ per degree of freedom; see Gaskell 1995 for details). This gives a projected maximum velocity ($v \sin i$) of 5340 km s^{-1}. It is probably safe to assume that the orbits have been circularized. The synchrotron self-Compton model and relativistic beaming considerations give an estimate of i, the inclination of the 3C 390.3 jets to the line of sight (Ghisellini et al. 1993). The ratio of optical to radio core flux supports such estimates (Wills & Brotherton 1995; Wills 1996). For 3C 390.3, Ghisellini et al. get $i \sim 29°$. This gives a true orbital velocity of 11,000 km s^{-1}. The minimum radius of

---

[12] The plot of the *ratio* of fluxes of the blue and red peaks versus time looks smoother than either individual light curve. I think the smoothness is because the errors of calibrating the fluxes in the standard way with respect to the narrow-line region lines cancel out when the ratio is taken.

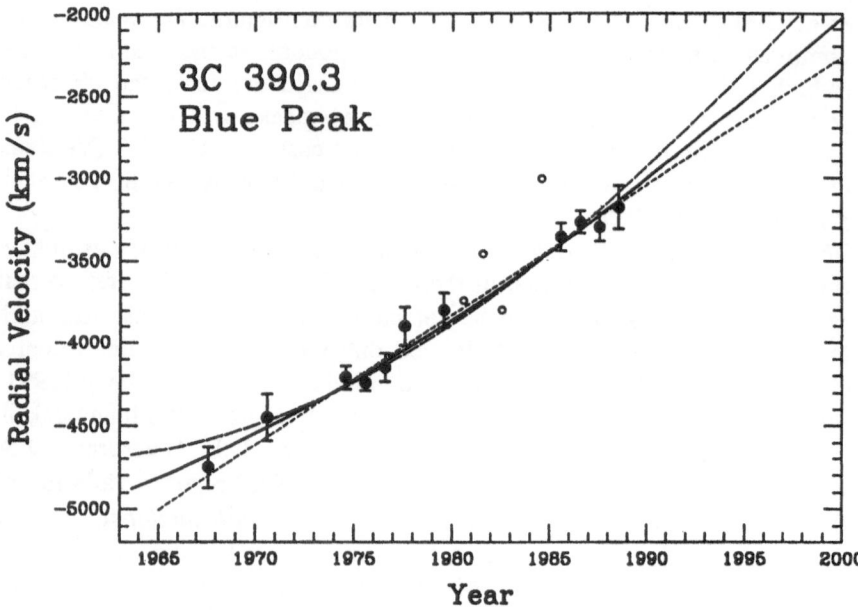

Fig. 1 Radial-velocity curve for the blue displaced broad peak of Hβ in 3C 390.3. The curves are for periods of 210 yr, 300 yr and an infinite period (from Gaskell 1995).

the orbit is therefore at least one light-year. The ratio of displacements of the peaks is 2:1 so the minimum masses of the two central engines are 4.4×10^9 and 2.2×10^9 solar masses. These are slightly higher than the mass predicted by the Koratkar & Gaskell (1991b) mass-luminosity relationship, but in good agreement with the more reliable estimate of 2.4×10^9 solar masses, from HST observations, for the mass of the CE in M 87 (Ford et al. 1994). They are also consistent with masses of remnants deduced from quasar counts (Soltan 1982) and the masses of 10^9 solar masses for the dark central objects in M 104 (Kormendy 1988) and NGC 3115 (Kormendy & Richstone 1992).

The spin axes of the central engines will undergo geodetic precession. We can derive the minimum precession period from the derived masses and orbital size. Equation 8 of Begelman, Blandford & Rees (1980) gives a period of 4×10^5 yrs.

For a given orbital velocity, the radius of the orbit and the period will both vary linearly with the sum of the masses. On the other hand, for orbits of approximately the same size, the period will increase with the square root of the mass while the orbital velocity will decrease with the square root of the mass. Since the observed separations of peaks are similar, we would expect the less luminous sources to show the most rapid radial velocity changes, if they do indeed have lower central-engine masses.

Note that the size and mass deduced are the same in a disk plus hot-spot model.

6.2 BLR Kinematics

As was pointed out by the late Michael V. Penston, if BLR cloud motions are gravity dominated, a supermassive binary does not give a line profile that is simply the sum of the profiles around two separate central engines. This is because the low-velocity gas in the peaks would have to arise from far out (see footnote 3 of Cheng, Halpern & Filippenko 1989). If the SMB model is correct, it has important implications for BLR cloud motions: they have to be sub-virial. Since non-gravitational forces such as radiation pressure and winds are already known to be significant (indeed, they are the dominant force for broad absorption line clouds in quasars), this is not unreasonable. If we ignore drag forces, ignore the extended mass distribution of the stars, and consider only radiation pressure in addition to gravity, the clouds will be subject to a net inverse-square law force which is less than the force of gravity. The velocities at a given distance will therefore be lower than the virial velocities. The masses Koratkar & Gaskell (1991b) derived from BLR cloud motions will be too small. Mathews (1993) has gone further and included the effects of drag forces. He shows that masses have been underestimated by factors of $\sim$ 10–20.

6.3 Implications for Reverberation Mapping

The separation deduced for the components of 3C 390.3 is not much greater than the BLR sizes in comparable objects (Koratkar & Gaskell 1991a,b; see Peterson 1994 for a recent review of the many new results), so, while each component of the BLR will be dominated by the variations of its corresponding central engine, the light from the brightest one may well strongly influence the BLR of the other. Thus, when a 3C 390.3 object goes into a "low state" (*i.e.*, the brightest continuum source fades), *both* BLR components could fade, but the time delay will be different. The behavior of 3C 390.3, as reported by Veilleux & Zheng (1991), is consistent with this.

6.3 Jet Structure

Curving radio jets, the motivation of Begelman, Blandford & Rees (1980) to propose SMBs, continue to provide strong support for the idea (Hunstead et al 1984; Roos 1988; Roos, Kaastra & Hummel 1993; Conway & Murphy

184

1993; Conway & Wrobel 1995). Any reader in doubt about precessing radio jets in quasars is referred to the beautiful radio maps of the PKS 2300-189 jets (Hunstead et al. 1984)[13]. To quote Begelman, Blandford & Rees, "...the massive binary seems to be the only way to set this [quasar jet precession] in a dynamical context". Precessing quasar jets are common. Hutchings, Price & Gower (1988) estimate that 30% of jet sources show evidence for precession. There is also at least one good example of helical *optical* emission, NGC 3516 (Veilleux, Tully & Bland-Hawthorn 1993), which also has a clear single displaced BLR emission-line peak (Wanders et al. 1993).

The determination of a precession period from jet structure is difficult, but there are many estimates in the literature. These are derived by fitting precessing jet models to helical structure in radio maps. References for more than a dozen can be found in Roos (1988) and Lu (1990). They find a correlation between the precession period and the radio and optical luminosities. I believe, however, that the selection effects in this need careful consideration.

Begelman, Blandford & Rees (1980) give formulae for the orbital and precession periods in the SMB model. As noted above, the minimum precession period for 3C 390.3, based on the parameters derived from the optical observations, is typical of precession periods deduced from radio maps. The precession periods are, of course, much longer than orbital periods. It is interesting that the shortest *precession* periods, as deduced from radio maps (for 3C 273 and 3C 345) are ~ 10^3 yrs – only slightly longer than the *orbital* period I get for 3C 390.3 from its radial-velocity curve. If these short precession periods are correct, then *very* rapid orbital motion might be seen in the optical spectra – perhaps on a timescale of only a year. This, and the usual temporal undersampling of quasar spectra, could produce some of the changes seen in some line profiles.

Before too much comparison of the radio and optical results is carried out, one should note that some caution is necessary in comparing samples. This is because different selection effects are at work in the two cases. Only motion with a long period produces obvious changes in jets. Only rapid motion produces readily separated line peaks in optical spectra. Precession is most obvious in jets close to our line of sight; orbital motion will be most obvious when jets are in the plane of the sky.

6.4 VLBI Cores

At least *some* SMBs should produce double radio cores. It should be remembered that the majority of quasars are *not* radio-loud, so the chances are that only *one* of the central engines will be a strong radio source. The

[13] Optical spectra of PKS 2300-189 show double displaced peaks.

185

unambiguous cases of double radio cores have fairly wide projected separations: Abell 400 cD = 3C 75 (Owen et al. 1985) and Mrk 266 (Mazzarella et al. 1988) both have separations of 8 kpc; NGC 3256 has a projected separation of 1.4 kpc. The nuclei in these sources are also roughly equal in luminosity, so some selection effect is going on. There will be many more cases of *unequal* radio luminosities. These are either lost because of limited dynamic range or written off as "background" sources. The three double cores just mentioned must represent the early stages of mergers. By contrast, the binaries selected by optical spectroscopy will have separations of 1 pc (or less, for the lowest mass cases). Some of these could possibly be detected with VLBI.

Double VLBI cores should be fairly rare. Radio-loud quasars are a minority of quasars. Radio core emission is also believed to be highly anisotropic because of relativistic beaming. We choose to observe radio cores because at least one of the central engines is radio-loud. Even if *both* members of the binary are radio-loud, chances are that the other one will not be beamed in our direction. Since VLBI maps have limited dynamic range, many weak emission features currently written-off as "noise" could prove to be second central engines. Radio-loudness probably also only has a limited lifetime, so both central engines might not be radio-loud at the same time.

The "classic" symmetric parsec-scale double with equal strength "mini-lobes" (see Conway et al. 1994 for a recent description of parsec-scale doubles) almost certainly has nothing to do with SMBs. Most of the asymmetric ones are probably also unrelated. However, a binary radio core is one possible explanation of the unusual VLBI structure of the displaced BLR peak quasar 4C 39.25. VLBI maps of 4C 39.25 show both stationary components *and* superluminal motion (Shaffer et al. 1987, Alberdi et al. 1993).

There is no case known of *active* double kpc-scale jets coming from separate cores with parsec-scale separation (the beautiful pairs of jets in 3C 75 start at least 8 kpc apart). There are several possible explanations of this. Perhaps they exist but are rare and we have simply not found or recognized a case yet. Perhaps the separate jets merge into one under the guiding of the combined magnetic field structure. The existence of misdirected "fossil jets" (see next section) suggests that perhaps pairs of jets from double cores are rare and that we might have just missed them.

6.5 Fossil Jets

In their maps of the kpc-scale structure of some bright FR II sources, Black et al. (1992) point out some cases where there have been "drastic changes of orientation in the past 10^8 yrs." 3C 223.1 and 3C 403 are good examples. These remarkable changes of orientation are not possible from a single massive rotating object. There are, however, two possible ways of getting such changes in the SMB model. The first is for the two components in an SMB to have differently oriented rotation axes. The first one emits a jet along its axis for some period of time and then the other emits a jet much later. The second possibility is for the jet to be emitted by just the most massive engine in the binary. When the members of the binary finally coalesce there can be an abrupt change in rotation axis.

6.6 Mergers

It has been known for a long time that quasar host galaxies have undergone interactions and mergers (see, for example, the review by Hutchings, 1983). Since SMBs probably arise from mergers, should we expect to see more signs of interactions around 3C 390.3 objects compared with non-3C 390.3 quasars? This is a question worth investigating. It is already known that Seyferts which show greater signs of interactions have stronger radio-cores and total radio-power than Seyferts showing less signs of interaction (Giuricin et al. 1990b). In doing this sort of comparison it should be remembered that SMBs can last a Hubble time. A recent interaction might only have recently fueled an SMB that actually formed much earlier.

Mere interaction of systems alone enhances activity, as does the presence of a close companion. Some quasars are isolated and undisturbed. Even in these cases, however, a merger might have taken place *with a small companion galaxy* (Gaskell 1985). This is very possible because of the shear numbers of such galaxies. Since dwarf galaxies can also have massive central objects (*e.g.*, M32 - Tonry 1987), even an encounter with a small galaxy could produce a SMB.

6.7 Orientation Effects

The ratio of radio-core to radio-lobe flux density (R) is a well-established indicator of the orientation of the spin axis of a quasar's central engine to the line of sight (see Antonucci 1993, 1996; Wills 1996). Wills & Brotherton (1995) have shown that the ratio of fluxes from the radio and optical cores (R_V) is an even better indicator of orientation (see also Wills 1996). Wills & Browne (1986) established that there is a strong correlation between line width and R. The correlations with R_V are even stronger (Brotherton 1995). These correlations imply that BLR motions are perpendicular to the jet axis. Wills & Browne (1986) interpret this in terms of a disk model of the BLR but it is also consistent with SMBs being responsible for the extra broadening in the non-face-on quasars. Indeed *the Wills & Browne correlation might be mostly due to the orbital motion of the SMBs*. Whether this is true or not, clearly, the deficiency of 3C 390.3 objects in core-dominated sources is an orientation effect. As already discussed in section 2, the differences in continuum and NLR properties are naturally explained as orientation effects.

While the unified picture of core-dominated versus lobe-dominated quasars is well accepted, there is no such agreement over why some quasars are radio-quiet. Nonetheless, orientation effects also play a role in the optical properties of radio-quiet quasars. Seyfert 1 galaxies (low-luminosity radio-quiet quasars showing a BLR) are preferentially seen face-on, while Seyfert 2 galaxies (ones where only the extended NLR is seen) are not (Keel 1980). There is good evidence that at least some Seyfert 2's are quasars seen from the side. Imaging, spectropolarimetry and statistics show that there is an "ionization-cone" with an opening angle of $30 - 60°$ (see Antonucci 1993). Because of the selection bias towards objects with a strong UV excess, most radio-quiet quasars are probably ones oriented so that the observer is inside that cone. This explains the similarities between the optical spectra of *core-dominated* radio-loud quasars and UV-excess-selected radio quiet quasars (Baldwin, Wampler & Gaskell 1989). I hypothesize therefore that *the reason why the displacements of peaks in radio-quiet quasars are smaller than in radio-loud quasars is primarily because of orientation effects*, and is not something fundamental related to whether a quasar has a strong jet or not (this is the opposite of what I proposed in Gaskell 1983b). Under this hypothesis, we see the best cases of 3C 390.3 objects in lobe-dominated quasars because extended radio lobes are a good way of finding quasars that do not have their radio jets aimed at us.

Testing this hypothesis will be tricky since we do not have a convenient orientation indicator like R or R_V available for radio-quiet quasars. The hypothesis does lead to four predictions about where the most spectacular cases of 3C 390.3 objects will be found:

1. *They will be found among quasars with the broadest lines.* This has already been shown to be the case (see table 3).
2. *They will be found among quasars with edge-on host galaxies.* IC 4329A provides support for this hypothesis since it is one of the most edge-on active galaxies known (De Zotti & Gaskell 1984).
3. *They will be found among quasars with a steep Balmer decrement.* The Balmer decrement correlates with host galaxy orientation (De Zotti & Gaskell 1984), but this is an additional check of the orientation hypothesis because the jet axis is not necessarily lined up with the axis of rotation of the host galaxy. IC 4329A has a very steep Balmer decrement. For the radio-quiet quasars in the Stirpe (1990) sample, the Balmer decrement for the quasars showing single or double displaced peaks is indeed steeper than for those

not showing such peaks[14]. The median Hα/Hβ ratios are 3.76 and 3.31 respectively. The one-tailed significance of the difference is 95%.

4. *They will be found in quasars with steep optical spectra.* A quasar's UV excess also correlates with galaxy orientation (Cheng, Danese & De Zotti 1983; De Zotti & Gaskell 1984). Again, this will be an extra check because the axis of the central engine is not necessarily aligned with the galaxy.

Except for the second of these predictions, they all obviously hold true already for radio-loud 3C 390.3 objects (the second does not hold because the radio-loud quasars are mostly in elliptical or disturbed galaxies).

6.8 Polarization

The polarization of the lines and continuum of quasars almost certainly occurs in scattering electrons or dust beyond the BLR. Since radio-quiet 3C 390.3 quasars are predicted to have their axes inclined to the line of sight, their broad lines should have higher polarization on average than the lines of non-3C 390.3 quasars. In both radio-loud and radio-quiet 3C 390.3 quasars, the low-velocity core will probably have a lower polarization since it is believed to come from further out. This is not a unique prediction of the SMB model. Since, in the SMB model the two peaks come from slightly different places it is possible that the polarizations will differ slightly, but this depends on the size of the scattering region. As already mentioned, Antonucci, Hurt & Agol (1995) have found significant line polarization in 3C 390.3 itself. Their results are probably consistent with the predictions of the SMB model.

6.9 Distribution of Extended Low Velocity Gas

In connection with the predictions of the orientation hypothesis it is worth noting that Simkin, Su & Schwarz (1980) found the FWZI of Hβ to be correlated with the inclination of the host galaxy while the FWHM of Hβ was not (see their figure 4*a*). This needs confirming with a larger data set, but the difference in the behavior of the FWZI and FWHM is easily explained if we remember that there is always an undisplaced peak due to gas surrounding the system. The undisplaced peak strongly influences the

---

[14] One inexplicable result deserves mention. Three of the Stirpe quasars had uncertain profile classifications and so were left out of this analysis. Curiously, they all proved to have the flattest Balmer decrements! Only one other classified profile approached them in flatness of the decrement. The two-tailed significance of this is over 99.98%

FWHM; the FWZI is free of this influence. The lack of variation of the width of the peak with inclination suggests that it has a more spherically symmetric distribution.

Wills & Browne (1986) studied FWHM(Hβ) versus R. In the SMB model there are two BLR components: the broad component associated with the binary central engine and the narrower undisplaced component associated with gas further out. The SMB model predicts that the broad component will show the strongest correlation with orientation. Thus the FWZI, or something similar, will show the strongest correlation with R. In table 6 of Brotherton (1995) the significance of the correlation of R_V (the best orientation indicator) with the full width of Hβ at the 75% level has a Spearman rank correlation coefficient of 0.45 (99.9% significance). For the full width of Hβ at the 25% level, a measure closer to the FWZI, the correlation coefficient rises to 0.61 (99.9999% significant). This strongly confirms the idea that it is the *broad base* which is orientation dependent. Brotherton et al. (1994), from a study of object-to-object variations of UV lines, identify two components: a "very broad line region" (VBLR), and an "intermediate line region" (ILR). Brotherton (1995) also identifies this component as the cause of object-to-object variations in Hβ. I believe the unshifted peak in 3C 390.3 objects is the ILR.

6.10 Are All Quasars Binaries?

There are certainly quasars that do not require binary central engines (quasars with single unshifted peaks and no signs of precession of their jets), but whether all quasars are binaries is a question worth exploring. The fraction of quasars showing some sort of displaced peak is already over 50% (see tables 1 and 2). Sulentic (1989) found that 50% of broad Balmer lines show a detectable displacement with respect to the rest frame of their host galaxies. My own (unpublished) study of radio-loud quasars gives an identical result. In both cases the threshold for detection of motion is about 200 km s^{-1}. Face-on binaries will not show double displaced peaks and, even if the system is inclined to the line of sight, there will be times in the orbit when the system looks single lined. The latter will happen for 3C 390.3 itself sometime in the years 2012 – 2029. If we make allowance for orientation effects and lower the detection threshold, the observed 50% implies that *essentially all BLRs are moving relative to their host galaxies*. This is particularly true when we consider that selection effects lead to a disproportionate number of both radio-loud and radio-quiet quasars being face-on.

There is one additional mechanism I should mention that could cause *modest* motion relative to the host galaxy without needing a binary. This is the motion of galactic cores relative to the center of mass of the galaxy. Miller & Smith (1992) discovered this in N-body simulations and they illustrate it in the movie accompanying their paper. The core of the galaxy wanders around erratically at a few hundred km s^{-1} with a quasi-period of a million years or more. It is easy to move the nucleus around because the restoring force is small (S.J. Aarseth, private communication). Any nuclear motion will damp out slowly because the damping time is long (S. D. Tremaine, private communication). Many galaxies are observed to have off-center nuclei (see Miller & Smith 1992 for references). The off-center nucleus mechanism could be responsible for the small velocity shifts seen in quasars while SMBs could be responsible for the large shifts. If this is the case, I have overestimated the number of SMBs.

6.11 Formation and Evolution of SMBs

Since Begelman, Blandford & Rees (1980) and Roos (1981) there has been a lot of work on the formation and evolution of SMBs and the production of gravitational radiation when they merge. Discussing this is beyond the scope of this paper, but I refer the interested reader to the review article by Valtonen & Mikkola (1991) for a discussion of the "few-body problem" aspects and to more recent work by Makino et al (1993), Zamir (1993), Xu & Ostriker (1994), Makino & Ebisuzaki (1994) and Governato & Maraschi (1994). This work confirms the essential details of the Begelman, Blandford & Rees picture. Finally, Wilson & Colbert (1995) have evoked the *merger* of SMBs as a possible explanation of radio-loudness/radio-quietness. In their model, however, the *radio-loud* quasars are the ones where the massive objects have merged to produce a *single* central object with greater angular momentum. In some ways this is the opposite of the model discussed here.

6.12 The Binary – Jet Connection

If we consider the members of "the jet set", the various classes of objects showing jets (for overviews, see Kundt 1987, 1996), they are all binaries. This is obvious for X-ray binaries, galactic superluminals, symbiotic stars and SS 433. Consideration of double star statistics shows that proto-stellar jet sources will mostly be binaries. 53% of solar type stars in the field are binaries (Duquennoy & Mayor 1991). Close binaries will already have coalesced. This is borne out by studies of young clusters. In these there is an excess of short-period binaries (see their figure 8b). Reipurth & Zinnecker (1993) have carried out a CCD search for *visual* binaries among pre-main sequence stars. They estimate binary frequencies of 60 - 90% (see their figures 8*a,b*). Obviously the binary frequency is high in pre-main sequence stars. There are some protostellar objects where double pairs of jets are seen, but these are very wide systems (the jets would not be resolved if they were not). The general question of jet production in very close binaries needs to be considered

If quasars with jets are also binaries *then perhaps all jet sources are binaries!* Does this mean anything? We believe that galactic jet sources are binaries because we need the companion to provide the supply of material for the compact object. It is normally thought that a single central object in quasars can be fueled with no difficulty, but Roos (1981, 1988 and references therein) has discussed how a binary can enhance the fueling of a central engine in a quasar. Perhaps many (all?) quasars are binaries because a binary has a high fueling rate. For quasars, the most spectacular examples of the 3C 390.3 phenomenon have strong jets, but I have argued that this could just be an orientation effect and not necessarily anything related to jet production.

7. Conclusions

I have argued that both optical spectroscopy and radio observations independently give strong evidence for the existence of supermassive binaries in many or even all quasars. The main predictions of the supermassive binary picture such as radial velocity variation and orientation dependence seem to be borne out, but more work on many aspects of the picture is still needed. The production of supermassive binaries through mergers is something we know *must* have happened and, indeed, still is happening. We need to understand the consequences.

Acknowledgments. I would like to express my appreciation to Wolfgang Kundt for the invitation to speak at this workshop and for all the effort he put into it. The workshop and its predecessor in Erice in 1986 have been a stimulus for my thinking in a number of areas. I am grateful to Ski Antonucci for sending an advance copy of his Arp 102B polarization results. I would also like to thank Sandy Faber for some useful discussion. Partial support for this work was provided by NASA through grant number AR-05796.01-94A from the Space Telescope Science Institute, which is operated by the Association of Universities for Research in Astronomy, Inc., under NASA contract NAS5-26555.

References

Alberdi, A. et al. 1993, A&A, 271, 93

Alloin, D., Boisson, C. & Pelat, D. 1987, A&A, 200, 17

Anderson, K.S. 1974, ApJ, 189, 195

Antonucci, R.R.J. 1993, ARA&A, 31, 473

Antonucci, R.R.J. 1996, in W. Kundt (ed.), Jets from Stars and Active Galactic
    Nuclei, Springer, Berlin

Antonucci, R.R.J., Hurt, T. & Agol, E. 1995, ApJ (Letters), in press

Barnes, J. 1993, ApJ, 393, video segment 2

Baldwin, J.A., Wampler, E.J. & Gaskell, C.M. 1989, ApJ, 338, 630

Begelman, M.C., Blandford, R.D. & Rees, M.J. 1980, Nature, 287, 307

Black, A.R.S., Baum, S.A., Leahy, J.P., Perley, R.A., Riley, J.M. & Scheuer,
    P.A.G. 1991, MNRAS, 256, 186

Blumenthal, G.R. & Mathews, W.G. 1975, ApJ, 198, 517

Boksenberg, A. & Netzer, H. 1977, ApJ, 212, 37

Brotherton, M.S., Wills, B.J., Francis, P.J. & Steidel, C.C. 1994, ApJ, 430, 495

Brotherton, M.S. 1995, ApJ, in press

Burbidge, E.M. & Burbidge, G.R. 1972, ApJ, 163, L21

Capriotti, E.R, Foltz, C.B. & Byard, P. 1979, ApJ, 230, 681

Cheng, F.Z., Danese, L. De Zotti, G. & Lucchin, F. 1984, MNRAS, 208, 799

Chen, K., Halpern, J.P. & Filippenko, A.V. 1989, ApJ, 339, 742

Chen, K. & Halpern, J.P. 1990, ApJ, 354, L1

Clavel, J. & Wamsteker, W. 1987, ApJ, 320, L9

Collin, S. & Dumont, A.M. 1989, A&A, 213, 29

Collin-Souffrin, S., Dyson, J.E., McDowell, J.C. & Perry, J.J. 1988, MNRAS,
    232, 539.

Collins, G.W., II, 1980, Ap&SS, 73, 355

Conway, J.E. & Murphy, D.W. 1993, ApJ, 411, 89

Conway, J.E., Myers, S.T., Pearson, T.J., Readhead, A.C.S., Unwin, S.C. & Xu,
    W. 1994, ApJ, 425, 568

Conway, J.E. & Wrobel, J.M. 1995, ApJ, 439, 98

Crenshaw, D.M. & Blackwell, J.H. Jr. 1990, ApJ, 358, L37

De Zotti G. & Gaskell, C.M. 1984, A&A, 147, 1

Disney, M.J. 1973, ApJ, 181, L55

Done, C. & Krolik, J.H. 1996, ApJ, in press

Duquennoy, & Mayor 1991, A&A, 248, 485

Dumont, A.M. & Joly, M. 1992, A&A 263, 75

Ekers, R.D., Fanti, R., Lari, C. & Parma, P. 1978, Nature, 276, 588

Eracleous, & Halpern, J.P. 1994, ApJS, 90, 1

Eracleous, M., Livio, M., Halpern, J.P. & Storchi-Bergmann 1995, ApJ, 438, 610

Ferland, G.J., Rees, M.J., Longair, M.S. & Perryman, M.A.C. 1979, MNRAS,
    187, 67P

Ferland, G.J., Netzer, H. & Shields, G.A. 1979, ApJ, 229, 274

Filippenko, A.V. 1985, ApJ, 289, 475

Ford, H.C., Harms, R.J., Tsvetanov, Z.I., Hartig, G.F., Dressel, L.L., Kriss, G.A., Bohlin, R.C., Davidsen, A.F., Margon, B. & Kochhar, A.K. 1984, ApJ, 435, L27

Gaskell, C.M. 1981, ApJ, 251, 8

Gaskell, C.M. 1982, ApJ, 263, 79

Gaskell, C.M. 1983a, ApJ, 267, L1

Gaskell, C.M. 1983b, in "Quasars and Gravitational Lenses"; 24th Liege Astrophysical Colloquium, p. 473.

Gaskell, C.M. 1985, Nature, 315, 386

Gaskell, C.M. 1987, in Astrophysical Jets and Their Engines, W. Kundt (ed.), Reidel, Dordrecht, p 29.

Gaskell, C.M. 1988a, ApJ, 325, 114

Gaskell, C.M. 1988b in H.R. Miller & P.J. Wiita (eds.), Active Galactic Nuclei, Springer, Berlin, p. 61

Gaskell, C.M. 1994, in P.M. Gondhalekar, K. Horne, & B.M. Peterson (eds.), Reverberation Mapping of the Broad-Line Region in Active Galactic Nuclei, Astronomical Society of the Pacific Conference Series, Vol. 69, San Francisco, p. 111.

Gaskell, C.M. 1995, ApJ, submitted

Gaskell, C.M. & Sparke, L.S. 1986, ApJ, 305, 175

Ghisellini, G., Padovani, P., Celotti, A., & Maraschi, L. 1993, ApJ, 407, 65

Giuricin, G., Mardirossian, F. & Mezzetti, M. 1990a, ApJS, 72, 551

Giuricin, G., Mardirossian, F., Mezzetti, M. & Bertotti, G. 1990b in R. Wielen (ed.) Dynamics and Interactions of Galaxies, Springer, Berlin, p. 477

Goad, M.R. et al 1996, in preparation

Golombek, D., Miley, G.K. & Neugebauer, G. 1998, AJ, 95, 26

Governato, F., Colpi, M. & Maraschi, L. 1994, MNRAS, 271, 317

Grandi, S.A. & Osterbrock, D.E. 1978, ApJ, 220, 783

Halpern, J.P. & Eracleous, M. 1994, ApJ, 433, L17

Halpern, J.P. & Filippenko, A.V. 1988, Nature, 331, 46

Hunstead, R.W., Murdoch, H.S., Condon, J.J., Phillips, M.M. 1984, MNRAS, 207, 55

Hutchings, J.B. 1983, PASP, 95, 799

Hutchings, J.B., Price, R & Gower, A.C. 1988, ApJ, 329, 122

Keel, W.C. 1980, AJ 85, 198

Keel, W.C. 1985, in J.S. Miller (ed.) Astrophysics of Active Galaxies and Quasi-Stellar Objects, University Science Books, Mill Valley, p. 1

Komberg, 1968, Sov. AJ, 11, 4

Koratkar, A.P. & Gaskell, C.M. 1991a, ApJS, 75, 719

Koratkar, A.P. & Gaskell, C.M. 1991b, ApJ, 370, L61

Korista, K. et al 1995, ApJS, 97, 285

Kormendy, J. 1988, ApJ, 335, 40

Kormendy, J. & Richstone, D. O. 1992, ApJ, 393, 559

Kundt, W. 1987, in Astrophysical Jets and Their Engines, W. Kundt (ed.), Reidel, Dordrecht, p. 1

Kundt, W. 1988, Ap&SS, 149, 175

Kundt, W. 1996, in W. Kundt (ed.), Jets from Stars and Active Galactic Nuclei, Springer, Berlin, 1

Lu, J.J. 1990, A&A, 229, 424

Lynden-Bell, D. 1969, Nature, 262, 649

Lynds, R. 1968, AJ, 73, 888

Makino, J. & Ebisuzaki, T. 1994, ApJ, 436, 607

Maoz, D, Netzer, H., Mazeh, T, Beck, S., Almoznio, E., Leibowitz, E., Brosch, N., Mendelson, & Laor 1991, ApJ, 367, 493

Marziani, P., Calvani, M. & Sulentic, J.W. 1992, ApJ, 393, 658

Marziani, P., Sulentic, J.W., Calvani, M., Perez, E., Moles, & Penston, M.V. 1993, ApJ, 410, 56

Mathews, W.G. 1982, ApJ, 258, 425

Mathews, W.G. 1993, ApJ, 412, L17

Mathews, W.G. & Capriotti, E. R. 1985 in J.S. Miller (ed.) Astrophysics of Active Galaxies and Quasi-Stellar Objects, University Science Books, Mill Valley, p. 185

Mathews, W.G. & Wampler, E.J. 1985, PASP, 97, 596

Mazzarella, J.M., Gaume, R.A., Aller, H.D. & Hughes, P.A. 1988, ApJ, 333, 168

Miley, G.K. & Miller, J.S. 1979, ApJ, 228, L55

Miley, G.K., Neugebauer, G. & Soifer, B.T. 1985, ApJ, 293, L11

Miller, J.S. & Peterson, B.M. 1990, ApJ, 361, 98

Miller, R.H. & Smith, B.F. 1992, ApJ, 393, 508

Norris, R.P. & Forbes, D.A. 1995, ApJ, 446, 594

Oke, J. B. 1987, in J.A. Zensus & T.J. Pearson (eds.), Superluminal Radio Sources, Cambridge Univ. Press, Cambridge, p. 267

Osterbrock, D.E. 1977, ApJ, 215, 822

Osterbrock, D.E., Koski, A.T. & Phillips, M.M. 1975, ApJ, 197, L41

Osterbrock, D.E., Koski, A.T. & Phillips, M.M. 1976, ApJ, 206, 898.

Osterbrock, D.E. & Shuder, J. M. 1982, ApJS, 49, 149

Owen, F.N., O'Dea, C.P., Inoue, M. & Eilek, J.A. 1985, ApJ, L85

Perez, E., Penston, M.V., Tadhunter, C., Mediavilla, E., & Moles, M. 1988, MNRAS, 230, 353

Peterson, B.M. 1994, in Reverberation Mapping of the Broad-Line Region in Active Galactic Nuclei, P.M. Gondhalekar, K. Horne, & B.M. Peterson, Astronomical Society of the Pacific Conference Series, Vol. 69, San Francisco, p. 1.

Peterson, B.M., Korista. K., & Cota, S. 1987, ApJ, 312, L1

Reipurth, B., & Zinnecker, 1993, A&A, 278, 81

Rokaki, E. in P.M. Gondhalekar, K. Horne, & B.M. Peterson (eds.), Reverberation Mapping of the Broad-Line Region in Active Galactic Nuclei, Astronomical Society of the Pacific Conference Series, Vol. 69, San Francisco, p. 257

Roos, N. 1981, A&A, 104, 218

Roos, N. 1988, ApJ, 334, 95

Roos, N., Kaastra, J.S. & Hummel, C.A. 1993, ApJ, 409, 130

Rosenblatt, E.I., Malkan, M.M., Sargent, W.L.W. & Readhead, A.C.S. 1994, ApJS, 93, 73

Sandage, A.R. 1966, ApJ, 145, 1

Saslaw, W.C., Valtonen, M.J. & Aarseth, S.J. 1974, ApJ, 190, 253

Shaffer, D.B., Marscher, A.P., Marcaide, J. & Romney, J.D. 1987, ApJ, 314, L1

Shuder, J.M. 1981, ApJ, 244, 12

Shuder J.M. 1982, ApJ, 259, 48

Simkin, S.M., Su, H.J. & Schwarz, M.P. 1980, ApJ, 23, 404

Smith, H.E. 1980, ApJ, 241, L137

Soltan, A. 1982, MNRAS, 200, 115

Stirpe, G. 1990, A&AS, 85, 1049

Stirpe, G., de Bruyn, A. G. & van Groningen, E. 1988, A&A, 200, 9

Sulentic, J.W. 1989, ApJ, 343, 54

Sulentic, J.W., Calvani, M., Marziani, P. & Zheng, W. 1990, ApJ, 335, 15

Tonry, J.L. 1987, ApJ, 322, 632

Valtonen, M. & Mikkola, S. 1991, ARA&A, 29, 9

Veilleux, S., Tully, R.B. & Bland-Hawthorn, J. 1993, AJ, 104, 1318

Veilleux, S. & Zheng, W. 1991, ApJ, 377, 89

Yee, H.K.C. 1980, ApJ, 241, 984

Wanders, I. et al. 1993, A&A, 269, 39.

Wanders, I., Goad, M.R., Korista, K.T., Peterson, B.M., Horne, K., Ferland, G.J.,
    Koratkar, A.P., Pogge, R.W. & Shields, J.C. 1995, ApJ, 453, L87

Wamsteker, W., Alloin, D., Pelat, D., & Gilmozzi, R. 1985, ApJ, 295, L33

Wanders, I. et al. 1993, A&A, 269, 39

Whitmire, D.P. & Matese, J.J. 1981, Nature 293, 722

Woltjer, L. 1959, ApJ, 130, 38

Wills, B.J. 1996, in W. Kundt (ed.), Jets from Stars and Active Galactic Nuclei,
    Springer, Berlin.

Wills, B.J. & Brotherton, M.S. 1995, ApJ, 448, L81

Wills, B.J., Brotherton, M.S., Fang, D., Steidel, C.C. & Sargent, W.L.W. 1993,
    ApJ, 415, 563

Wills, B.J. & Browne, I. W. A. 1986, ApJ, 302, 56

Xu, G. & Ostriker, J.P. 1994, ApJ, 437, 184

Zamir, R. 1993, ApJ, 403, 278

Zheng, W. 1992, ApJ, 385, 127

Zheng, W., Binette, L. & Sulentic, J.W. 1990, ApJ, 365, 115

Zheng, W., Veilleux, S., & Grandi, S. A. 1991, ApJ, 381, 418

Accretion and Jet Power

Beverley J. Wills

Department of Astronomy, University of Texas at Austin, Texas, 78712

Abstract. In the first of a series of three lectures we discuss ways of measuring the power available to feed the jets in powerful FR II radio sources. For unobscured radio-loud QSOs we present evidence that this power is directly related to the UV-optical luminosity, or probably more accurately, to the power radiated through processes of accretion in a strong gravitational potential. It has been suggested on theoretical grounds that powerful radio jets are a necessary component of the central engine. It then follows, from the similarity of the optical-UV power output, spectral energy distribution, and emission-line spectra of radio-loud and radio-quiet QSOs, that radio-quiet QSOs have the same power available to feed jets as do radio-loud QSOs. This then leaves us with the puzzle of why we do not see the powerful jets in radio-quiet QSOs.

1 Introduction

The enormous luminosity of QSOs[1] is believed to arise from accretion of gas through a large gravitational potential, over distance scales of ~ 0.001 pc, giving rise to the characteristic optical-ultraviolet-X-ray spectral energy distributions with $T \sim 10^4 - 10^5$ K (the Big Blue Bump). Material is carried inwards in an accretion disk. Conservation of angular momentum requires the transport of angular momentum and energy outwards. One popular idea for achieving this (Königl & Kartje 1995, but see Kundt & Gopal-Krishna 1980) is that the disk, consisting of ionized plasma, is threaded by magnetic field lines that are twisted by the disk's rotation. Plasma flows out along the fields lines that then leave the disk perpendicular to the surface, allowing the plasma to be accelerated into collimated, relativistic outflows along the disk axis. Thus jets regulate the transport of angular momentum and energy; in this hypothesis they are a necessary component of the central engine, enabling the accreting fuel to spiral inwards.

The optical-UV-X-ray spectral energy distributions are remarkably similar for radio-loud and and radio-quiet QSOs (RLQs and RQQs). So also are the strong, broad emission-line spectra that arise in the surrounding high velocity gas of the broad line region (BLR), and the narrow-line spectra that

---

[1] 'QSO' refers to all luminous AGN ($L \gtrsim 10^{11}$ $L_\odot$, $H_0 = 100$ km s^{-1} Mpc^{-1}, and QSOs are called 'radio-loud' if $F_{5GHz}/F_{4400\,\text{Å}} \gtrsim 10$, where F is the rest frame flux-density in mJy; such strong radio emission is assumed to indicate powerful radio jets.

arise from lower-velocity gas of the narrow line region (NLR) at greater nuclear distances (~ 1 pc – Kpc); both the BLR and NLR are ionized by the 0.01 – 1 keV photons from this central continuum source. These similarities strongly suggest the same type of central engine in RLQs and RQQs.

Even if we do not know the exact processes by which jets are formed, there are two empirical approaches that suggest that the power that feeds the jets is directly related to the power radiated by the central engine in the accretion process. We discuss these in §2 and §3.

2 Jet Power, and Luminosity Relationships

Two relations together show that jet power and accretion power (assumed to be represented by power radiated by QSO's Big Blue Bump) are strongly coupled – one between emission-line and optical-continuum luminosity, and the other between luminosity of emission lines and extended radio emission.

Yee (1980) and Shuder (1981) have shown that emission-line luminosities (e.g., Hα) are closely proportional to the luminosity of the non-stellar, featureless continuum at 4800 Å rest wavelength, over a range of more than 10^5 in luminosity, for a heterogeneous assortment of narrow and broad line Seyfert and radio galaxies, RQQs and RLQs. They show that this relationship is consistent with photoionization of the emission-line regions by an extrapolation of the observed non-stellar continuum.

Rawlings & Saunders (1991) report a close proportionality between the total kinetic power of the Kpc to Mpc-scale jets and narrow-line luminosity, for a complete sample of 3CR FR II radio galaxies. Despite their careful, detailed calculations of kinetic jet power, the result is not very different if simply the radio luminosity is used instead, provided any beamed core emission is excluded. The completeness of the sample is important, to demonstrate that the proportionality is not simply induced by the bias towards selecting sources of increasing radio and optical luminosity at higher redshift. Such an unbiased sample is not available for RLQs; however, if we do include radio and narrow emission-line data for FR I radio galaxies, and RLQs with z< 1, this relation extends over a range of 10^4 in luminosity.

The first relation, between emission-line and continuum luminosity, shows that the ionizing continuum is closely related to the observed optical continuum, and that both are measured at least roughly, by the emission-line luminosity. This is useful because, in orientation-dust Unified Schemes, FR II radio galaxies are RLQs with the central engine hidden or partially hidden from the observer by a dusty torus or inner galaxy disk whose axis is parallel to the jet direction. In this case, even though the RLQ continuum and broad line region may be obscured, much of the extended gas is still illuminated and so narrow-line luminosity can apparently be used as a measure of the ionizing continuum and hence the accretion power. The second relation, between emission-line and extended radio luminosity, therefore shows that the kinetic power of the jet is directly related to the EUV-optical luminosity – the accretion power. Moreover, Rawlings & Saunders argue that the central engine channels a significant fraction of power into the jets compared with that radiated by accretion; this high efficiency implies a massive, spinning object that both powers the jets and controls the accretion rate.

Impressive as these relations are, the scatter in them is almost an order of magnitude. This scatter could be accounted for entirely by the uncertainties in determining jet power and by how well line luminosity measures the ionizing continuum. Significant scatter is certainly expected from variations in gas covering, optical depth to ionizing photons, dust reddening and obscuration. In other words, the true relationship between jet power and accretion power may actually be much tighter.

3   Radio Core-Dominance Relationships

It is conventional to use the core dominance, R, as a measure of the Doppler-boosting of radio synchrotron emission arising from relativistic flows at the base of the jet. R is defined as the ratio of flux-density of the compact radio source coincident with the QSO nucleus, to the (assumed isotropic) flux-density in the extended radio lobes, measured in the QSO rest frame. The nuclear flux density is typically measured with VLA resolutions of $\sim 1''$, and the nuclear source is usually unresolved at these resolutions. If the luminosity in the extended lobes is a good measure of the power available to feed the nuclear jets, and the bulk velocities at the base of the jet are similar from one RLQ to another, then R should be an indicator of the angle ϕ, of the jet to the line-of-sight. On the simplest beaming hypotheses, R may range over factors of $\sim 10^4 - 10^5$, for blazars with jets within a few degress of the line of sight (small ϕ) to radio-galaxies with jets near the sky-plane ($\phi \sim 90°$).

Jets depend on synchrotron losses for their visibility, and there is strong evidence that their radio emission beyond several tens of Kpc from the nucleus depends increasingly on interactions with the host galaxy or intergalactic environment (Bridle et al. 1994). Their emission may be beamed even on scales of many Kpc. Also, they represent jet power averaged over millions of years. For these reasons extended radio luminosity may not be a good measure of present jet power, and so we looked for another way to normalize the beamed core flux-density (see Wills & Brotherton 1995 for further discussion and references). We found that the use of optical luminosity for UV-excess FR II RLQs thought to have low reddening and negligible synchrotron contribution, significantly improved two relationships – those of R with jet angle and R with width (FWHM) of the broad Hβ emission line. The jet angle, ϕ is determined completely independently from measurements of superluminal motion and limits on inverse Compton scattered X-ray flux. These relationships are compared in Figs. 1 and 2.

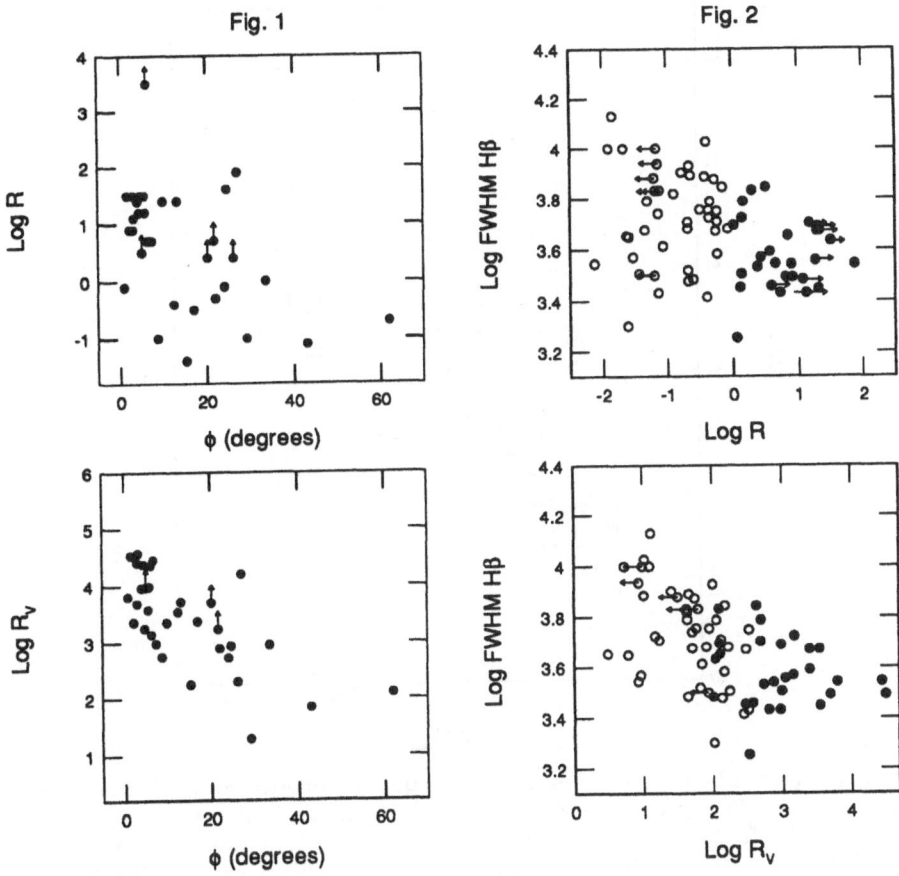

Fig. 1. Radio core-dominance vs. ϕ, the angle of the beam to the line of sight. The 2-tailed probability that these variables are unrelated decreased from 10^{-2} to 10^{-5} by using a rest-wavelength V-band luminosity to normalize core-dominance R.

Fig. 2. Radio core-dominance vs. the width of the broad H β line. Here, the 2-tailed probability that these variables are unrelated decreased from 4×10^{-3} to less than 10^{-5} by using R_V instead of R.

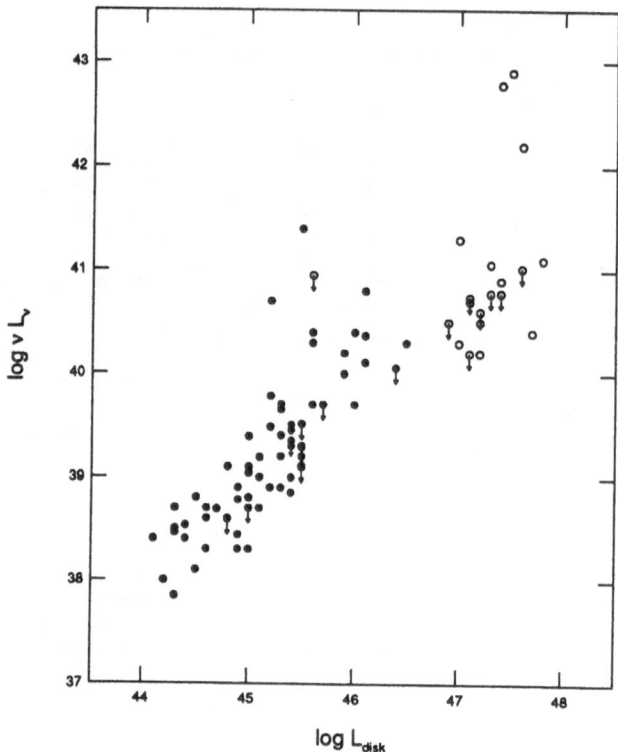

Fig. 3. Total radio luminosity vs. disk (accretion) luminosity, for radio-quiet QSOs. Data are for PG QSOs (the complete z < 0.5 sample is shown as filled circles, higher redshift QSOs are shown as open circles). The L_{disk} values are from Falcke et al. (1995), with radio luminosities updated using Kellermann et al. (1994). Many previous upper limits on νL_ν have become real measurements, showing even more convincingly the strong radio-UV luminosity relation.

4 Radio Quiet QSOs

Miller et al. (1992) and Falcke et al. (1995) show that there is a strong emission line vs. optical-UV continuum luminosity relationship for a complete, optically-selected PG sample, as also found for RLQs, showing a strong link between the observed and ionizing (EUV) continua. More importantly, despite the 100 times weaker radio luminosity, this complete sample of predominantly RQQs shows a closely proportional relationship between total 5GHz luminosity and accretion luminosity (Miller et al. 1993; Falcke et al. 1995). Fig. 3 shows this strong relationship from the paper by Falcke et al., updated with more recent radio luminosities from Kellermann et al. (1994), just for the PG QSOs. The relationship also suggests a narrow range of jet power to accretion power – providing further support for a real dichotomy between RLQs and RQQs.

We suggest that investigations of this dichotomy via the distribution of radio loudness for QSOs should be made in the radio luminosity vs. optical-UV luminosity plane rather than, as has sometimes been done, in the single radio-luminosity dimension, or even the single dimension of the ratio of radio luminosity to optical-UV luminosity. [Perhaps it would be better to calculate radio jet power instead of radio luminosity (Rawlings & Saunders 1991), and a bolometric luminosity to represent accretion power, instead of optical-UV luminosity.]

5 Summary and Discussion

Previously-discovered relationships among emission line, optical continuum and extended radio luminosities indicate a proportionality between jet power and accretion power in high luminosity (FR II) sources. The improved correlations between core-dominance and jet angle, and between core-dominance and line width, when the beamed jet luminosity is normalized by the optical continuum, suggest an even stronger link on sub-parsec scales.

The similarity of the optical-to-EUV luminosities and spectra indicates the same accretion mechanism for the central engine of RLQs and RQQs – this is also suggested by the similar relationship between extended radio luminosity and optical luminosity, albeit at very different radio luminosities. These similarities, together with the conclusion that this relationship is a strong proportional one for QSOs with powerful radio jets, suggest that the same power is potentially available to feed radio jets in RQQs. It is therefore a real puzzle that 90% of QSOs are radio-quiet.

The conclusion that radio jet power depends directly on ultraviolet-optical luminosity may seem to contradict the established belief that there is a very wide spread in extended radio luminosity for a given optical luminosity. It now takes on new meaning to ask whether there is a real dichotomy in radio loudness between RLQs and RQQs. If the central engines are identical, the differences in (unbeamed) jet brightness must have an extrinsic cause and therefore be dependent on the environment beyond sub-parsec scales. The question then becomes one of the distribution of environment conditions, and whether there is a discontinuity in the environmental conditions between RLQs and RQQs. An emission line approach to this will discussed in the next chapter.

References

Bridle A. H., Hough D. H., Lonsdale C. J., Burns J. O., Laing R. A. (1994): Deep VLA Imaging of Twelve Extended 3CR Quasars. AJ **108**, 766–820

Emmering R. T., Blandford R. D., Shlosman I. (1992): Magnetic Acceleration of Broad Emission Line Clouds in Active Galactic Nuclei. ApJ **385**, 460–477

Falcke H., Malkan M. A. and Biermann P. L. (1995): The jet-disk Symbiosis II. Interpreting the radio/UV Correlations in Quasars. A&A **298**, 375–394

Kellermann K. I., Sramek R. A., Schmidt M., Green R. F., Shaffer D. B. (1994): The Radio Structure of Radio-loud and Radio-quiet Quasars in the Palomar Bright Quasar Survey. AJ **108**, 1163–1177

Königl A., Kartje J. F. (1994): Disk-Driven Hydromagnetic Winds as a Key Ingredient of Active Galactic Nuclei Unification Schemes. ApJ **434**, 446–467

Kundt W., Gopal-Krishna (1980): Extremely Relativistic Electron-positron Twin-jets form Extragalactic Radio Sources. Nature, 288, 149–150

Miller P., Rawlings S., Saunders R., Eales S. (1992): A Spectrophotometric Study of BQS Quasars. MNRAS **254** 93–110

Miller P. Rawlings S., Saunders R. 1993: The Radio and Optical Properties of the z < 0.5 BQS Quasars. MNRAS **263**, 425–460

Rawlings S., Saunders R. (1991): Evidence for a Common Central Engine Mechanism in all Extragalactic Radio Sources. Nat **349**, 138–140

Shuder J. M. (1981): Emission-line–Continuum Correlations in Active Galactic Nuclei. ApJ **244**, 12–

Wills B. J., Brotherton M. S. (1995): An Improved Measure of Quasar Orientation. ApJ **448**, L81–L84

Yee H. K. C. (1980): Optical continuum and emission-line luminosity of active galactic nuclei and quasars. ApJ **241** 894–

Jets and QSO Spectra

Beverley J. Wills and M. S. Brotherton

Department of Astronomy, University of Texas at Austin, Texas, 78712

Abstract. QSOs' emission lines arise from highest velocity ($\sim 10^4$ km s^{-1}), dense gas within ~ 0.1 parsec of the central engine, out to low-velocity, low-density gas at great distances from the host galaxy. In radio-loud QSOs there are clear indications that the distribution and kinematics of emission-line gas are related to the symmetry axis of the central engine, as defined by the radio jet. These jets originate at nuclear distances < 0.1 pc — similar to the highest-velocity emission line gas. There are two ways we can investigate the different environments of radio-loud and radio-quiet QSOs, i.e., those with and without powerful radio jets. One is to look for optical-UV spectroscopic differences between radio-loud and radio-quiet QSOs. The other is to investigate dependences of spectroscopic properties on properties of the powerful jets in radio-loud QSOs. Here we summarize the spectroscopic differences between the two classes, and present known dependences of spectra on radio core-dominance, which we interpret as dependences on the angle of the central engine to the line-of-sight. We speculate on what some of the differences may mean.

1 Introduction

The optical-ultraviolet emission-line profiles and intensity ratios are remarkably similar for radio-loud QSOs (RLQs) and radio-quiet QSOs[1] (RQQs), for a wide range of luminosities (Baldwin et al. 1995). The broad emission lines arise predominantly by photoionization of the broad-line region (BLR), a region of dense gas within ~ 0.1 parsec of the central engine. Lower velocity (~ 500 km s^{-1}) gas of the narrow-line region (NLR) arises at distances from parsecs to kiloparsecs. The BLR and NLR are not single, homogeneous regions: the highest-velocity, highest-density gas ($\sim 10^{12-13}$ cm^{-3}) — the very broad line region (VBLR) — occurs closest to the central engine and produces the broadest emission lines ($\sim 10^4$ km s^{-1}). These are typically blueshifted by ~ 1000 km s^{-1} with respect to the systemic (NLR) redshift. Gas of the ILR — intermediate in velocity, density, and distance, between the VBLR and NLR — produces profiles of ~ 2000 km s^{-1} width. Much of the diversity in line ratios and profiles can be reproduced by a combination of spectra from the VBLR, ILR and NLR (Brotherton et al. 1994b).

---

[1] 'QSO' refers to all luminous AGN (L$\gtrsim 10^{11}$ L$_\odot$, H$_0 = 100$ km s^{-1} Mpc^{-1}. A radio-loud QSO is one having F$_{5GHz}$/F$_{4400} \gtrsim 10$, where F is the rest frame flux-density in mJy. Such strong radio emission is assumed to indicate powerful radio jets. For a short 'course' on emission lines, see Netzer (1990), and reviews to appear in the Proceedings of IAU Colloquium No. 159, *'Emission Lines in Active Galaxies'* (Shanghai, 1996).

The powerful, synchrotron-emitting radio jets arise within VBLR distances, and often extend hundreds of Kpc from the nucleus. Thus, differences in the optical-ultraviolet spectra of the BLR between the RLQs with powerful jets, and the RQQs of 1000 – 10,000 less radio luminosity, should give clues to differences in material that fuels or is expelled by the central engine, and how this might be related to the formation and collimation of the jets. Differences in the optical-ultraviolet spectra of the lower-speed gas might reflect jet interaction with material in the inner regions of the host galaxy, up to many Kpc distant — by means of entrainment, shock excitation, and photoionization by beamed high-energy synchrotron emission.

The spin axis of the central engine is indicated by the position angle of the base of the jet, and by the radio core dominance that indicates the angle of the jet to the line-of-sight (see the chapter, 'Accretion and Jet Power'). Thus dependence of spectral properties on core dominance provides a statistical way to investigate axisymmetric structure of the gaseous environment in RLQs, necessary for the interpretation of radio-loud – radio-quiet differences.

First we summarize spectroscopic differences between RLQs and RQQs, then summarize some spectroscopic dependences on radio core-dominance. We suggest a picture for future testing, which is consistent with the present observations.

2 Radio-Loud – Radio-Quiet Spectroscopic Differences

Despite the great similarity of RLQ and RQQ spectra, differences are being recognized as a result of improved data over a wide spectral range, for appropriate samples of QSOs. We present results that derive from our own and other investigations, using ground-based spectrophotometry in the optical-infrared, and ultraviolet spectroscopy with the Hubble Space Telescope, emphasizing those that are most simply interpreted without detailed statistical analysis. Even the strongest relations involve more than two variables, and multivariate analyses of carefully defined samples are needed. Some of this work is in progress.

2.1 The Fe II – [O III] Relation

One of the strongest correlations is the inverse correlation between the strength of blended optical Fe II emission (between Hγ and Hβ, and 5150 – 5300 Å), and the strength of [O III] λ5007 — the strongest optical line from the NLR. This was nicely presented by Boroson & Green (BG, 1992) for Seyfert 1 galaxies and QSOs from the PG survey — a predominantly radio-quiet sample. Fig. 1 illustrates typical spectra of the Hβ region, and in Fig. 2, the inverse relation is shown by comparing equivalent widths, including BG data for RQQs (open symbols), but adding data for RLQs from the investigation by Brotherton (1995; filled circles). Two blazars were excluded, because of their strong, variable synchrotron continua. The samples cover comparable optical luminosities. There is a striking difference between the RLQs and RQQs: the RLQs appear clustered towards the weak-Fe II, strong-[O III] end of the relation.

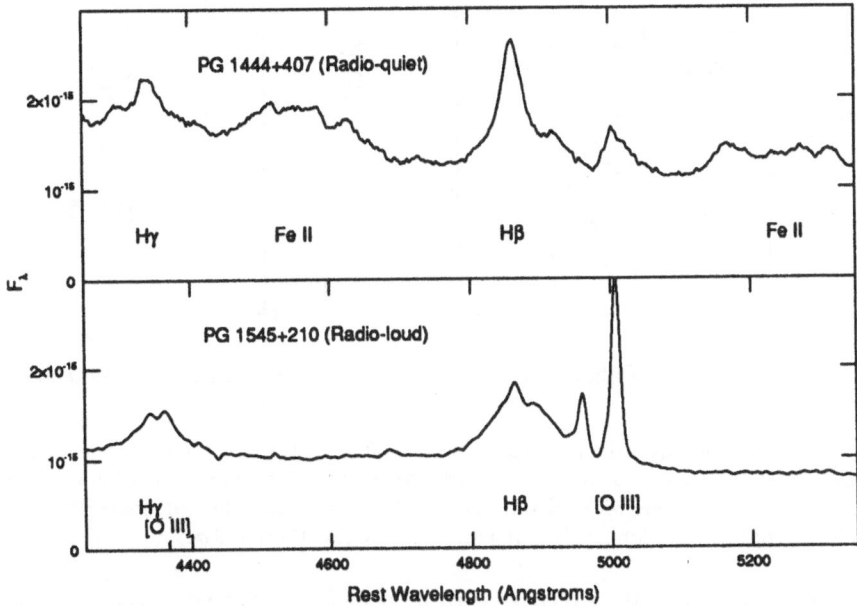

Fig. 1. The Hβ–[O III] λ5007 region in a typical radio-loud and typical radio-quiet QSO. For the radio-quiet QSO, Fe II blends contribute at all wavelengths in this region, and [O III] λ5007 is barely visible. For the radio-loud QSO Fe II blends are weaker, and stronger NLR emission of [O III] λ5007 and [O III] λ4363 is present.

The [O III] is representative of NLR emission. The strong optical Fe II emission must arise in regions of high density and high optical depth to the ionizing continuum, and the line widths of blended Fe II appear to be similar to other broad lines, suggesting an origin in the BLR. However, problems in explaining the great strength of Fe II optical emission have led to suggestions of a different source of excitation from the standard ultraviolet lines (Ly α, C IV λ1549, C III] λ1909). This problem is probably related to the general 'energy budget' problem, where the observed ultraviolet continuum does not appear strong enough, in general, to produce the strengths of low ionization lines . Assuming that the Fe II emission does in fact arise in the BLR, the inverse correlation can be explained by object-to-object differences in covering of the ionizing continuum by the BLR gas. The greater the BLR covering, the greater is the shielding of the more distant NLR from ionizing photons — a possibility also considered, but fleetingly, by BG.

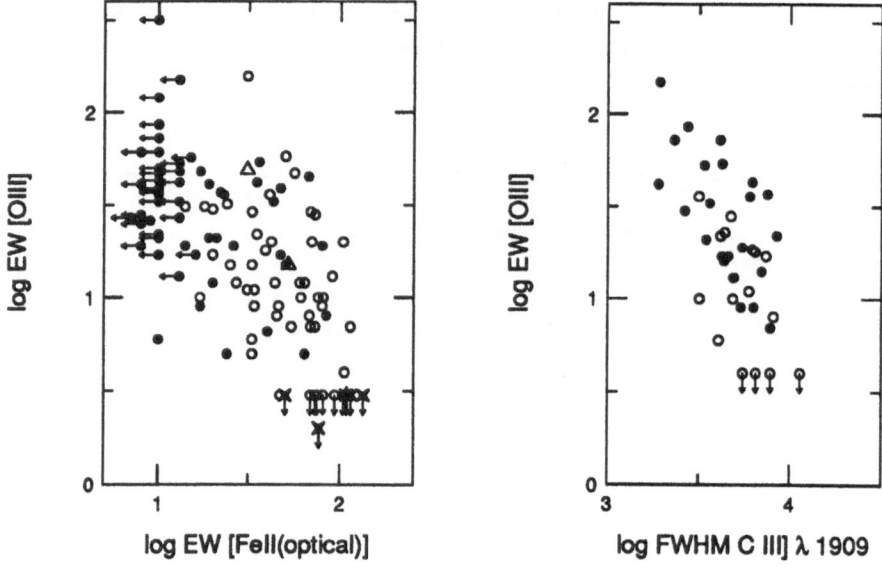

Fig. 2. (left) The equivalent width of [O III] λ5007 vs. the equivalent width of optical Fe II. Open circles are for RQQs from BG, filled circles for RLQs (Brotherton 1995), open triangles for high-ionization BAL QSOs, and stars for low-ionization BAL QSOs. Upper limits for EW[Fe II] are 1.5 σ; for EW [O III], 3 σ.

Fig. 3. (right) The EW of [O III] λ5007 vs. the width (FWHM) of the C III] λ1909 feature. Data are from Brotherton (1996). The symbols are the same as for Fig. 2.

2.2 Broad Absorption Lines and Associated Absorption

Broad absorption lines (BALs) — troughs of absorption extending between ∼5,000 km $^{-1}$ and 25,000 km s^{-1} blueward of the corresponding emission line peak — are recognizable in ∼10% of QSOs, but have never been seen in RLQs (Stocke et al. 1992). Associated absorption is narrow, intrinsic absorption occurring near (within a few hundred km s^{-1}) or blueward of the emission line peak. Associated absorption is common in BAL QSOs and occurs in other RQQs. It is also common in RLQs — probably occurring more often in lobe-dominated RLQs. Two classes of BAL QSO are distinguished — high-ionization BAL QSOs with C IV, N V and Ly α BALs, and low-ionization BAL QSOs, which show, in addition to the high-ionization BALs, BALs of Mg II, Al III, and probably Fe II and Na I. Under current debate is the hypothesis that BAL QSOs and other RQQs are the same — the observed difference depending on whether or not intrinsic broad absorbers happen to lie along our line-of-sight to the nucleus. This hypothesis receives strong support from the general similarity of their broad-emission-line ultraviolet spectra (Weymann et al. 1991).

However, low-ionization BAL QSOs do show some other spectral differences. One of these is illustrated in Fig. 2 (stars and open triangles represent low- and high-ionization BAL QSOs, including one [O III] $\lambda 5007$ upper limit) — nearly all the low-ionization BAL QSOs have very weak [O III] and are super-Fe II emitters. Nearly all known low-ionization BAL QSOs show significant scattering polarization and significantly reddened continua. The PG QSOs are selected by UV-excess and are therefore biased against being dust-reddened; however, many QSOs selected by high infrared luminosity (the IRAS ultraluminous QSOs) show high scattering polarization and reddened continua – and a significant fraction show low-ionization BALs (based on small numbers, as yet), and/or belong to the rare class of super-Fe II emitters. The strong Fe II is also significantly associated with narrow Hβ from the BLR, and softer X-ray spectra (0.3 – 2 kev) (Laor et al. 1994, for PG QSOs; Boller et al. 1995, & Grupe et al. 1995, for soft X-ray ROSAT AGN). The polarization, reddening, and BALs can be understood as line-of-sight effects; the Fe II emission, Hβ width and X-ray slope are not as simply interpreted. We therefore do not suggest that the entire [O III]–Fe II anticorrelation is a line-of-sight effect; there is probably more than one physical cause.

Recently BALs and narrower associated absorption have been linked with both warm and cool X-ray edge absorption, placing the absorbing material within or just beyond BLR distances from the nucleus (e.g. Mathur et al. 1995).

2.3 Stronger ILR Emission in Quasars?

Fig. 2 shows that radio-loud QSOs tend to have stronger emission from the NLR compared with the BLR. The cause may be related to the alignment between jets and extended narrow-line emission, resolved on scales of kpc, and between radio power and the strength and width of [O III] emission in radio galaxies (e.g. Baum & Heckman 1989; Whittle 1992). Narrower, but stronger, BLR lines of C IV $\lambda 1549$ and C III] $\lambda 1909$ in RLQs compared with RQQs suggest a greater contribution from ILR gas (Brotherton et al. 1994a; Francis et al. 1993), and this is further supported by a link between the ILR and NLR gas, being investigated by Brotherton (1996), and illustrated by the inverse correlation between [O III] strength and C III] $\lambda 1909$ width (Fig. 3).

Hypotheses to explain the enhanced emission from lower-velocity gas in RLQs include jet-shocked gas, jet-induced star-formation, and ionization by beamed ultraviolet synchrotron emission.

2.4 BLR Line Asymmetries

There are significant statistical differences in line profiles:

- C III] λ1909 and C IV λ1549 are narrower in RLQs than in RQQs.
- The C IV line generally has stronger red than blue wings in RLQs, but the blue wing is often stronger than the red in RQQs (Wills et al. 1993, 1995).
- The Hβ broad line also often has stronger red wings in RLQs, with the RQQs showing similar frequency of red and blue asymmetries (e.g. Sulentic 1989, Corbin 1993; BG92).

Systematic line asymmetries imply radially flowing gas with obscuration. The most likely obscuration related to the emission-line region seems to be either dust within the unilluminated backside of gas clouds, or an emission region within an obscuring torus. In these cases the above line asymmetries suggest that RQQs, and perhaps all QSOs, have greatest inflow in the innermost BLR, but that in RLQs, additional outflow occurs, often producing redshifted line wings.

Another likely contributor to some line asymmetries is blending by other emission lines, for example, Fe II in the red wing of C IV λ1549. In some QSOs with strong Fe II emission or BALs, the λ1909 feature is dominated by blended Fe III emission (Hartig & Baldwin 1986; Baldwin et al. 1996), and may well account for the larger width of this 'C III]' feature in RQQs.

3 Radio Core-dominance Relationships for RLQs

The previous chapter (Accretion and Jet Power) presented evidence that, for RLQs, core-dominance is a measure of orientation of the jet axis to the line-of-sight, and we retain that interpretation here.

3.1 Line Widths & Asymmetries

The width of the broad Hβ line (FWHM) is inversely correlated with core-dominance and we have suggested that this is the result of viewing predominantly planar motions that are perpendicular to the jet axis: in core-dominated RLQs the radial velocities are smaller, in lobe-dominated RLQs, up to $\sim$8,000 km^{-1}, and in broad-line radio galaxies where we may be viewing the central QSO at even higher inclination, widths up to $\sim$20,000 km^{-1} are found (Wills & Browne 1986; Wills & Brotherton 1995; see also the previous chapter, *Accretion and Jet Power*). This dependence is weaker for the C IV λ1549 line, which has a greater contribution from ILR gas than does Hβ (Brotherton 1995).

Core-dominated RLQs have stronger red wings for C IV λ1549 (Wills et al. 1995; see Barthel, Tytler & Thomson 1990). For a higher-redshift sample, Corbin (1991) found a trend in the opposite sense, but this was a marginal result. On the other hand, for Hβ, it is the lobe-dominated RLQs that often show strong red wings (e.g. Fig. 1; BG; Brotherton 1996). For examples of Hβ

line asymmetries, see Fig. 6 in BG. Again, line asymmetries imply radial flow plus obscuration, so these correlations suggest axial flow of high-ionization gas, and radial flow of low-ionization gas in a plane perpendicular to the jet axis.

3.2 Associated Absorption and Reddening

Excluding from consideration those core-dominant RLQs (blazars) with steep, beamed IR – ultraviolet synchrotron continua, there is a strong trend for the most lobe-dominant sources to show steeper optical-ultraviolet continua. This has been seen for the 3CR sample (see the lobe-dominated sources in Smith & Spinrad 1980), especially when the broad-lined radio galaxies are included, and is shown more quantitatively for the 408 MHz Molonglo Quasar Sample (Baker & Hunstead 1995). Baker & Hunstead find the same reddening trend in the Balmer decrements.

There is a corresponding trend for increased numbers of associated absorption systems in lobe-dominated RLQs (Aldcroft et al. 1994, Wills et al. 1995).

All these trends suggest increasing concentrations of low-ionization material at the largest angles to the jet axis, some of which must be close to the active nucleus (see X-ray absorption results, §2.2).

3.3 Fe II and [O III] Emission

The larger Fe II(optical)/[O III] and Hβ/[O III] ratios are seen in the most core-dominated RLQs (Jackson & Browne 1991), and the ultraviolet 'Little Blue Bump', composed of blended Fe II emission and Balmer continuum, shows the same trend in the Molonglo Quasar Sample (Baker & Hunstead 1995). Jackson and Browne interpreted this trend in terms of axisymmetric Fe II line emission. Perhaps the Fe II-emitting region occurs near the inner edge of a dusty torus and is therefore shielded from the observer more readily at higher inclination angles than other broad lines produced nearer the center.

The stronger red wings on the C IV line for core-dominated RLQs, mentioned above, could be low-optical-depth ultraviolet Fe II emission, related to the greater strength of Fe II (optical) in core-dominated RLQs.

4 Summary and Discussion

Despite the great similarity between the spectra of RLQs and RQQs, the following observational differences are statistically very significant:
In RLQs lower-velocity (NLR, ILR) emission lines are more prominent, and C IV and Hβ emission lines are asymmetric with stronger red wings. While there is a wide dispersion in properties, RQQs have, on average, stronger

Fe II emission, and ~10% show strong, broad-absorption troughs of high- and low-ionization gas – a phenomenon that is apparently unique to RQQs.

Two apparently independent relations account for much of the diversity in the optical-ultraviolet spectra:

(i) For RLQs, with increasing core-dominance, emission lines are narrower, Fe II emission is stronger, and C IV more often has stronger red wings. With decreasing core-dominance, Hβ more often has stronger red wings, and the liklihood of associated absorption and reddening increases.

(ii) For RQQs, there is an inverse relation between the strengths of Fe II and [O III] $\lambda5007$ emission, with the strong-Fe II–weak [O III] QSOs being associated with low-ionization BALs, reddening and polarization.

We note that the properties accounting for the radio-loud – radio-quiet differences are those involved in these two relations.

If we interpret line profiles as velocity profiles (rather than as blended emission), and assume dust exists on the unilluminated sides of photoionized gas regions and in a torus, then we can interpret the observations as follows: For RLQs, increasing radio core-dominance means smaller inclination of the rotation axis to the line-of-sight. Kinematics, line emission, and the distribution of dust and absorbing gas are axisymmetric. Broader emission lines at higher inclinations means greater velocities in a plane perpendicular to the axis. Red wings on C IV for core-dominant RLQs imply high-ionization axial outflow, and red wings on Hβ for lobe-dominated RLQs imply lower-ionization planar outflow. At higher inclinations (edge-on view), high-density, low-ionization Fe II-emitting gas, being located at the inner edge of the purported dusty torus, is partially shielded from view, and the line-of-sight to the nucleus is more likely to pass through dust and outflowing associated absorbers. Perhaps Fe II-rich grains, having the highest evaporation temperatures, exist at the inner edge of the torus, producing Fe II-rich gas.

Comparison with the profiles of RQQs suggests that the outflow of emission-line gas is unique to RLQs. Blueshifted VBLR emission seen in C IV in RQQ may be common to all QSOs (in RLQs, this is difficult to determine because of the presence of stronger red wings), indicating higher-ionization, higher-velocity inflowing gas closer to the central engine. This inflow may be related to accretion.

It has not been determined whether distribution of nuclear gas in luminous RQQs is symmetric about a rotation axis, although the existence of several examples of alignment between weak radio structure and ionization cones in some quite luminous Seyfert galaxies suggests that this might be the case (NGC 1068, Antonucci 1993). BAL material does in some instances cover less than 4π sr, so the appearance of BAL QSOs must depend on orientation. It would be important to understand whether the distribution of BAL gas is related to any structure axis.

What underlies the Fe II – [O III] anti-correlation? We suggested that decreasing covering of the ionizing continuum by dense, thick, Fe II-emitting gas could simultaneously reduce Fe II emission and allow escaping photons to ionize lower-density, more-distant gas, increasing [O III] emission. The characteristics of line-of-sight dusty material and low-ionization high-velocity outflowing BAL gas dominate increasingly with increasing Fe II and decreasing NLR ([O III]) strengths. If all RQQs have BALs then orientation is important in determining BAL characteristics. If this relation were one of orientation, then the different locations of RLQs and RQQs in Fig. 2 would require them to be intrinsically different. In the next chapter we suggest ways in which the inner-galaxy environment may explain the observed radio-loud – radio-quiet differences.

References

Aldcroft,T., Bechtold,J., Elvis,M. (1994): Mg II Absorption in a Sample of 56 Steep-Spectrum Quasars. ApJS **93**, 1

Antonucci, R. R. J. (1993): Unified Models for Active Galactic Nuclei and Quasars ARAA **31**, 473-521

Baker, J. C., Hunstead, R. W. (1995): Revealing the Effects of Orientation in Composite Quasar Spectra. ApJL **452**, L95–L98

Baldwin, J. A., Ferland, G. J., Korista, K., Verner, D. (1995): Locally Optimally Emitting Clouds and the Origin of Quasar Emission Lines. ApJ **455**, 119

Baldwin, J. A., Ferland, G. J., Carswell, R. F., Phillips, M. M., Wilkes, B. J., Williams, R. E. (1996): in preparation. ApJ **000**, 000

Barthel, P. B., Tytler, D. R., Thomson, B. (1990): Optical Spectra of Distant Radio Loud Quasars. I - Data: Spectra of 67 Quasars. AAS **82**, 339

Baum, S. A., Heckman, T. (1989): Extended optical line emitting gas in powerful radio galaxies - What is the radio emission-line connection? ApJ **336**, 702

Boller, T., Brandt, W. N., Fink, H. (1995): Soft X-ray Properties of Narrow-Line Setfert 1 Galaxies. AA, in press

Boroson T. A., Green R. F. (1992): The Emission Line Properties of Low-redshift Quasi-Stellar Objects. ApJS **80**, 109–135 (BG)

Brotherton, M. S., Wills, B. J., Steidel, C. C., Sargent, W. L. W. (1994a): Statistics of QSO Broad Emission Line Profiles. II. The C IV λ1549, C III] λ1909, and Mg II λ2798 Lines. ApJ **423**, 131–142

Brotherton, M. S., Wills, B. J., Francis, P. J., Steidel, C. C. (1994b): The Intermediate Line Region of QSOs. ApJ **430**, 495–504

Brotherton, M. S. (1995): The Profiles of Hβ and [O III] λ5007 in Radio-Loud Quasars. ApJ, in press

Brotherton, M. S. (1996): A Comparison of Hβ and [O III] λ5007 in Radio-Loud and Radio-Quiet QSOs. ApJ, to be submitted

Corbin, M. R. (1991): The Emission Line Properties of Steep Radio Spectrum Quasars. ApJ **375**, 503–516

Corbin, M. R. (1993): On the Relation of the X-ray and Emission Line Properties of Low-redshift QSOs. ApJ **403**, L9-L11

Francis, P. J., Hooper, E. J., Impey, C. D. (1993): The Ultraviolet Spectra of Radio-Loud and Radio-Quiet Quasars. AJ **106**, 417

Grupe, D., Beuermann, K., Mannheim, K., Reinsch, K., Thomas, H.-C., Fink, H. H. (1995): Properties of Bright Soft X-ray-Selected ROSAT AGN. AA, in preparation

Hartig, G. F., Baldwin, J. A. (1986): The Emission-line Regions in Broad Absorption Line Quasars. ApJ **302**, 64

Jackson, N., Browne, I. W. A. (1991): Optical Properties of Quasars. II - Emission-line Geometry and Radio Properties. MNRAS **250**, 422

Laor, A., Fiore, F., Elvis, M., Wilkes, B. J., McDowell, J. C. (1994): The Soft X-ray Properties of a Complete Sample of Optically Selected Quasars. 1: First Results. ApJ **435**, 611

Mathur, S., Elvis, M., Singh, K. P. (1995): Strong X-Ray Absorption in a Broad Absorption Line Quasar: PHL 5200. ApJ **455**, 9

Netzer, H. (1990): AGN Emission Lines. in Active Galactic Nuclei, eds. T. J.-L. Courvoisier & M. Mayor (Springer-Verlag: Heidelberg)

Smith, H. E., Spinrad, H. (1980): Spectrophotometry of Faint, red 3C QSO Candidates. ApJ **236**, 419

Stocke, J.T., Morris, S. L., Weymann, R. J., Foltz, C.B. (1992): The radio properties of the Broad Absorption Line QSOs. ApJ 396, 487–503

Sulentic, J. W. (1989): Toward a Classification Scheme for Broad-Line Profiles in Active Galactic Nuclei. ApJ **343**, 54

Weymann, R. J., Morris, S. L., Foltz, C. B., Hewett, P. C. (1991): Comparisons of the Emission-line and Continuum Properties of Broad Absorption Line and Normal Quasi-Stellar Objects. ApJ 373, 23–53

Whittle, M. (1992): Virial and Jet-induced Velocities in Seyfert Galaxies. III. Galaxy Luminosity as Virial Parameter. ApJ **387**, 121

Wills, B. J., Brotherton, M. S., Fang, D., Steidel, C. C., Sargent, W. L. W. (1993): Statistics of QSO Broad Emission-Line Profiles. I. The C IV λ1549 Line and the λ1400 Feature. ApJ **415**, 563–579

Wills B. J., Brotherton M. S. (1995): An Improved Measure of Quasar Orientation. ApJ **448**, L81–L84

Wills B. J., Browne, I. W. A. (1986): Relativistic Beaming and Quasar Emission Lines. ApJ **302**, 56

Wills B. J., Thompson, K. L., Han, M., Netzer, H., Wills, D., Baldwin, J. A., Ferland, G. J. Browne, I. W. A., Brotherton M. S. (1995): Hubble Space Telescope Sample of Radio-loud Quasars: Ultraviolet Spectra of the First 31 Quasars. ApJ **447**, 139

An Interpretation of Radio-Loud - Radio-Quiet QSO Differences

Beverley J. Wills

Department of Astronomy, University of Texas at Austin, Texas, 78712

Abstract. Here we speculate on what observations are telling us about the difference between radio-loud and radio-quiet QSOs. The observations are (i) the relation between ultraviolet-optical luminosity and 'jet power', (ii) the dependences of emission and absorption line spectra, and the spectral energy distribution, on radio core-dominance, assumed to be an indicator of orientation, (iii) the spectral differences between radio-loud and radio-quiet QSOs, and (iv) the inverse relation between the strength of broad, blended Fe II multiplets and [O III] λ5007, and the apparently-related association between Fe II strength, reddening, broad absorption lines, and scattering polarization. We present and discuss a picture in which there are two main variables: (i) the inclination of the plane of the host galaxy to the axis of the inner jet (the central engine's rotation axis), and (ii) the angle of the line-of-sight to this rotation axis. The radio-loud QSOs are those with jets aiming away from the plane of the host galaxy.

1 Introduction

Some hypotheses proposed to explain why ~90% of QSOs[1] are radio quiet include (i) an evolutionary phenomenon where radio-loudness is a short-lived phase in the existence of all QSOs, or a series of short-lived phases (Schmidt 1970), (ii) the result of differences in mass concentration in the host galaxy nucleus (Heckman 1983), (iii) the result of fundamental angular momentum differences (Wilson & Colbert 1995), (iv) the result of poorly-collimated sub-relativistic wind in radio-quiet QSOs (RQQs) (Boroson, Persson & Oke 1985).

Some hypotheses simply discuss conditions under which jets might form, but do not attempt to explain all known differences between radio-loud QSOs (RLQs) and RQQs. We take a different approach, by first examining the relation between ultraviolet-optical luminosity and jet power (see the chapter, *Accretion and Jet Power*). There, we concluded that jet power (represented approximately by unbeamed radio power) is directly related to the Big Blue Bump luminosity, for RLQs and RQQs, while the radio luminosity is a factor of ~1000 less in the RQQs. Then we argued that the generally great similarity of the Big Blue Bump, non-synchrotron X-ray emission, and the emission

---

[1] 'QSO' refers to all luminous AGN ($L \gtrsim 10^{11}\ L_{\odot}$, $H_0 = 100$ km s^{-1} Mpc^{-1}. A radio-loud QSO is one having $F_{5GHz}/F_{4400} \gtrsim 10$, where F is the rest frame flux-density in mJy. Such strong radio emission is assumed to indicate powerful radio jets.

line spectra implied a very similar central engine mechanism, independent of radio–emission. This, together with the relations between unbeamed radio and Big Blue Bump luminosity, led us to the hypothesis that the central engines of RQQs and RLQs (fueling, accretion, and power available to generate a jet) are essentially identical. There is some theoretical support for this, but no single hypothesis is clearly favoured.

2   Direct Interpretation of Observations

Apart from the radio emission, observed differences in photon energy distribution, emission-line spectra and absorption spectra may lead to clues concerning the collimation and propagation of luminous radio jets. In the previous chapter *Jets and QSO Spectra* we summarized the most significant relations between radio emission and these ultraviolet–optical properties, and suggested a consistent picture for the nuclear gas.

For RLQs, we noted that there are dependences of emission-line profiles and line strengths on the jet inclination, implying that emission regions are symmetric about the jet (rotation) axis — in particular BLR gas velocities are larger perpendicular to the jet. Dependences of profile asymmetry on inclination imply axisymmetric obscuration and an axisymmetric velocity field. Thus we suggested that dust shielded from the central ionizing continuum by BLR gas is responsible for revealing high-ionization axial (polar) outflow in core-dominant RLQs, and low-ionization equatorial outflow in lobe-dominant RLQs. This equatorial outflow ties in with associated absorption outflows and increased reddening in lobe-dominant RLQs. The low-latitude reddening is probably associated with hot AGN dust as well as the interstellar medium of the host galaxy. The inner edge of a dusty torus, at the evaporation radius for iron-rich grains, may shield from our view Fe II rich gas produced there, explaining why Fe II blends appear weaker in lobe-dominant RLQs.

These interpretations for RLQs conjure up the dusty-torus model for RQQs proposed by Weymann et al. (1991), in which BAL clouds are ablated from the surface of the torus and accelerated, by thermal wind and radiation pressure. BAL QSOs are just those where the high-velocity outflow lies along the line-of-sight. However, we have little evidence, so far, for axisymmetry in RQQs.

Further observational differences between RLQs and RQQs are seen in the inverse relation between the strengths of [O III] $\lambda5007$ emission from NLR gas at many pcs to Kpcs from the center, and the high-velocity Fe II emission from the BLR. RLQs lie at the strong [O III] – weak Fe II end of this relation, and the BAL QSOs at the weak [O III]– strong Fe II end. In this sense the BAL QSOs represent 'extreme RQQs'. We favour an explanation for this relation in terms of relative covering by dense, high-speed, Fe II-rich gas. The lower the covering by dusty low-ionization BLR gas, the more photons are able to escape to ionize the more-distant, low-density, NLR gas.

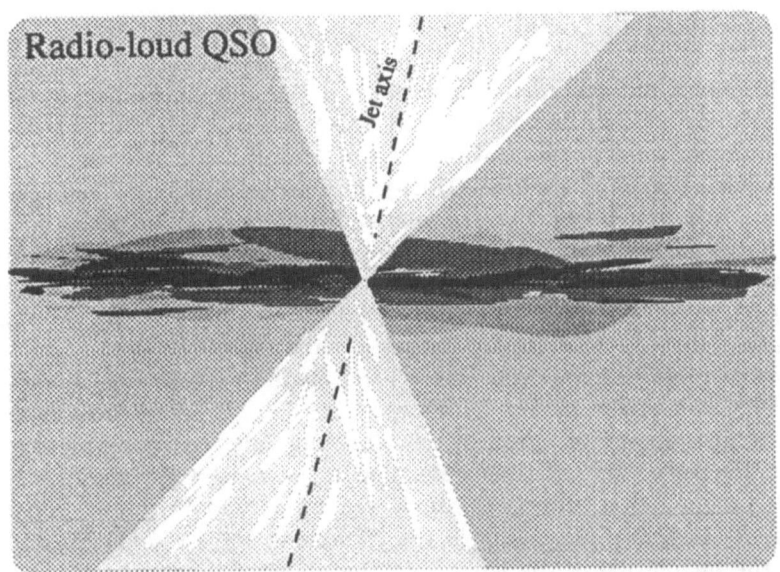

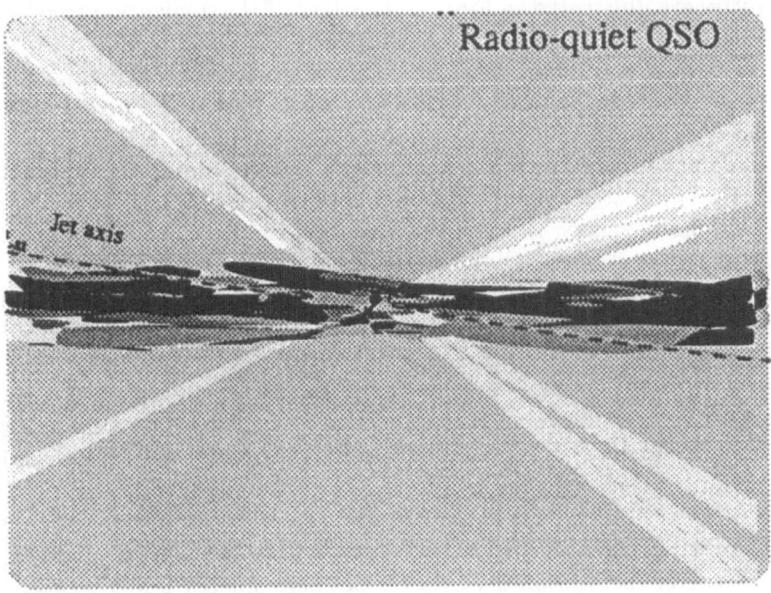

Fig. 1. A hypothesis for the different ultraviolet-optical properties of radio-loud and radio-quiet QSOs. The RLQs have jets, and ionization cones formed by the shadow of a dusty torus, in a direction away from the highest concentrations of low-ionization gas and dust in a galaxy, whereas the opposite is true for radio-quiet QSOs. In a planar, axisymmetric, distribution of broad emission line (BLR) gas, RLQs' ionizing photons excite gas out of the plane, to distances of several Kpc. In radio-quiet QSOs, the NLR gas is partly shielded from ionizing photons by high-optical-depth broad line gas (Fe II emitting), that is ablated from heated grains near evaporation temperature. We show a warped torus whose inner regions are perpendicular to the radio axis.

3   A Possible Model

The ultraviolet–optical differences between RQQs and RLQs can be summarized as follows. Compared with RLQs, RQQs appear to be associated with more emission from low-ionization gas (Fe II) and show evidence for more line-of-sight reddening and high-velocity outflowing (BAL) gas. The presence of a hot, dusty, environment with low- and high-ionization (BAL) outflows along the line-of-sight therefore has something to do with *lack* of powerful radio emission. If radio-loud and radio-quiet central engines are so similar, why should a line-of-sight effect be so important? It must be that RQQs' central engines are located in a similar nuclear environment, but one in which the observer is more likely to view the central engine through dusty, low- and high-ionization outflows.

We suggest that, for QSOs of the same Big Blue Bump luminosity, the same power is available to feed jets, that a 'jet axis' exists in both RLQs and RQQs, and that the observations can be accounted for by two main variables – one, the inclination of the jet axis to the line-of-sight, and the other, the inclination of the jet axis to the plane of the host galaxy. These possibilities are illustrated in Fig. 1. The edges of the 'ionization cone' within which the NLR can be excited are defined by the shadow of a dusty, inner torus. In several well-observed, but low-luminosity cases, the cone is fairly symmetric about the jet axis (N4261, Circinus: Urry & Padovani 1995). In the radio-loud case the jet axis is, in some observations of FR II radio galaxies, perpendicular to the plane of the host galaxy (Ekers & Simkin, and references therein; Heckman et al. 1985), or a dust lane, or even parallel to the rotation axis of the host galaxy or extended emission-line gas. In RQQs the inner regions may be symmetric about the jet axis, but the outer 'torus' may warp to match the galaxy plane; a synchrotron photon and high-energy particle beam may pound dense gas and dust in the inner regions, ablating Fe II-rich gas that shields the NLR. The range of possible inclinations of the jet axis to the plane of the galaxy could be quite large, depending on the vertical thickness of dense material near the nucleus. This geometry would determine the relative numbers of RLQs and RQQs. The geometry need not even be as simple as illustrated, for example, in the case of merging galaxies.

Our proposed picture could explain differences in inclination dependence. A wider range of viewing angles available for the inner parsec could explain the jet–observer inclination-dependence of axisymmetric emission regions in RLQs compared with RQQs. RLQs' 'illumination cones' are well-defined by the dusty torus and are free of the Galactic plane. Still, grazing views of the torus can result in associated absorption and reddening. Reddening can also result from a low-latitude view of the galactic plane, especially if the jet–cone axis is tilted towards the observer and the galactic plane. Jet–observer inclination dependence for RQQs is less easily defined because galactic obscuration is important, and the 'torus' geometry may be more complex. Greater nuclear dust-covering may result in stronger Fe II emission. The inverse Fe II–[O III] relation — increasing [O III] and decreasing Fe II emission – could result as the angle between the jet–cone axis and the galactic plane increases, illuminating more, distant, low-density NLR gas. This also explains the weaker Fe II and stronger NLR emission in RLQs.

The dustier, low-ionization absorption environment seen towards RQQs may not be only a jet–galactic plane inclination dependence. It has been commonly thought that powerful radio-loud AGN occur in elliptical galaxies, and the radio-quiet AGN occur in spirals (Hutchings et al. 1989). This would tie in nicely because spiral galaxies are thought to be, more often, richer in gas and dust. This host-galaxy dichotomy has also been suggested to relate to the nuclear and host galaxy mass (Heckman 1983) or angular momentum (Wilson & Colbert 1995) hypotheses for the generation and maintainance of powerful radio jets, and it would be necessary to measure all three parameters (mass, angular momentum, and dusty, low-ionization environment) to disentangle these hypotheses. We note that elliptical galaxies can contain significant amounts of dust. There are apparently RQQs in elliptical hosts (Disney et al. 1995), and radio-loud sources in spiral galaxies, although tidal tails produced by mergers that are thought to fuel AGN, could be mistaken for spiral arms. Are RQQs ever found in ellipticals? Hubble Space Telescope imaging may provide the answer.

Why the lack of powerful jets in RQQs? We suggest that this has something to do with the inner jet being within the galactic plane – perhaps lack of collimation as a result of greater mass densities, or perhaps related to orientation of the jet and galactic magnetic fields. Radio cores may appear weaker as a result of absorption by highly-ionized nuclear plasma. It may be a problem that light relativistic fluid jets are likely to propagate unimpeded through the galactic plane (Leahy, this workshop).

Tests of such a picture could be to investigate, by radio and optical-ultraviolet imaging, polarimetry, and spectroscopy, the relative orientation of some RQQs' weak jets, possible illumination cones, and host galaxy orientation. One could look for absorption in the nuclear, radio-core spectrum, and investigate its possible relation to absorption seen in the optical, ultraviolet and X-ray regions.

Having found significant differences between the spectra of RLQs and RQQs, one could question our original assumption of the similarity of the central engines of RLQs and RQQs. However, we might argue that the greatest differences we find are probably in gas thought to exist at least $\sim$ 1 pc from the central engine.

References

Disney, M. J., Boyce, P. J., Blades, J. C., Boksenberg, A., Crane, P., DeHarveng, J. M., Macchetto, F., Mackay, C. D., Sparks, W. B.;Phillipps, S. (1995): Interacting Elliptical Galaxies as Hosts of Intermediate-redshift Quasars. Nature **376**, 150

Ekers, R. D., Simkin, S. M. (1983): Radio Structure and Optical Kinematics of the cD Galaxy Hydra A (3C 218) ApJ **265**, 85

Heckman, T. M. (1983): Radio Emission and the Masses of Elliptical Galaxies. ApJ **273**, 505

Heckman, T. M., Illingworth, G. D.,Miley, G. K., Van Breugel, W. J. M. (1985): The kinematics of stars and gas in radio Galaxies. ApJ **299**, 41

Hutchings, J. B., Janson, T., Neff, S. G. (1989): What is the Difference Between Radio-loud and Radio-quiet Quasi-Stellar Objects? ApJ **342**, 660

Urry, C. M., Padovani, P. (1995): Unified Schemes for Radio-Loud Active Galactic Nuclei PASP **107**, 803

Wilson, A. S., Colbert, E. J. M. (1995): The Difference between Radio-Loud and Radio-Quiet Active Galaxies. ApJ **438**, 62 Toward a Classification Scheme for Broad-Line

Jochen Eislöffel addressing Beverley Wills

Jets in Gamma-Bright AGN:
Constraints on Reprocessing Mechanisms

Stefan J. Wagner

Landessternwarte, Königstuhl, Heidelberg, Germany

Abstract. Several dozen active galaxies have been detected by instruments sensitive to hard gamma-ray radiation. All of these objects are radio-loud AGN and share several other characteristics in the low-energy part of their spectra. The gamma-ray emission varies on time-scales from days to years and is characterized by a power-law spectrum. In order to determine the broad-band energy distributions of these objects, a series of simultaneous multifrequency campaigns has been arranged. Apart from providing the low-frequency properties during those states when the sources were bright enough in the gamma-ray range to be detected, such campaigns provide clues to the degree of correlation between variations in different energy regimes. Several cases were found which indicate that the optically thin part of the synchrotron radiation is closely connected to the gamma-ray range. This sets constraints on the reprocessing mechanisms responsible for the gamma-ray emission.

In addition to studies in the GeV regime, ground-based Cherenkov telescopes allow investigations of the gamma-ray properties in the TeV regime. One such source, the BL Lac object Mrk 421 has been found to emit variable TeV radiation which might be related to the variable emission seen in the optical regime.

1 Introduction

In 1991 the Compton Gamma-Ray Observatory (CGRO) was launched. It carried on board four instruments, covering the energy range from a few 10 keV up to a few 10 GeV. The highest energies are studied with the EGRET instrument, a spark chamber operating in the energy range above 30 MeV. Its wide field of view (0.6 sr) observes different regions of the sky for typically 2 weeks. Within the ~ 80 pointings during the first 2.5 years of operation it detected ~ 50 AGN, five pulsars and GRBs, as well as ~ 60 unidentified sources.

The most complete list of EGRET sources has been compiled by Thompson et al. (1996). Von Montigny et al. (1995) describe the gamma-ray characteristics and (non-simultaneous) broad-band energy distributions of the gamma-bright AGN. Models explaining the gamma-ray emission in terms of inverse-Compton (IC) scattered radiation or hadron-initiated cascades have been reviewed e.g. by Schlickeiser (1996).

All of the gamma-bright AGN are flat-spectrum radio sources, either belonging to the group of radio-selected BL Lac objects or flat-spectrum Quasars (Eckart et al., 1987). All of the sources show polarized power-law continua in the optical range. The low-energy (radio through optical) spectra are dominated by synchrotron radiation. In this energy range all of the sources vary on fairly short time-scales, most of them belong to the class of Intraday Variables (IDV sources) as described by Wagner and Witzel (1995). Most of the radio emission is compact and many of the sources show superluminal velocities in VLBI campaigns. In three sources superluminal motion was discovered as a result of following-up the detection of gamma-ray emission (PKS 0420-014 (Wagner et al., 1995a), PKS 0528+134 (Zhang et al., 1994), and PKS 1633+382 (Barthel et al., 1995)). All of the above characteristics are generally attributed to a relativistic jet pointing at small angles to the line-of-sight of the observer. It is therefore most likely that the gamma-ray emission can be attributed to the relativistic jet as well.

2 Gamma-ray observations

Among the characteristics of the gamma-ray emission described by von Montigny et al. (1995), variability is striking. It is the most important one for the discussion below. Almost all of the sources detected so far vary in the energy range above 70 MeV. The only sources where variability could not be proven are the faint ones detected only in the deepest pointings (with other observations of the same region of sky giving only upper limits). Rapid variability can be studied during an individual observation only, if the average flux is large enough to yield sufficiently high statistics within individual bins of 3 to 5 days. In more than 50 % of these cases the sources were found to be variable on such short time-scales. Examples are the early discovery of strong emission from 3C 279 which showed an increase of a factor of 2 within a few days and a subsequent decay which was even faster (factor of two within 24 h) (Kniffen et al., 1993), PKS 0528+134 (Hunter et al., 1993, Mukherjee et al., 1996), PKS 2251+158 (Hartman et al., 1993), and 1633+382 (Mattox et al., 1993). One of the important conclusions which can be drawn from the rapid variations is the compactness of the gamma-ray emitting volume. If the high gamma-ray luminosities emerge from a volume less than about one light-day across, the density of gamma-ray photons is so high that they should suffer extreme losses due to photon-photon pair production (e.g. McBreen, 1979). This self-absorption can be reconciled with the observed gamma-ray luminosities if the emitting plasma moves relativistically towards the observer and the high-energy photons are Doppler-boosted, similar to the radio emission. This is a more direct evidence of the gamma-ray emission being related to the relativistic jet of these sources.

In most cases the flux is too low to permit binning in sub-intervals, and light-curves can be obtained only by integrating the flux of the entire viewing periods. Structure-function analysis of such data shows that the characteristic time-scales are of the order 10 - 100 days, which is comparable to or shorter than the sampling time-scale. Apart from sources located in regions of the sky which are of particular interest, most objects are observed 1 - 3 times a year. This is illustrated in figure 1 which gives the light-curve and the structure-function of the (fairly well-sampled) source PKS 0208-512.

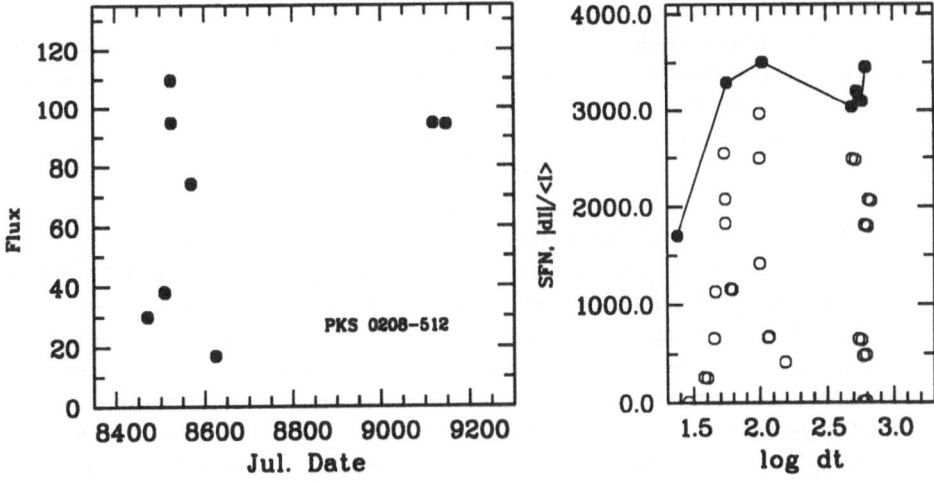

Fig. 1. Gamma-ray variations of PKS 0208-512. The left panel shows the light-curve obtained from EGRET observations (e.g. Thompson et al., 1996), the right panel illustrates the structure-function (full circles). The corresponding $\{\Delta\,I_{ij},\,\Delta\,t_{ij}\}$ pairs are shown in open circles to illustrate the sampling problem of the structure-function analysis.

Gamma-ray variability is too fast to derive the characteristics of the changes with the temporal sampling available with EGRET. This strongly limits the possibility to draw conclusions on the correlation between variations in the gamma-ray range ond other bands which occur on time-scales longer than a few weeks.

The range of variability is about an order of magnitude, almost comparable to the dynamic range between the brightest detections and the faintest levels providing secure detections. It is unclear, therefore, whether the steady, quiescent level of gamma-ray flux has been detected in any but the brightest sources. Most *detections* of AGN in the high-energy gamma-ray range seem to mark a gamma-ray *flare* of that particular source.

3 Low-frequency properties during gamma-bright states

The variability of the gamma-ray emission implies that broad-band characteristics have to be studied simultaneously in order to carry out quantitative comparisons to the gamma-ray properties. Apart from long-term monitoring campaigns which include certain fractions of the gamma-bright AGN (such as the Michigan cm monitoring program (Aller et al., 1996), or the Metsahovi mm monitoring program (Valtaoja et al., 1996)), dedicated campaigns have been set up to carry out simultaneous observations (e.g. Marscher, 1996; Reich et al., 1993; Wehrle et al., 1996). Many of these campaigns focussed on radio observations. In the following, optical observations are discussed. Optical data have the advantage of not being subject to self-absorption. Since optical synchrotron radiation is assumed to be optically thin, any direct correlation of optical and gamma-ray activity would only be subject to delays of the gamma-emission introduced by transport and reprocessing.

During high states of gamma-ray emission, all sources show enhanced levels of optical brightness. While this observation is limited by an incomplete sample and insufficient and inhomogeneous sampling, we have not found a single case of enhanced gamma-ray activity during 'quiescent' optical states. The optical light-curves are usually sampled sufficiently well to define a quiescent level. An example of this long-term monitoring is shown in figure 2, comparing the optical and gamma-ray light-curves of PKS 0420-014. While both light-curves are undersampled, the brightest gamma-ray state occurs exactly at the time of the brightest optical state. Likewise, slightly enhanced gamma-ray states are related to periods of lower activity (still enhanced with respect to the quiescent level).

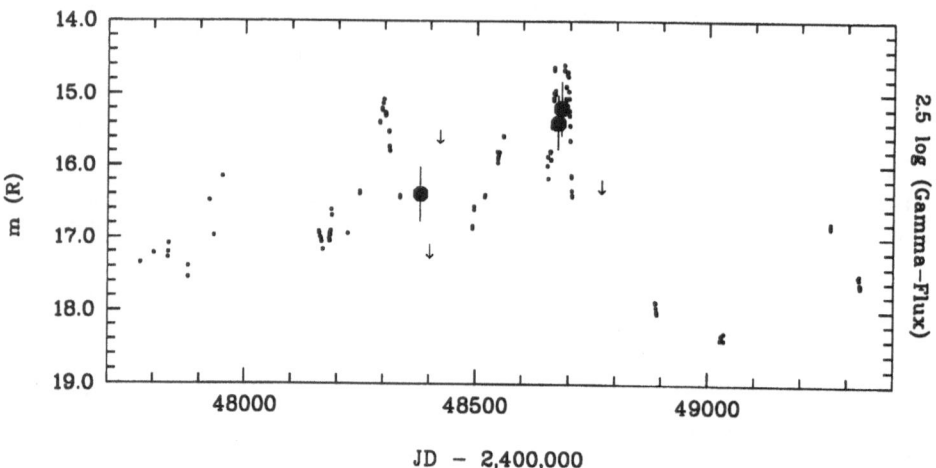

Fig. 2. Gamma-ray emission of PKS 0420-014 (full circles) and optical variations (small circles) during 1991 - 1994 (see Wagner et al., 1995a).

While there is in general no indication that very rapid optical activity in BL Lac objects and flat-spectrum quasars occur preferentially during bright states of long-term optical light-curves (Wagner and Witzel, 1995) we have found that high states of gamma-ray activity often correspond to very fast changes of the optical flux. An example is given by von Linde et al. (1993), who found rapid

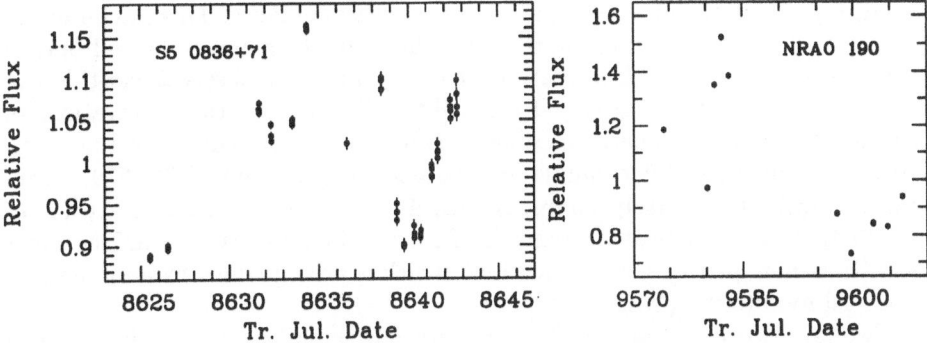

Fig. 3. Optical flares of NRAO 190 and 0836+71. The two R-band light-curves were obtained during epochs when the sources were bright in the gamma-ray range (though not bright enough to study variability on short time-scales).

changes coinciding with a high gamma-ray state of S5 0836+71, the most distant gamma-ray source known. Although this object has been studied for intraday variability in several campaigns (e.g. Wagner et al., 1990) it was hardly found to show such rapid variations as during the high gamma-ray state in January 1992. Other examples are S5 0954+65 (Mukherjee et al., 1995) and NRAO 190 (McGlynn et al., 1994, 1996). The optical light-curves of those sources, as measured by the Heidelberg monitoring group are shown in figure 3. Objects which are detected by EGRET whenever they are observed at very similar luminosities (such as the BL Lac source S5 0716+714) show these very rapid variations permanently (Heidt and Wagner, 1996) indicating a very high duty cycle of both gamma-ray activity and intraday variability.

In a few cases not only the total flux but also the polarization characteristics have been monitored. They also show rapid changes during epochs of high gamma-ray activity. These changes in polarized flux are not directly correlated to changes in total flux (Wagner et al., in preparation).

All of these characteristics illustrate that it might be misleading to directly interpret non-simultaneous broad-band energy distributions. Even a "simultaneously" derived flux density such as a single photometric measurement taken during a fortnight of gamma-ray observation might fall short in representing the *average* flux density which would correspond to the measurements above 70 MeV. The changes in Stokes parameters further illustrate that even well-sampled light-curves of total flux may not be sufficient to reveal the full information necessary for quantitative understanding of the IC scattered contribution to the gamma-ray emission.

4 Correlated flares

The gamma-bright AGN can be defined fairly well on the basis of their properties at lower frequencies. This suggests that the emission processes of gamma-radiation are closely linked to the synchrotron characteristics. This can be tested quantitatively by comparing the time variability in the gamma-ray and optical ranges. Due to the sampling pattern of the gamma-ray observations this can be done best for fast flares (rapid variability). For most of the observations of fast gamma-ray flares mentioned above, there were hardly any optical observations. A more successful example are the early observation of 3C 279, where some similarity to (sparsely sampled) optical data are discussed by Hartman et al. (1996). A less significant gamma-ray flare was found in 3C 279 in December 1994 (Hartman, private communication), which is accompanied by a minor flare in optical monitoring (Wagner, 1996b).

Two additional events have been recorded with sufficient sampling both in the optical and in the gamma-ray ranges. PKS 1406-076 flared in January 1993 (figure 4) in both bands (Wagner et al., 1995b). A similar event was observed in PKS 1622-293 in June/July 1995 (Mattox et al., 1995).

Fig. 4. PKS 1406-076 exhibited a gamma-ray flare (open circles) and an outburst of optical synchrotron radiation (filled symbols) during January 1993 (see Wagner et al., 1995b).

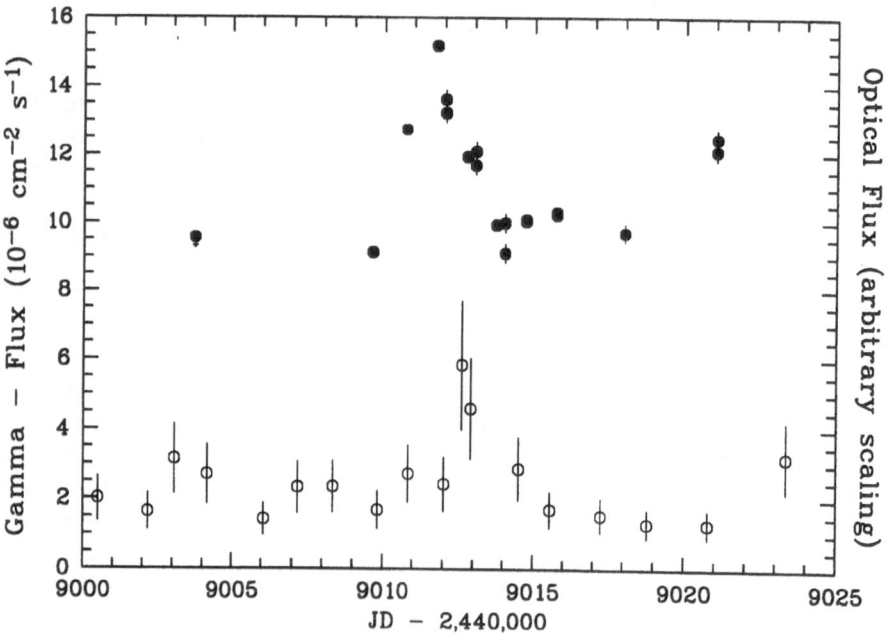

PKS 1406-076 has not been studied extensively prior to its detection by EGRET. The limited monitoring carried out after the observations shown in figure 4 suggest that the flaring rate in this object is fairly low. The January 1993 outburst is the only one recorded in about 100 days of observations spread over a two year interval. Likewise, the gamma-rays have not yet varied again in a similar way. This reduces the probability of the coincidence being by chance (physically unrelated outbursts of optical and gamma-ray radiation happening within the same three day interval) to below 10^{-6}.

Except for clues on the radiation processes, the correlated flare confirms the correct identification of the gamma-ray source. The statistical evidence of gamma-bright sources being correctly identified with flat-spectrum radio sources is very high (Mattox et al., in preparation). The large number of yet unidentified EGRET sources may indicate the presence of a second major population among gamma-bright sources which has a large scale-height or a nearly isotropic distribution on the sky. In view of the large error boxes of the gamma-ray sources any individual identification has to rely on unambiguously correlated activity.

5   Constraints on reprocessing mechanisms

The optical and the gamma-ray outbursts both last for about 2 ± 1 days. If the radiation recorded in one of the energy bands is reprocessed from the other, this reprocessing has to occur in a volume which is not much larger than the emitting volume to avoid the flare being spread out in time. The optical flare precedes the gamma-ray flare by one day. This suggests that either the gamma-rays are reprocessed synchrotron radiation or that the increase in both frequency ranges is triggered independently by a primary process with the optical radiation responding faster.

Mannheim and Biermann (1992) suggested that the decay products of relativistic hadrons give rise to flares of electromagnetic radiation. With photon energies corresponding to the energies of the secondaries decaying in an originally highly relativistic hadronic shower, one would expect photons of higher photon energies leading those of lower energies - contrary to what is observed in PKS 1406-076. The temporal sequence cannot rule out the possibility, however, that the optical flare marks the trigger of the onset of the hadronic shower (e.g. if the optical flare were associated with the injection and acceleration of the primary particles). The (delayed) gamma-ray flare will then still be related to the decaying primary particles. Apart from energetic constraints (e.g. Schlickeiser, 1995) such a scenario would require very special tuning to explain the shapes of the flares (Wagner, 1996a). Likewise, the model of Bednarek and Kirk (1995) also suffers from the problem of producing *delayed* optical radiation.

The Synchrotron-Self-Compton (SSC) mechanism is one of the most popular models which has been put forward to predict and explain the gamma-ray emission of AGN (Marscher, 1980, Maraschi, et al., 1992). In a strict sense, SSC models require the very electrons that radiate the synchrotron emission to upscatter the soft photons instantaneously and in situ and cannot explain the lag between the peaks in the two frequency ranges. IC scattering may instead occur downstream of the site of synchrotron emission. It is very likely in this case, however, that the distribution function of the radiating particles changes during the optical flare.

Several models have been suggested which involve IC scattering of external photons by the relativistic electrons within the jet. These external photons may either originate from the accretion disk (e.g. Dermer et al., 1992), the broad-line region (e.g. Sikora et al., 1994) or the dusty torus (Wagner et al., 1995a). The simplest interpretation of the tight correlation of optical variability and gamma-flares suggests that the optical flare provides the photons which are scattered up in energy. In this case the seed photons would have to emerge from a site that can produce very luminous flares of synchrotron radiation. The only site within an AGN for such a flare that has been discussed in any detail is the jet. The seed photons would then not be genuinely external but would be reflected back into the jet. The delay between the optical and the gamma-ray flare would correspond to the time required for the scattering part of the jet to reach the gamma-photosphere. If the seed photons do not originate from within the jet, such a model would be very constrained by the requirement to reproduce the IDV characteristics (Wagner and Witzel, 1995) which generally call for relativistic beaming. The studies by Begelman et al. (1994) and Sikora et al. (1994) may tie IDV and high gamma-luminosity within one model.

A variant of the above suggestions would be two-fluid models (see e.g. Kundt and Gopal-Krishna, 1980, Sol et al., 1989). Even in a simple jet model the particle distribution varies along and across the jet. Treated as a multiple set of fluids, synchrotron-flares which originate in localized sections of the jet will be IC scattered by other parts of the jet where electron densities, local magnetic field structures and particle distribution functions are different, invalidating attempts to model the scattering within straightforward SSC scenarios. Electron-Proton beams may also give rise to the gamma-ray emission through annihilation (Henri et al., 1993). As mentioned by Henri et al., optical synchrotron radiation may accompany the gamma-ray flare. The delay between the optical and gamma-ray burst would be caused by the high gamma-ray opacity at the onset of the flare.

Many of the models discussed above have illustrated one specific reprocessing scenario. More detailed studies are needed which take all of these scenarios into account since all of them contribute to the gamma-ray emission to some extent (i.e. a certain degree of SSC, scattering of external photons, etc. is unavoidable). The occurrence of correlated variations suggests that most of the reprocessing is dominated by a single mechanism (otherwise the response in the gamma-ray range would be diluted very much). On the other hand, the characteristics described in sections 2-4 illustrate that the full set of observables must be monitored with high precision and dense sampling to be modeled quantitatively.

6 TeV flares in Mrk 421

As shown above, the rapid flares seen in several gamma-bright AGN show up also at lower photon energies, indicating that the IDV variations are correlated throughout the entire spectrum up to the GeV range. At even higher energies the photon fluxes get extremely low and sufficiently high photon statistics can be

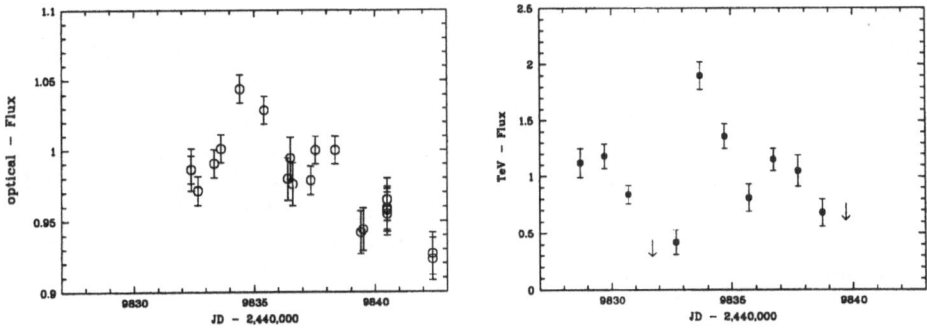

Fig. 5. The optical flares of Mrk 421 and simultaneously recorded TeV emission (compiled from Buckley et al., 1995).

achieved only by the atmospheric Cherenkov technique. The Whipple telescope which has a maximum efficiency at energies of a few TeV had reported significant detections of the nearby BL Lac object Mrk 421 during several observing runs (e.g. Kerrick et al., 1994). During some of these observations the source was flaring in a manner similar to the variations seen among the brighter EGRET detected sources. While being an EGRET-detected source as well, the GeV fluxes of Mrk 421 have always been too low to permit investigations of variability on short time-scales. On longer time-scales the source has been fairly steady and showed little variability in the GeV range.

Mrk 421 is different from other EGRET-detected sources also in being a low-luminosity source at all frequencies. The synchrotron spectrum extends up into the X-ray regime where high (flux and spectral) variability has been detected with several X-ray instruments (see e.g. Wagner and Witzel, 1995). The amplitude of variations is largest close to the high-energy cutoff of the synchrotron spectrum and significantly lower in the optical range. Within the context of IC models for the GeV emission, which identify the GeV emission as scattered optical radiation, one expects that the variation within the gamma-range follow the trend of increasing amplitude with photon energy. This explains the high degree of variation of Mrk 421 at TeV energies and its low degree of variations in the GeV regime. Likewise, this assumption predicts correlated variations in the TeV regime and at lower frequencies.

We carried out optical monitoring in parallel to the TeV observations by the Whipple group in April 1995. The Whipple team reported variations in Buckley et al. (1995), reproduced in the right hand panel of figure 5. As shown in the left hand panel of fig. 5, optical variations with a similar pattern but much reduced amplitude have been recorded. The sampling pattern is not perfect and the campaign lasted only for 15 days but the degree of correlation is very high, indicating that the non-thermal radiation in this source is indeed correlated over at least 13 decades of photon energy (Takahashi et al., 1996, Wagner et al., 1996).

Acknowledgements This work was supported by the Deutsche Forschungsgemeinschaft (SFB 328). The author thanks Bob Hartman and John Mattox for cooperation concerning the gamma-ray data, Holger Bock, Anke Heines, and Martin Kümmel for much work on the optical data, Reinhard Schlickeiser and Alan Marscher for discussions, and Wolfgang Kundt for organizing this interesting workshop.

References

Aller, M. et al., 1996, in *IAU Symp. 175, Extragalactic Radio Sources*, C. Fanti et al. (Eds.), Kluwer, in press

Buckley J., Quinn J., Weekes T., Catanese M., Carter-Lewis D.A., et al., 1995, IAUC 6167

Barthel P.D., Conway J.E., Myers S.T., Pearson T.J., and Readhead A.C.S., 1995, ApJ 444, L21

Bednarek, W. and Kirk, J.G., 1995, A&A 294, 366

Dermer C.D., Schlickeiser R., and Mastichiadis A., 1992, A&A 256, L27

Eckart, A., Witzel, A., Biermann P.L., Johnston, K.J., Simon, R., Schalinski, C.J., and Kühr, H.: 1987, A&A Supp. 67, 121

Hartman R.C., Bertsch D.L., Dingus B.L., Fichtel C.E., Hunter S.D., et al., 1993, ApJ Lett. 407, 41

Hartman R.C. et al., 1996, ApJ, in press

Henri, G., Pelletier, G., and Roland, J., 1993, ApJ 404, L41

Heidt, J. and Wagner, S., 1996, A&A 305, 42

Hunter S.D., Bertsch D.L., Dingus B.L., Fichtel C.E., Hartman R.C., et al., 1993, ApJ 409, 134

Kerrick A.D., Akerlof, C.W., Biller, S.D., Buckley, J.H., Cawley, M.F., et al., 1995, ApJ Lett. 438, 59

Kniffen D.A., Bertsch D.L., Fichtel C.E., Hartman R.C., Hunter S.D., et al., 1993, ApJ 411,133

Kundt, W. and Gopal-Krishna, 1980, Nature 288, 149

Mannheim, K., and Biermann, P.L., 1992, A&A 253, L21

Maraschi, L., Ghisellini, G., and Celotti, A., 1992, ApJ 397, L5

Marscher, A.P., 1980, ApJ 235, 3865

Marscher, A., 1996, in *Quasars and AGN: High Resolution Radio Imaging*, M. Cohen and K. Kellerman (Eds.), Proc. Nat. Acad. Sci., in press

Mattox J.R., Bertsch D.L., Chiang J., Dingus B.L., Fichtel C.E., et al., 1993, ApJ 410, 609

Mattox, J.R., Wagner, S.J., McGlynn, T.A., Malkan, M., Schachter, J.F., Sreekumar, P., 1995, *IAUC*, 6179, 6181

McBreen, B., 1979, A&A 71, L19

McGlynn T.A., Mattox J.R., Vestrand W.T., Dingus B.L., Bertsch D.L., et al., 1994, IAUC 6061

McGlynn T.A., Hartman R.C., Aller M.F., Filippenko A.V., Marscher A.P., et al., 1996, ApJ, submitted

Mukherjee R., Aller H.D., Aller M.F., Bertsch D.L., Collmar W., et al., 1995, ApJ 445, 189

Mukherjee R. et al., 1996, ApJ, in press

Reich, W. et al., 1993, A&A 273, 65

Schlickeiser, R., 1996, A&A Supp., in press

Sikora M., Begelman M.C., and Rees M.J., 1994, ApJ 421, 153

Sol, H., Pelletier, G., and Asséo, E., 1989, MN 237, 411

Takahashi, T. et al., 1996, ApJ, in press

Thompson, D. et al., 1996, ApJ Supp., in press

Valtaoja, E. et al., 1996, A&A Supp., in press

von Linde J., Borgeest U., Schramm K.-J., Graser U., Heidt J., et al., 1993, A&A 267, L23

von Montigny C., Bertsch D.L., Chiang J., Dingus B.L., Esposito J.A., et al., 1995, ApJ 440, 525

Wagner, S.J., 1996a, in *17th Texas Symposium on Relativistic Astrophysics*, H. Böhringer et al. (Eds.), New York Acad. Sci., 759, p.526

Wagner, S.J., 1996b, A&A Supp., in press

Wagner S.J. and Witzel A., 1995, ARAA 33, 163

Wagner, S.J., Sanchez-Pons, F., Quirrenbach, A., and Witzel, A., 1990, A&A 235, L1

Wagner S.J., Camenzind M., Dreissigacker O., Borgeest U., Britzen S., et al., 1995a, A&A 298, 688

Wagner, S.J., Mattox, J.R., Hopp, U., Bock, H., Heidt, J., et al., 1995b, ApJ Lett., 454, L97

Wagner, S.J. et al., 1996, ApJ, in press

Wehrle, A. et al., 1996, A&A Supp., in press

Zhang, Y.C. et al., 1994, ApJ 432, 91

Spectral Evolution Along the Jets of M 87 and 3C 273

Klaus Meisenheimer, Martin Neumann, Hermann-Josef Röser

Max-Planck-Institut für Astronomie
Königstuhl 17, D−69117 Heidelberg, Germany

1 Motivation and outline of our project

The emission of extragalactic radio sources is entirely dominated by synchrotron radiation from highly relativistic particles (electrons, positrons?) moving through magnetic fields. The smoothness of the continuous spectrum generated by this radiation process limits the accuracy with which physical parameters can be inferred from observational quantities: For instance, radio images of the relevant Stokes parameters ($I(\nu)$, $Q(\nu)$, $U(\nu)$) typically span a frequency range of about a factor of 10 and thus provide little more than the morphology of the source, its magnetic field structure (projected onto the plane of sky), and in some fortunate cases an indication of spectral changes. The latter could in principle represent an intrinsic physical change – *i.e.* a change in particle spectrum. In practice, however, the variations of the observed synchrotron spectrum due to changes in the effective magnetic field strength ($B_\perp$, perpendicular to the line of sight) may equally well modify the observed spectral index of curved spectra. More details about the radio plasma can be derived in those cases where multi-epoch radio maps reveal proper motions of the radio pattern and thus allow some conclusion about one of the essential parameters: the flow velocity of the radio plasma.

In any case, the smoothness of the synchrotron continuum asks for a multi-frequency approach: Only if the spectrum is observed over many decades in frequency can one hope to disentangle magnetic field changes from variations in the particle spectrum. This puts those sources into the limelight which are detected over a wide range, *e.g.* not only in the radio but also in the NIR/optical frequency band. Since *jets* are the prime movers of powerful radio sources, the key objects for our understanding of radio sources are that handful of objects in which the radio jet can be traced up to optical frequencies (and beyond).

Due to their large angular extent ($> 10''$) and their high surface brightness at all frequencies, the most promising jets are those of M 87, 3C 273 and PKS 0521−36. Leaving aside the latter (for which only poor radio maps are available) we concentrate on a detailed investigation of the jets in M 87 and 3C 273. Our observational program aims at a determination of the – spatially resolved –

overall spectrum. To this end we have employed optical, near-infrared, radio[1], and X-ray observations[2] with the best resolution attainable from the ground ($\sim 1''$ for optical maps, $0''.1$ for radio maps) and HST observations (optical resolution $0''.1$). HST images at $\lambda = 300$ and $600\,\mathrm{nm}$ of the jet in 3C 273 have been obtained in Cycle 5. Together with new high-fidelity VLA and MERLIN maps they will allow us to study the jet of 3C 273 on a scale of 200 pc, sufficiently close to the ground-based resolution of M 87 (100 pc) for a direct comparison.

Here we summarize our results from combining ground-based optical work with radio maps at matching resolution.

2 The synchrotron spectrum of the jet in M 87

Our first analysis of the spatially resolved radio-optical synchrotron spectrum of the jet in M 87 was based on photometry in three optical bands (B,R,I) plus one radio frequency (15 GHz, from Owen *et al.* 1989). The results of this study are already published (Biretta & Meisenheimer 1993, Meisenheimer *et al.* 1996a).

We found that the overall synchrotron spectrum is characterized by a rather flat radio-optical power-law $S_\nu \sim \nu^{-0.65}$ which cuts off steeply at frequencies $\nu_c \simeq 2 \times 10^{15}\,\mathrm{Hz}$. This indicates that the underlying electron spectrum has a sharp high-energy cutoff around $\gamma_c \equiv E_c/m_e c^2 \simeq 10^6$. Due to the limited frequency range of the optical data ($\nu_B/\nu_I \lesssim 2$) the pointwise determination of ν_c had to rely on the assumption that the spectral *shape* of the spectrum remains constant along the jet. Meanwhile we have obtained NIR-maps (H,K) of M 87 which in combination with the optical data ($\nu_B/\nu_K \simeq 5$) allow a much more accurate determination of the spectral shape (Neumann 1995, Neumann *et al.* 1996b). Thus we are able to fit synchrotron model spectra at every point *along* the jet (no significant variations were found *across* the jet). The best fits where obtained with a model spectrum assuming a sharp high-energy cutoff in the electron spectrum, although the data do not rule out a smoother high-energy cutoff (*cf.* Meisenheimer *et al.* 1996a for details of the model spectra).

Fig. 1 summarizes our new results on the spectral evolution along the jet of M 87. The cutoff frequency ν_c shows a global decrease from 3×10^{15} to $\simeq 7 \times 10^{14}\,\mathrm{Hz}$ modulated by oscillations by a factor of 2. Note that the peaks and dips of ν_c exactly match those observed in the brightness distribution along the jet. The power-law slope remains perfectly constant at $\alpha_{PL} = -0.660$, at least for the brightest parts of the jet ($9''$ to $21''$ from the core). The small oscillations at the position of bright knots are not significant but rather result from the limited accuracy with which the radio maps could be aligned to the optical data (better than $0''.05$!). We are not certain whether the variation from $\alpha_{PL} = -0.69$ to $\alpha_{PL} = -0.64$ observed in the inner jet ($3''$ to $9''$ from the core) is real or introduced by residuals in removing the starlight of the galaxy M 87.

This extremely accurate determination of the pointwise spectrum along the jet fully justifies our earlier suggestion that the *shape* of the synchrotron spec-

---

[1] In collaboration with R. Conway, R. Davis (Jodrell Bank) and R. Perley (NRAO).
[2] With H. Fink and others (MPE Garching).

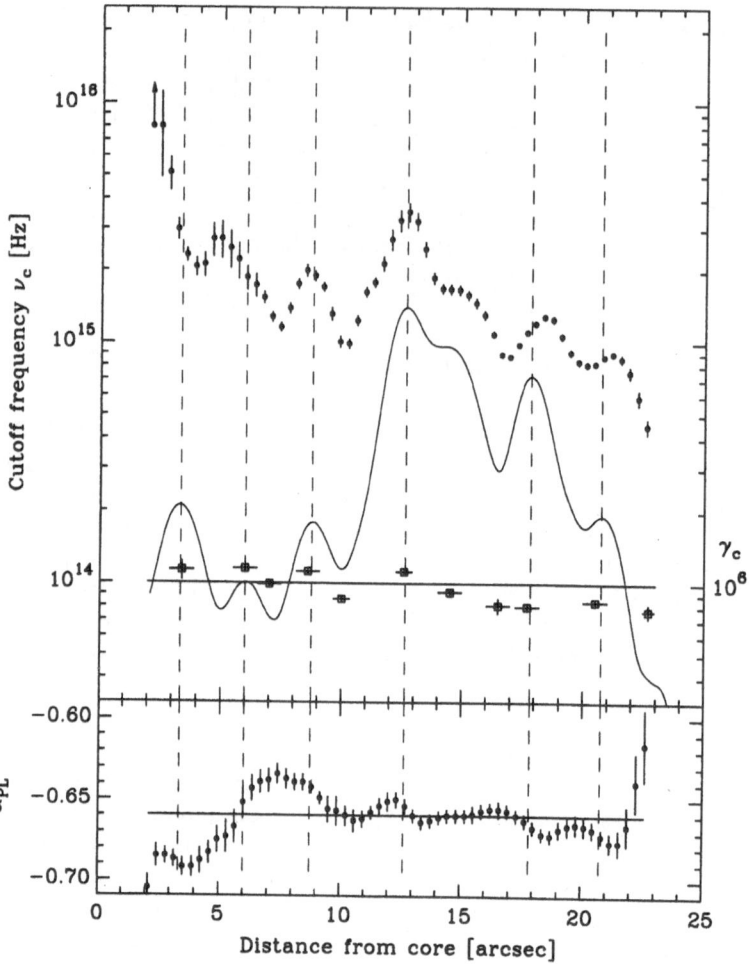

Fig. 1. Fit parameters of the model spectra along the jet in M 87. In the upper panel we show the cutoff frequency ν_c and the brightness at 15 GHz (logarithmic scale, thin line). In the lower panel the radio-optical power-law index α_{PL} is shown. The values of γ_c ($\square$, refer to labels on the right) were calculated assuming a minimum-energy magnetic field. The positions of prominent knots are indicated by dashed vertical lines.

trum remains completely unchanged along the jet (Biretta & Meisenheimer 1993). The spectral variations observed (*e.g.* of the optical spectral index) are solely due to changes in ν_c along the jet. Moreover, using a minimum-energy estimate of the local magnetic field in selected knots and inter-knot regions we derive a maximum energy γ_c which remains constant at $\gamma_c = 10^6$ within ±15%. We regard this as convincing evidence that the energy spectrum of the electrons (positrons) remains unchanged along the entire jet. The observed variations in ν_c only reflect the variations of the local field $B_\perp$ (perpendicular to the line of sight) according to standard synchrotron theory:

$$\nu_c = 1.26 \, 10^{15} \, \frac{B_\perp}{30 \, \text{nT}} \left(\frac{\gamma_c}{10^6} \right)^2 \, [\text{Hz}]. \tag{1}$$

The typical synchrotron loss scale for an isotropic pitch-angle distribution, *i.e.* $\langle \sin \theta \rangle = 1/\sqrt{2}$,

$$x_{1/2} = \frac{c\tau_{1/2}}{\sqrt{2}} = 41 \left(\frac{B_\perp}{30 \, \text{nT}} \right)^{-2} \left(\frac{\gamma_c}{10^6} \right)^{-1} \, [\text{pc}] \tag{2}$$

would predict that losses of 15% should occur on scales of $< 8 \, \text{pc}$ ($\leq 0\rlap{.}{''}1$ at the distance of M 87). This is two orders of 10 shorter than the length of the jet and at least $10\times$ smaller than the typical knot to inter-knot distance. Possible solutions to this dilemma will be discussed in a separate contribution about "Particle Acceleration in Extended Radio Sources" (see Meisenheimer, these proceedings).

3 The spectrum of the jet in 3C 273.

Although being of similar angular extent as the jet of M 87, the linear size of the much more distant jet of the quasar 3C 273 ($z = 0.158$, *i.e.* $3.7 \, \text{kpc}/''$ for $H_0 = 50 \, \text{km/s/Mpc}$) is at least 80 kpc, that is $> 40\times$ longer than that of M 87. The typical separation of $1''$ between knots in the 3C 273 jet makes it impossible to separate the spectra of the inter-knot regions from that of the knots with our ground-based optical data ($1\rlap{.}{''}3$ FWHM resolution). Therefore, we have to limit our analysis of the jet's spectrum to 11 components into which the jet can be decomposed (see Röser & Meisenheimer 1991). In fact, the analysis has to be restricted even further by summing up the flux from components A1+A2, B2+B3, and D2+H3, which are too close together to obtain an unambiguous decomposition into individual components. Leaving aside also component B1 (the flux of which is severely affected by overspill from the brighter knots A2 and B2) we end up with 7 independent positions for the multi-frequency spectra. Due to its optical faintness and since the spectrum peaks around 10^{14} Hz no unambiguous spectral fits could be derived from radio and optical data alone. However, employing the new NIR camera MAGIC at the Calar Alto 3.5 m telescope we recently succeeded in obtaining a K-band map of the jet with sufficient S/N-ratio (Neumann *et al.* 1996a) to attempt a spectral analysis of the 7 components.

The spectra displayed in Fig. 2 show that along the *"optical jet"* ($12''$ to $20''$ from the core) the result resembles that of M 87: The spectral *shape* is conserved but the cutoff frequency ν_c shifts. The high-frequency cutoff is much gentler than in M 87, such that optimum fits require both a smooth cutoff in the energy distribution (as *e.g.* predicted by 1st-order Fermi acceleration) and a break by $\Delta q = 1$ in the electron spectrum $n(\gamma) \sim \gamma^{-q}$ around $\gamma_b/\gamma_c = 0.2$ ($\nu_b/\nu_c = 0.04$).

Beyond the brightest optical knot (DH3 = D2+H3) which marks the end of the *optical jet*, however, a dramatic change in jet spectrum occurs. From the best-fit parameters (see Fig. 3) we conclude that the spectrum of the dominant

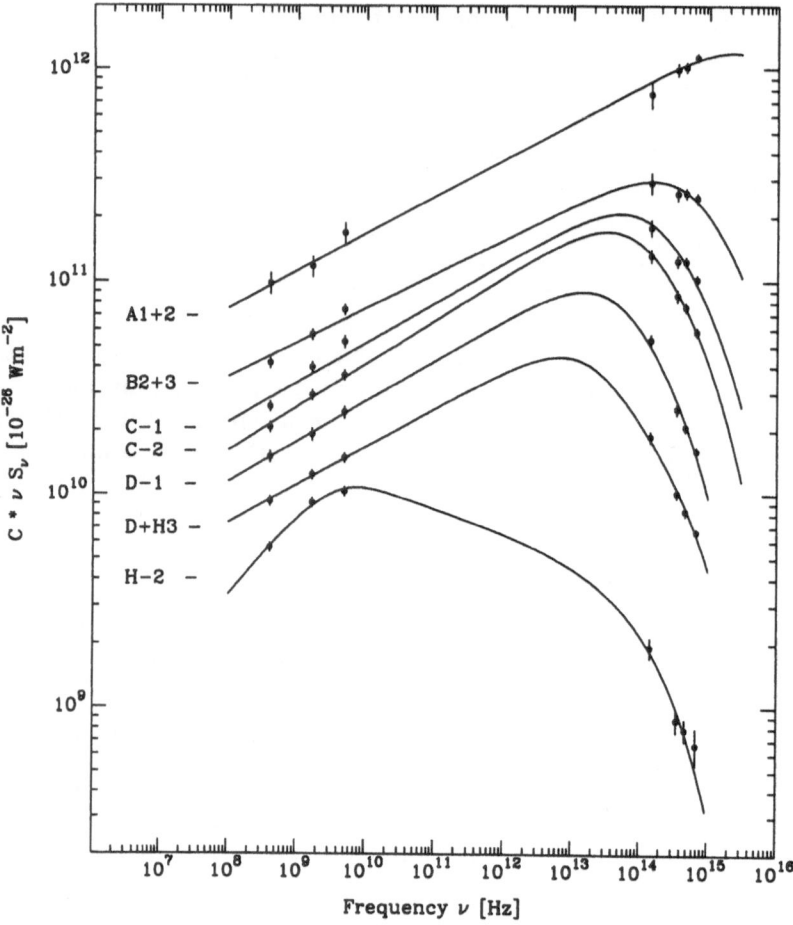

Fig. 2. Synchrotron spectra of 7 independent components in the jet of 3C 273. In order to avoid confusion, we scaled the observed spectra by $C = 1$ (component H-2), $C = 2$ (D+H3), $C = 11$ (D-1), $C = 40$ (C-2), $C = 70$ (C-1), $C = 100$ (B2+3), and $C = 380$ (A1+2). The contribution of the (optically quiet) backflow (Röser *et al.* 1996) is roughly subtracted from the radio flux of components A1+2 to D-1.

radio hot spot H-2 is characterized by a rather flat low-frequency spectrum $S_\nu \sim \nu^{-0.6}$ which steepens by the standard value $\Delta\alpha = 0.5$ around $\nu_b = 3\,\mathrm{GHz}$.

This spectrum exactly fits the hot spot model proposed by Meisenheimer & Heavens (1986). Since, in addition, a recent comparison of HST images with radio maps (Röser *et al.*, in prep., see also Fig. 3 in Bahcall *et al.* 1995) confirms the model prediction of a positional offset of $\Delta z \simeq 0\farcs2$ between radio and optical peak, we are rather confident that the basic ideas of that model are correct: The strong terminal shock of the jet is located near the upstream edge of H-2. Here efficient particle acceleration takes place (either by diffusive shock acceleration or a similar process). The balance of acceleration gains and synchrotron losses limits

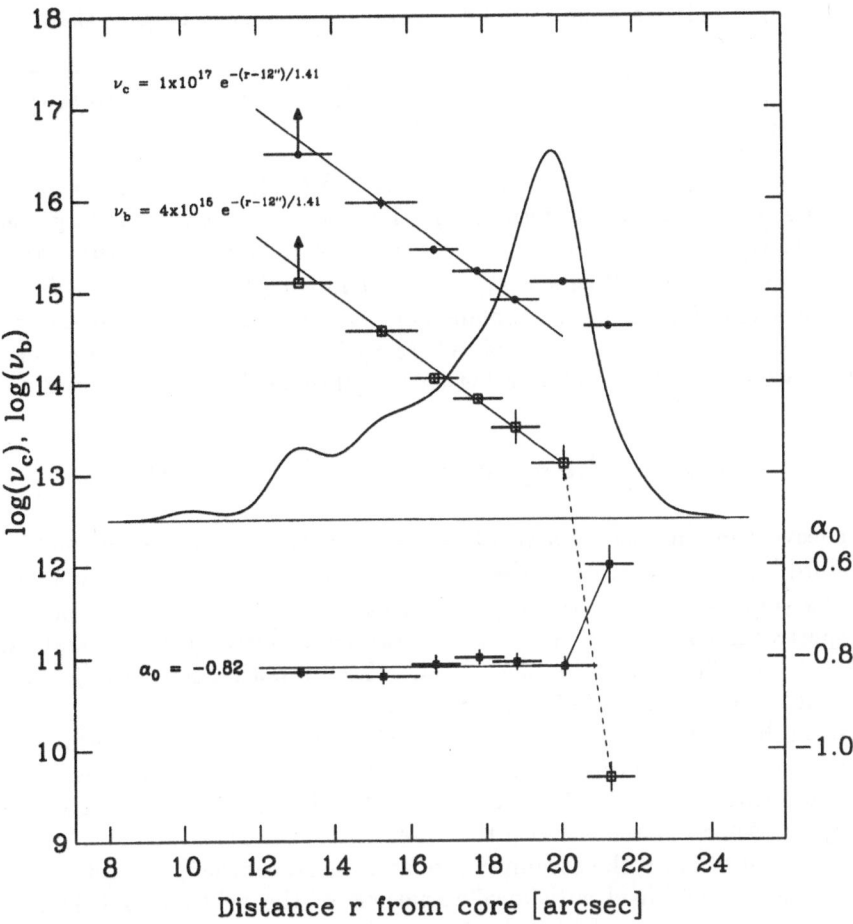

Fig. 3. Evolution of the spectral parameters along the jet of 3C 273. For the cutoff frequency ν_c (•) and the break frequency ν_b (□) refer to labels on the left. Positional reference is provided by the K-band brightness along the jet (linear scale, thin line). For the low-frequency power-law index α_0 refers to labels on the right.

the maximum energy γ_c. Further downstream, the particles suffer synchrotron losses according to eqn. (2), shifting γ_c to lower and lower energies. This results in the observed overall spectrum with a break by $\Delta\alpha$ at a frequency ν_b which corresponds to the maximum energy of electrons at the *downstream end* of the hot spot. A self-consistent model (that is: minimum-energy field $B_{m.e.} \simeq B_\perp$, derived from losses) requires that the plasma flows with about $0.1\,c$ away from the shock.

The interpretation of the spectral evolution along the *optical jet* is much less certain. Assuming a minimum-energy magnetic field, the exponential decline in ν_c corresponds to γ_c decreasing from about 10^7 to 10^6 between knots A and D. This conclusion remains valid even for a relativistic jet flow (where both ν_c and

$B_{m.e.} \equiv [f_{m.e.}S_\nu]^{2/7}$ are modified by a Doppler factor $D \neq 1$) since the intrinsic

$$\gamma_{c,0} = \sqrt{\frac{\nu_{c,0}}{1260\,\mathrm{Hz}}\left(\frac{B_{m.e.,0}}{30\,\mathrm{nT}}\right)^{-1}} = \sqrt{\frac{\nu_{c,obs}D^{-1}}{1260\,\mathrm{Hz}}30\,\mathrm{nT}\,[f_{m.e.}S_{\nu,obs}D^{-3+\alpha}]^{-2/7}}$$

differs only by a factor $D^{0.04}$ from that derived directly from the *observed* values of ν_c and $B_{m.e.}$ according to formula (1). Thus we get an extremely large energy loss scale $\lambda_{1/2} \equiv \frac{\gamma_c}{d\gamma_c/dz} \simeq 5\,\mathrm{kpc}$, about a factor of 100 longer than the maximum synchrotron loss scale (2) for a $\Gamma_{jet} = 10$ jet moving towards us. On the other hand, it seems impossible to explain both the decrease in ν_c and the rise in surface brightness along the jet of 3C 273 with a model like that proposed for M 87 in which a tight correlation between brightness S_ν and ν_c is found.

4 Comparison of the spectral evolution in both jets

If we leave aside the *radio hot spot* H-2 in 3C 273 (which can be explained by the standard hot spot models; *cf.* Meisenheimer *et al.* 1989) we find for both jets:

- The slope of the particle energy distribution (as inferred from the power-law spectral index between 10^{10} and $10^{14}\,\mathrm{Hz}$) remains remarkably constant along both jets. The power-law index of 3C 273 is somewhat steeper ($\alpha_{PL} = \alpha_0 = -0.82$) than that of M 87 ($\alpha_{PL} = -0.66$).
- In both jets the decrease of γ_c is at least 100 times slower than one would expect from synchrotron losses. Along the short jet of M 87 the loss rate is too small to be clearly determined, while for the jet in 3C 273 we derive $\lambda_{1/2} = 5\,\mathrm{kpc}$ (consistent with the lower limit in M 87).
- In the jet of M 87, the assumption of an (almost) unchanged electron energy spectrum, combined with local variations of the field strength leads to a consistent model explaining both the brightness distribution $S_\nu(z)$ and the run of the cutoff frequency $\nu_c(z)$ (Meisenheimer *et al.* 1996a). This is not possible for 3C 273 where brightness and ν_c seem to be anti-correlated.

Our finding that along both jets the spectral *shape* is preserved strongly argues for a common physical origin of the spectra. So we have to identify the physical parameter which could explain the opposite sense of the correlation between observed brightness and cutoff frequency, found in both jets.

A hint might be provided by the proper-motion studies: While the pattern speed in M 87 never exceeds mildly relativistic values[3] (Reid *et al.* 1989, Biretta *et al.* 1995), the VLBI jet in the inner 50 mas of 3C 273 clearly shows superluminal motion (*e.g.* Zensus *et al.* 1990). Apparently, the plasma flows with Lorentz factors $\Gamma_{jet} \gtrsim 10$. Thus it is appealing to assume that the difference between both jets is caused by relativistic effects dominating the jet in 3C 273 (this could also explain the steeper power-law spectrum, *cf.* Ballard & Heavens 1991). As shown above, our estimation of γ_c is little affected by relativistic

---

[3] The higher values found in knot D seem to reflect changes of the brightness pattern rather than proper motion.

beaming. We thus have to look into its effect on ν_c and S_ν. In order to get a decoupling of ν_c from S_ν we have to demand that γ_c *falls off* along the jet (decreasing Γ_{jet}!) while the Doppler factor D is *increasing* outwards. An obvious solution to this dilemma would be that γ_c is positively correlated with Γ_{jet}, while $S_{\nu,obs}$ depends on the Doppler factor D which may – depending on orientation – either decrease or increase with Γ_{jet}. Indeed we could demonstrate that an *ad hoc ansatz* $\gamma_c \sim d(u_{jet}\Gamma_{jet})/dz$ in a decelerating jet flow (with flow speed $c\,u_{jet} \equiv v_{jet}$) and a large angle between the axis of the kpc jet w.r.t. the line of sight ($\theta_{jet} \gtrsim 30°$, i.e. D decreasing with $\Gamma_{jet} \gtrsim 2$) could unify the observations of both jets (Meisenheimer *et al.* 1996b). The required misalignment of the kpc jet with the line of sight is also supported by the high radio polarization of H-2 (see *e.g.* Conway & Davis 1993).

Our *ansatz* does not rule out alternative explanations. Since it implies several clear predictions for the high-resolution HST images of 3C 273 (*e.g.* minor optical spectral index variations between knots and inter-knot regions), we are confident that our current analysis of the high-resolution data will tell us whether our *ansatz* is on the right track.

References

Bahcall, J.N. *et al.* 1995, *Astrophys. J.* **452**, L91

Ballard, K.R. and Heavens, A.F. 1991, *Mon. Not. R. astr. Soc.* **251**, 438

Biretta, J. A. and Meisenheimer, K. 1993, H.-J. Röser and K. Meisenheimer (eds.): *Jets in Extragalactic Radio Sources.* Lecture Notes in Physics Vol. **421**. Springer Heidelberg, p. 159.

Biretta, J. A., Zhou, F. and Owen, F. N. 1995, *Astrophys. J.* **447**, 582

Conway, R.G. and Davis, R.J. 1994, *Astron. Astrophys.* **284**, 724

Heavens, A.F., Meisenheimer, K. 1987, *Mon. Not. R. astr. Soc.* **225**, 335

Meisenheimer, K. and Heavens, A.F. 1986, *Nature* **323**, 419

Meisenheimer, K., Röser, H-J., Hiltner, P., Yates, M.G., Longair, M.S., Chini, R. and Perley, R.A. 1989a, *Astron. Astrophys.* **219**, 63

Meisenheimer, K., Röser, H.-J. and Schlötelburg, M. 1996a, *Astron. Astrophys.* in press.

Meisenheimer, K., Neumann, M., Röser, H.-J. and Conway, R.G. 1996b, submitted

Neumann, M. 1995, PhD Thesis, Universität Heidelberg.

Neumann, M., Meisenheimer, K., Röser, H.-J. and Stickel, M. 1995, *Astron. Astrophys.* **296**, 662

Neumann, M. Meisenheimer, K. and Röser, H.-J. 1996a, in preparation.

Neumann, M. Meisenheimer, K. and Röser, H.-J. 1996b, in preparation.

Owen, F.N., Hardee, P.E. and Cornwell, T.J. 1989, *Astrophys. J.* **340**, 698

Reid, M.J., Biretta, J.A., Junor, W. Muxlow, T.W.B and Spencer, R.E. 1989, *Astrophys. J.* **336**, 112

Röser, H.-J. and Meisenheimer, K. 1991, *Astron. Astrophys.* **252**, 458

Röser, H.-J., Conway, R.G., Meisenheimer, K. 1996, submitted to *Astron. Astrophys.*

Zensus, J.A., Unwin, S.C., Cohen, M.H. and Biretta, J.A. 1990, *Astron. J.* **100**, 1777

X-Ray Observations of Cen A

Stefan J. Wagner[1] and Stefan Döbereiner[2]

[1] Landessternwarte, Königstuhl, Heidelberg, Germany
[2] MPE, Garching, Germany

Abstract. We have observed the nearby radio galaxy Centaurus A (NGC 5128) with the ROSAT HRI. With a distance of about 3 Mpc, this object contains the closest of the three extragalactic Jets which have been studied in X-rays. We confirm the existence of separate knots of X-ray emission which are coincident with the radio-knots. The angular scale of 15 pc arcsec^{-1} now permits us to resolve the jet laterally even in the X-ray domain and to study the morphology of the jet in detail. We do not detect a counter-jet, but the HRI images reveal an X-ray filament which is associated with the shock front of the southern inner radio lobe (on the counter-jet side). This filament probably traces shock-heated gas.

In addition we detect diffuse emission from the host galaxy of the AGN which is partially absorbed by the cold matter of the prominent absorption band on the line of sight towards the nucleus of NGC 5128.

1 Introduction

Cen A is a large, prototypical radio-galaxy, which has been studied in great detail. It is the closest active galaxy to the Milky Way and permits studies of unsurpassed spatial resolution in many frequency bands. Unfortunately, the central part of this object is hidden behind a prominent dust band. This prohibits any investigation of the nucleus at optical and UV wavelengths and still severely affects studies in the adjacent IR and X-ray regimes.

Cen A does not only contain the closest extragalactic jet, but also one of the very few jets which have been detected at X-ray energies. The radio-emission on scales of tens of parsecs to kiloparsecs has been studied extensively (see e.g. Clarke et al., 1986, 1992). They have resolved many of the knots found in earlier investigations into complex substructure with steep spatial gradients of the radio spectral indices. This complex morphology suggests that the radio jet is hollow with most radio emission being emitted from filaments. The flow seems to have some helical component. It is collimated out to the prominent knot B, downstream of which it expands in a linear fashion until it hits knot G where the jet looks disrupted and the plasma enters the northern 'inner' radio lobe. While a symmetrically placed southern 'inner' radio lobe has comparable luminosity, there is no radio counter-jet. This asymmetry may be due to relativistic beaming (as suggested by the rapid variations of Cen A from mm wavelengths (Kellerman, 1974) into the gamma-ray regime (Kinzer et al., 1994)). Polarization studies imply the counter-jet side to be the far one (Sect. 5).

Near-IR studies by Joy et al. (1991) found an elongated feature which is spatially coincident with the radio-jet and conclude that the IR emission is thermal in nature. This conclusion is subject to the corrections for reddening, however.

Optical investigations found no direct emission from the synchrotron jet. At large distances from the nucleus, however, Morganti et al. (1992) discovered gas which is photoionized by a hard continuum. They suggest that this gas actually intercepts the jet and is photoionized by the unabsorbed (and probably beamed) nuclear radiation, indicating that Cen A is a misdirected blazar.

The most elaborate earlier X-ray studies have been carried out by Feigelson et al. (1981) using EINSTEIN HRI data. They showed that the X-ray emission is mostly due to a small number of knots which are often spatially coincident with radio knots.

2 X-ray Observations

The ROSAT HRI has a higher sensitivity and improved spatial resolution as compared to the EINSTEIN HRI detector. This enables a more detailed investigation of the X-ray properties. We supplemented our 68 ksec HRI observation with archival PSPC data and new optical/near-IR observations for this project. The HRI image is displayed in figure 1. The most prominent features are the X-ray jet and two diffuse bars of emission perpendicular to the jet. These features are the brightest parts of a diffuse component, centered on the nucleus which is intercepted by an absorption band. In addition, figure 1 reveals about one dozen point sources and a thin tangential filament of emission 5 arcmin south-west of the nucleus.

We present some of the results of our ROSAT HRI observations. A full discussion is given in Döbereiner et al. and Wagner et al., in preparation.

3 Diffuse Emission

The well-defined X-ray jet (Sect. 4), about 40 point-sources and the X-ray filament (Sect. 5) are superposed of a diffuse component, centered on the nucleus of the galaxy. This diffuse emission is intercepted by bands running perpendicular to the jet orientation. They perfectly trace the well-known absorption bands which are very prominent in optical images of NGC 5128. The absorption in the central parts prohibits the derivation of the radial luminosity profile of the diffuse emission. We cannot compare the X-rays directly with the optical light. The surface brightness of the diffuse emission outside the absorption bands is consistent with the hypothesis that the diffuse emission is due to the underlying stellar population of this giant radio-galaxy. We cannot exclude, however, that a fraction as high as 40 % may be due to additional interstellar X-ray gas. The scale length of the diffuse emission suggests that most of it lies within the dust band. We can therefore use the small-scale modulation of the X-ray intensity parallel to the X-ray jet (which is caused by the "foreground" absorption of the

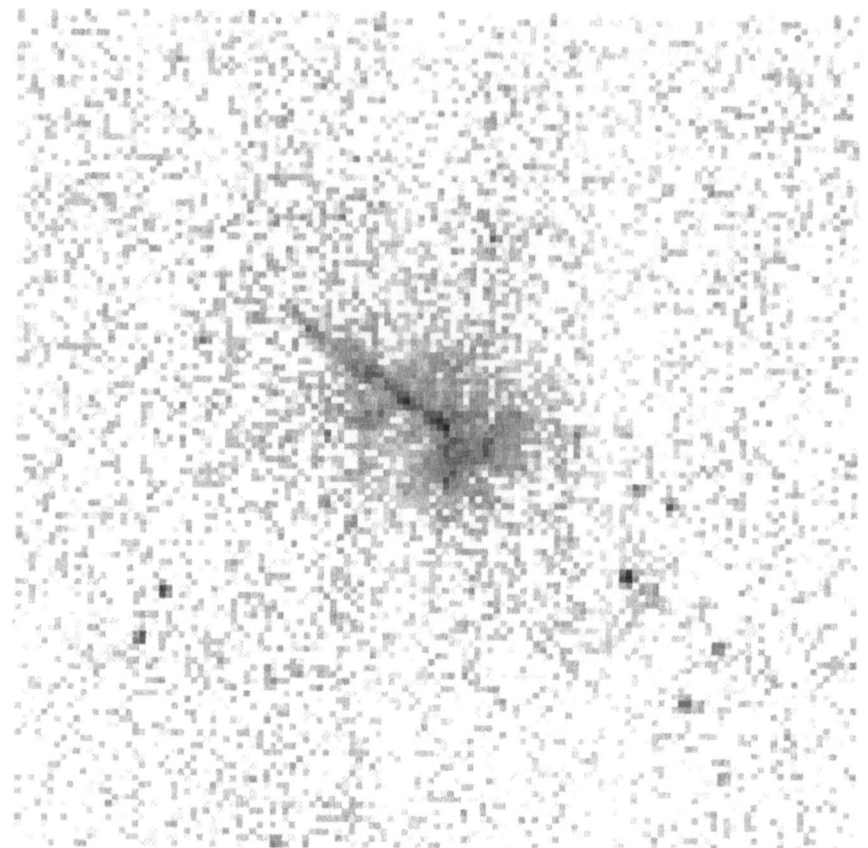

Fig. 1. ROSAT HRI image of Centaurus A (NGC 5128). A logarithmic gray scale is used. The box is 18' × 18'. NE is to the top-left. Apart from the jet extending 5 arcmin along PA = 53° diffuse emission along two ridges running perpendicular to the jet on either side of the nucleus are clearly visible. We find no emission from a counter jet but a faint filament extending tangentially over 1 arcmin on the counter jet side (close to the bright point source 5 arcmin southwest of the nucleus).

dust band) to constrain the amount of small-scale modulation along the ridge line of the jet which is due to this absorption rather than substructure of the X-ray emissivity of the jet.

4   The X-ray Jet

The jet is the most prominent feature in the X-ray image of Cen A which is shown in contours in Fig. 2. The morphology of the observed jet is subject to photoelectric absorption by the dust band. In order to study the true spatial variations we subtracted a smooth background component and corrected for small-scale

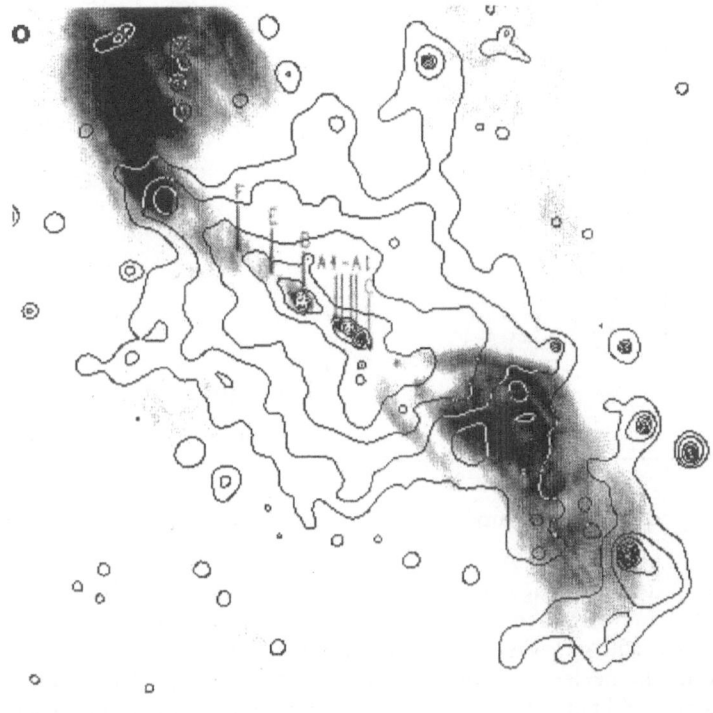

Fig. 2. X-ray emission from HRI observations of NGC 5128 (contours) superposed on a 18 cm radio map (greyscale; see Clarke et al., 1992). The knotted X-ray jet, the diffuse emission, intercepted by absorption bands running perpendicular to the jet orientation, and a tangential X-ray filament at the front-end of the southern radio lobe are clearly visible.

variation of the foreground absorption as described in Sect. 2. Uncertainties in this procedure may result in errors of the corrected images by about 15 % with a safe upper limit of 40 %. It is obvious that the separation of the X-ray jet into individual knots is not due to modulations of foreground extinction. Apart from the nucleus, knots A and B are the most prominent, knots E, F, and G are clearly visible. The X-ray emission thus coincides closely with the radio emission, as shown in Fig. 2. Knots C and D (labelled by Feigelson) have no radio counterparts. Since this part of the jet is still strongly affected by absorption, the detection of knots C and D is subject to details of the correction for foreground absorption.

The nucleus has a spatial profile which is consistent with the PSF. All of the remaining knots are extended along the jet, as demonstrated in Fig. 3. The knots B, E, and F are clearly extended perpendicular to the jet as well. Their projected diameters are about 50 pc each. The detailed morphology differs slightly from the structure seen in the 6 cm maps of Clarke et al., 1992. In addition to

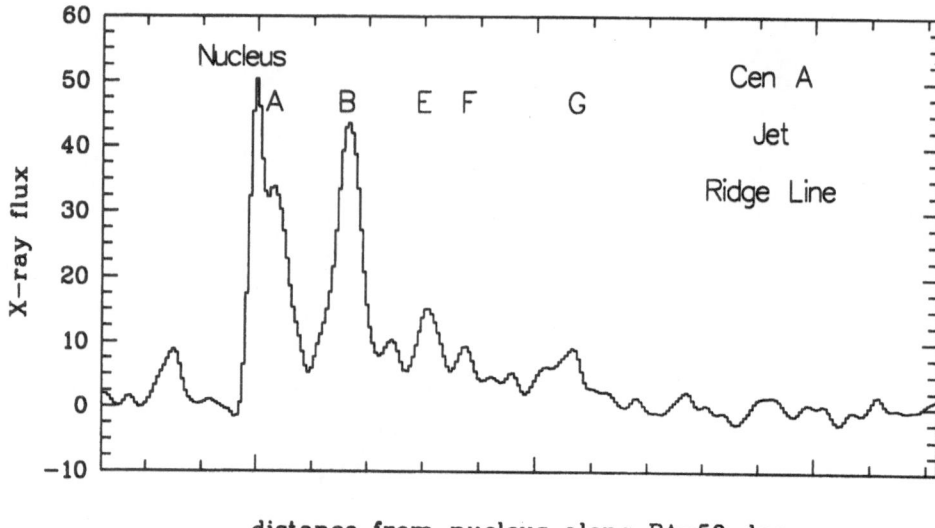

Fig. 3. X-ray intensity along the ridge line of the jet. The flux within a 5" wide strip was background subtracted and corrected for absorption using an average of the deviations from a de Vaucouleurs profile on either side of the jet bin as a tracer of the absorption. With the exception of the nucleus, all knots are extended along the jet direction. In addition to the well defined knots, diffuse X-ray emission from the jet is clearly visible.

the (resolved) knots we detect smooth X-ray emission along the jet with some indication of emission even beyond knot G (see table 1).

While the effects of foreground absorption on the morphology of the jet and the knots can be constrained to within a few 10 %, the corrections to the total luminosities are uncertain by at least an order of magnitude. To facilitate the comparison with the earlier studies of Feigelson, we computed luminosities assuming power-law spectra with a photon index of 1.7 and a galactic n_H of $10^{21} cm^{-2}$ for the all of the knots. In Tab. 1 we compare these fluxes to the synchrotron fluxes of Burns et al. 1983 and Clarke et al. (1992).

A detailed comparison with the results of Feigelson et al. (1981) demonstrates that the relative intensities of the different knots have changed. While the changes of the absolute luminosities remain insignificant (due to the large errors introduced in the corrections for absorption) differential changes are not affected. The variations provide an important constraint on the radiation mechanism. The HRI data do not contain spectral information. We cannot distinguish thermal and nonthermal models from the spectral shape. Feigelson et al. (1982) ruled out a dominant IC contribution. Thermal emission cannot be ruled out on the basis of the observed luminosities, but a close similarity between X-ray and radio morphology would be unlikely. The significant differences in the flux densities of the knots between our data and those of Feigelson et al. cannot be explained by any thermal model. Hence we conclude that the X-ray emission is due to synchrotron radiation.

Table 1. Flux densities of the prominent knots in the jet of Cen A

Knot	I (1 keV) [10^{38} erg/s]	L (6 cm) [Jy]
Nucleus	> 4.6	6.98
A1-A4	> 4.0	0.76
B	> 8.6	0.33
E	~ 2.4	0.11
F	~ 1.1	0.15
G	~ 1.5	2.28

5 The Southern Filament

We do not detect a counter-jet in the X-ray image. On the counter-jet side, however, we detect a filament which is extended perpendicular to the jet direction at a distance of 5.5 arcmin to the nucleus. This filament is 90" long and unresolved along its short axis. The reality of the feature has been confirmed by lower S/N detection in archival PSPC and EINSTEIN HRI data. The bright point source 5.5 arcmin southwest of the nucleus is spatially coincident with a distant early-type galaxy and most likely unrelated to the filament. This superposition makes it impossible however, to determine constraints on the X-ray spectrum of the filament from the PSPC data. As can be seen in Fig. 2, the filament traces the front end of the southern inner radio lobe. Clarke et al. (1992) used radio data to detect depolarization and Faraday rotation in the inner part of this southern lobe and concluded that the southern lobe is on the far side of the galaxy. The front end of the lobe has a large projected distance from the core and is not affected very much from the depolarization which is related both to the absorption band and to halo gas. As shown by Clarke et al. (1992; see their figure 5), the magnetic field is enhanced and oriented along the front. This clearly demonstrates that the magnetic field in this region is compressed. This compression is likely to be caused by shocks from the inner radio lobe propagating into the external medium. The close association of the X-ray filament with this regime of compressed gas suggests that the X-ray filament is due to shock heated gas (Wagner et al., in preparation). Assuming a temperature of 0.5 keV and galactic absorption only ($n_H = 10^{21} cm^{-2}$), the filament has an X-ray luminosity of 3±1 10^{38} erg/sec.

Acknowledgements SW acknowledges support from the Deutsche Forschungsgemeinschaft (SFB 328) and thanks Wolfgang Kundt for organizing this interesting workshop.

References

Burns J.O., Feigelson E.D., Schreier E.J., 1983, ApJ 273, 128

Clarke D.A., Burns J.O., Feigelson E.D., 1986, ApJ 300, L41

Clarke D.A., Burns J.O., Norman M.L., 1992, ApJ 395, 444

Feigelson E.D., Schreier E.J., Delvaille J.P., Giacconi R., Grindlay J.E., and Lightman A.P., 1981, ApJ 251, 31

Joy M., Harvey P.M., Tollestrup E.V., Selgren K., McGregor P.J., and Hyland A.R., 1991, ApJ 366, 82

Kellermann K.I., 1974, ApJ Lett. 194, 135

Kinzer R.L., Johnson W.N., Kurfess J.D., Strickman M.S., Grove J.E., et al., 1994, in *The Second Compton Symposium*, Eds. C. Fichtel, N. Gehrels, and J.P. Norris. AIP Conf. Proc. 304, 531

rganti R., Fosbury R.A.E., Hook R.N., Robinson A., and Tsvetanov Z., 1992, MNRAS, 256, 1P

Superluminal Sources

René C. Vermeulen

California Institute of Technology,
Astronomy 105–24, Pasadena, CA 91125, USA

Abstract: This review of the superluminal motion phenomenon in Active Galaxies takes a largely statistical approach. Predictions for the apparent velocity distribution under simple beaming models are presented and compared to the observations. The potential applications for tests of unification models and for cosmology (source counts, measurements of H_0 and q_0) are discussed. First results from a large homogeneous survey are presented. The data do not show compelling evidence for the existence of intrinsically different populations of galaxies, BL Lac objects, or quasars. β_{app} in the range $(1 - 5) h^{-1}$ occur with roughly equal frequency; higher values, up to $\beta_{\mathrm{app}} = 10 h^{-1}$, are rather more scarce than appeared to be the case from earlier work, which evidently concentrated on sources which are not representative of the general population. The β_{app} distribution suggests that there might be a skewed distribution of Lorentz factors over the sample, with a peak at $\gamma_b \approx 2 h^{-1}$ and a tail up to at least $\gamma_b \approx 10 h^{-1}$. There appears to be a clearly rising upper envelope to the β_{app} distribution when plotted as a function of observed 5 GHz luminosity; a combination of source counts and the apparent velocity statistics in a larger sample could provide much insight into the properties of radio jet sources.

1. Introduction

This review follows the presentation given at the conference on "High Resolution Radio Imaging of Quasars and AGN" (Vermeulen 1995) with only small adaptations, and deals mostly with the preliminary statistics of superluminal motion now emerging from a large homogeneous sample.

A wealth of information has been gathered on a select few sources, such as 3C 345 (Wardle et al. 1994, Zensus, Cohen, & Unwin 1995, Zensus 1996), 4C 39.25 (Alberdi et al. 1993), and BL Lac (Mutel et al. 1990, Mutel 1996). In such famous objects, intensive and prolonged monitoring has led to substantial insight into the jet kinematics, and a large, detailed body of data is available to confront hydrodynamical model calculations of jets and shocks. It is seen that in the inner few parsecs, the jets often have substantial curvature, and the moving radio knots show apparent velocity changes. In the case of 3C 345, the jet is thought to have a constant Lorentz-factor ($\gamma{\sim}10$), and to bend away from the line-of-sight by a few degrees in the inner parsecs (e.g. Zensus et al. 1995).

However, it is also attractive to be able to work with close to a hundred sources, even though these mostly have rather sparse data (for many objects the bare minimum of a snapshot image at two epochs). As long as the source selection criteria are well known, this approach has the virtue that population models can be constructed, involving a distribution of Lorentz factors, jet bending, pattern motions, acceleration, or whatever other complexity is thought to be indicated by the data. These models can be compared to the observed distribution of (superluminal) apparent velocities with the help of Monte Carlo simulations. Thus, as long as the data are obtained and used in a homogeneous fashion, the apparent velocity statistics of the sample reveal for the population as a whole complexities which are missed in individual objects.

In a strictly kinematic approach, velocities or upper limits should only be used if derived from an observed change in separation between two recognizable features in a radio source. This excludes values based only on the timing of the appearance of a new knot and/or a radio flare. The inferences drawn can then be compared to other indicators of relativistic motion and a small angle to the line of sight, such as variability (e.g. Teräsranta & Valtaoja 1994, Hughes, Aller, & Aller 1994) and a high brightness temperature (e.g. Readhead 1994, Daly 1996), a high ratio of core to extended radio luminosity (e.g. Orr & Browne 1982), detectable γ-rays (e.g. von Montigny et al. 1995, Barthel et al. 1995), a deficit of inverse-Compton X-rays (e.g. Ghisellini et al. 1993), or the shape of radio source count functions (e.g. Padovani & Urry 1992, Urry & Padovani 1995).

Throughout this review, $H_0 = 100h$ km s^{-1} Mpc^{-1} is used in Friedmann cosmology. Where appropriate the dependence of the results on h and q_0 is shown. Indeed, since superluminal radio sources can be observed over a wide range of redshifts, their statistics can contribute to a discrimination between different cosmological models (Cohen et al. 1988).

2. Randomly Oriented Sources

The model predictions in this review, based mostly on Monte Carlo simulations, follow the formalism and assumptions outlined in Vermeulen & Cohen (1994). In particular, the radio jets are assumed to be narrow and (in first instance) straight, to have random intrinsic orientations, and to have no relationship between the Lorentz factor and the intrinsic (isotropic) luminosity. Under these circumstances, it follows immediately from the solid angle available that many jets will be pointed near the plane of the sky, and show apparent velocities $\beta_{\mathrm{app}} = v_{\mathrm{app}}/c \approx 1$. There will be a modest fraction of superluminals, from jets at angles of $\theta \sim 1/\gamma$, and a small percentage of knots with $\beta_{\mathrm{app}} < 1$, in jets pointed almost straight towards us. This distribution of velocities is shown in Figure 1.

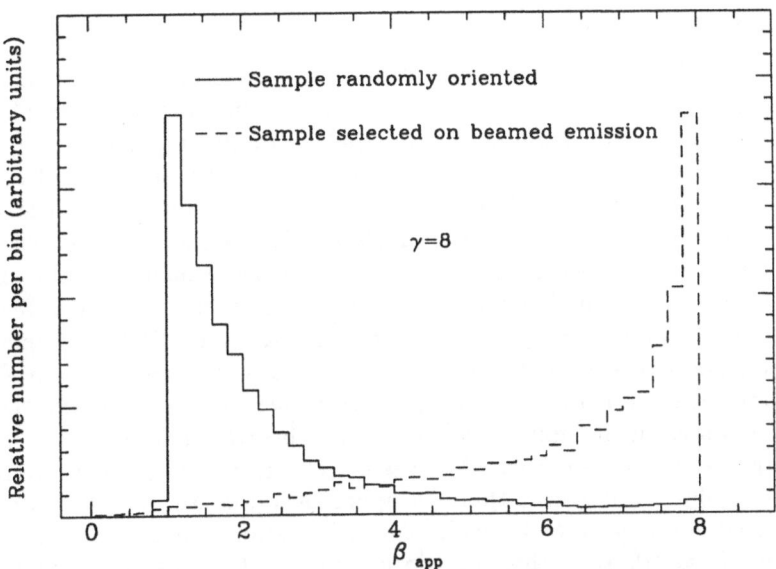

Fig. 1. Predicted apparent velocity distributions for a randomly oriented sample (solid line), and for a sample selected on Doppler boosted emission (dashed line). A single bulk and pattern Lorentz factor is assumed, $\gamma = 8$, but the value chosen does not affect the global shape of the predicted distributions.

2.1 Cosmological parameters

In principle, it should be possible to use a sample of randomly oriented relativistic jet sources with a range of redshifts, for example selected on their low-frequency radio emission, to measure both h and q_0, by using the predominant observed apparent velocity (or perhaps the lower limit, Pelletier & Roland 1988) as a standard velocity of c, or $2c$ if separating features from opposite jets are observed. No appropriate samples have currently been studied well enough. It would be even better if features with a known common origin separating from each other in oppositely directed relativistic jets can be found. By using the arm-length or perhaps brightness ratio in addition to the apparent motions, the jet velocity and angle to the line of sight can then be obtained separately, and a limit to h follows immediately. Some of the Compact Symmetric Objects (Readhead 1995) are promising in this regard; in particular, there is a detection of bi-directional motion in 1946+708 (Taylor, Vermeulen, & Pearson 1995).

2.2 AGN unification

Given values for h and q_0, motion statistics can also be used to test unification models. For example, a high fraction of superluminals in a sample would indicate jets oriented predominantly at small to moderate angles to the line of sight. Several groups are pursuing the apparent velocity statistics of complete samples of quasars selected at low frequency, which minimises orientation bias (see Hough 1994, and Zensus & Porcas 1987). This is hard work because many of the cores are quite weak (down to 1 mJy), in agreement with the expectation that their emission is Doppler beamed away from us (Scheuer & Readhead 1979). Indeed, the current statistics (Vermeulen & Cohen 1994) do uphold the trend of decreasing apparent velocities for sources with decreasing core fraction (R). However, there is much scatter (e.g. Vermeulen et al. 1993), and the trend is primarily seen in the upper envelope to the observed motions.

At the very largest measured values of core-dominance, which often occur in BL Lac objects, it seems that the apparent velocities may decrease again, as expected for jets viewed within the $1/\gamma$ beaming cone. The statistics on high-luminosity BL Lacs therefore do not require the assertion that they are intrinsically different from the core-dominated quasars (e.g. Gabuzda et al. 1994,Gabuzda 1995, Urry & Padovani 1995). The lower luminosity BL Lac objects probably should be compared to FR-I luminosity radio galaxies (Urry, Padovani, & Stickel 1991, Urry & Padovani 1995, Giovannini 1996) but while the apparent velocities in general are slower than in more powerful objects (see also §4.1, §4.4), there are not enough motions known to constrain this brand of unification.

3. Samples Selected on Beamed Emission

Again following earlier work (Vermeulen & Cohen 1994), the differential source counts are assumed to be a power law function of intrinsic flux density, with index 2.5. It is assumed that in a complete flux-limited sample at high radio frequency, selection of sources from a population of randomly oriented jets takes places through Doppler favoritism. Only flat-spectrum continuous jets are considered, for which Doppler boosting depends on the Doppler factor to the power 2.0; this could be rather different with other assumptions (e.g. Lind & Blandford 1985).

3.1 Upper β_{app} envelope

Assuming initially that the Lorentz factor associated with the observed moving radio knots (γ_p) is the same as that of the bulk flow (γ_b), then the width of the beaming cone ($\sin\theta \sim 1/\gamma_b$) is well matched to the angle at which the largest apparent motion occurs ($\sin\theta \sim 1/\gamma_p$). Thus, many fast motions are expected in samples selected on beamed emission. Figure 1 shows the apparent velocity distribution predicted for a sample selected on beamed emission with a single value of $\gamma_p = \gamma_b$ for all jets. It is in marked contrast to the distribution expected for randomly oriented jets. Most of the observed motions should now be slightly below the maximum possible velocity ($\beta_{app} = \beta_p \gamma_p$), which should be a sharply defined upper edge. The match between optimal beaming and large motion occurs for all relativistic jets, and when normalized to $\beta_p \gamma_p$, the shape of the β_{app} distribution function is nearly independent of the Lorentz factor.

3.2 Separate pattern velocities

If there were a ratio $r = \gamma_p / \gamma_b \neq 1$ between the Lorentz factors of the bulk flow and the radio knots (traveling or standing shocks, for example), Doppler beaming selection would favor angles inside ($r < 1$) or outside ($r > 1$) those at which the largest β_{app} occur. Thus, for both slow and fast patterns (compared to the bulk flow), there usually would be a less pronounced upper β_{app} envelope, and a larger fraction of relatively slower motions. Fortunately, with careful Monte Carlo modeling (Vermeulen & Cohen 1994) it will be possible to break the degeneracy between slow and fast pattern statistics by taking into account other measurements of the bulk Doppler factor, such as provided by the inverse-Compton X-ray deficit (Ghisellini et al. 1993) or the equipartition Doppler factor (Readhead 1994).

4. First Results from the CJ Survey

The combination of the Pearson-Readhead (Pearson & Readhead 1988) and Caltech-Jodrell Bank VLBI surveys (Polatidis et al. 1995, Taylor et al. 1994) yields a complete flux-limited sample of 293 flat-spectrum sources brighter than 0.35 Jy at 5 GHz. Second epoch observations, with a time interval of 2–3 years, are in progress, and will hopefully be completed in 1995. Apparent proper motions or useful upper limits are now available for 81 of these sources, after temporarily putting aside $\sim 1/3$ of the objects, in which the morphological changes, if any, are more ambiguous (mostly those with rather featureless or extended jets). The new motions were derived by fitting to the data at each epoch the relative positions and flux densities of a few Gaussian components with fixed shapes, determined in preliminary iterations. Where more than one motion could be measured in a source with multiple components, the value used is that between the two brightest features. The apparent velocities derived for both $q_0 = 0.5$ and $q_0 = 0.05$ are shown in Figure 2.

4.1 Galaxies, BL Lacs, and quasars

Ignoring upper limits and empty fields, the mean apparent velocity is slightly smaller for the galaxies ($N = 7$, $\langle \beta_{app} \rangle = 2.1$) and the BL Lacs ($N = 8$, $\langle \beta_{app} \rangle = 2.3$) than for the quasars ($N = 44$, $\langle \beta_{app} \rangle = 3.2$); all for $q_0 = 0.5$. This is in the sense expected for unification models. However, the KS test yields a probability of 23% that the galaxies and the quasars have the same β_{app} distribution, and a

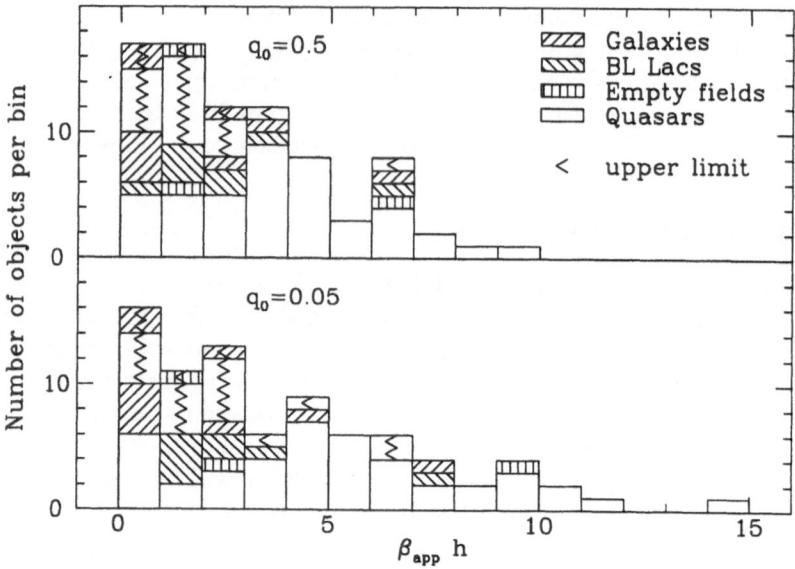

Fig. 2. The observed apparent velocity distribution for 81 objects in the homogeneous PR+CJ flat-spectrum sample, showing the scarcity of higher values, compared to earlier work and to predictions for beamed samples. There is no firm evidence for differences related to optical identification.

probability of 30% for the BL Lac objects and the quasars. With the full sample, stronger tests of unification models will be possible.

4.2 Scarcity of fast apparent motions

The upper cutoff is not nearly as sharp as predicted in the simplest model discussed in §3.1, and the apparent velocities do not cluster near the maximum; the same conclusion as derived from an earlier more heterogeneous group of 25 core-selected quasars (Vermeulen & Cohen 1994). There is a substantial fraction ($\sim$25%) of stationary features or upper limits. Then, if $q_0 = 0.5$ it seems that β_{app} in the range 1–$5\,h^{-1}$ occur with roughly equal frequency, with a tail (17%) of higher values, up to $10h^{-1}$. If $q_0 = 0.05$, the β_{app} distribution tapers off even more gently; most values are still below $10h^{-1}$.

This detailed distribution differs from that in the group analysed earlier (Vermeulen & Cohen 1994), of which, with $q_0 = 0.5$, 36% had $\beta_{app} = 5$–$10\,h^{-1}$, more than twice the fraction in the new sample. The scarcity in CJ of these somewhat faster motions is certainly not an artefact caused by undersampling in time. It seems that the superluminal quasars reported thus far in the earlier literature are not representative of the population as a whole. There is not an obvious correlation of β_{app} with flux density in the new sample, so other factors must be at work. Clearly, there was a bias in earlier work towards analysing and publishing fast superluminals. Furthermore, a lot of the sources previously studied are highly variable (many appear in the Variable Source Sample defined by Wehrle et al. 1992, for example), and there is already evidence for a correlation between variability at high frequency and a large apparent velocity (Teräsranta & Valtaoja 1994).

4.3 Plausible Lorentz factor distributions

Setting aside the upper limits, and adopting $q_0 = 0.5$ (the least extreme case), the β_{app} distribution can be reproduced if there is a wide range of Lorentz factors in the sample, with in particular a long tail to high values. As a numerical example for illustrative purposes only, if $h = 0.55$ then $\gamma_b \approx 4$ could be the peak of a skewed bell-shaped distribution spanning the range $\gamma_b \approx 2 - 18$. Interestingly, such a distribution is akin to that derived from largely independent data, involving radio source counts (Padovani & Urry 1992).

Well-fitting models can also be found in which the observed patterns have a different Lorentz factor than the bulk flow. For $q_0 = 0.5$ one would need either $r = \gamma_p/\gamma_b \approx 0.25$ or $r \approx 10.0$; intermediate values are ruled out, unless there is also a considerable range of γ_b values. Such low or high r are un-appealing; $r = 0.25$ because, with β_{app} up to $10\,h^{-1}$, it requires a rather high $\gamma_b \geq 40\,h^{-1}$ in the radio jets of all objects; $r = 10$ because, conversely, it requires that almost all objects have $\gamma_b \leq 1\,h^{-1}$, in contradiction with other evidence for substantial Doppler beaming. It seems that this high r case is akin to a recent incarnation of the light-echo models (Ekers & Liang 1990), in which relativistic motion is admitted, but Doppler beaming is not. While it is entirely plausible that pattern velocities do play a role, for example in causing apparently stationary patterns in relativistic jets, further evidence that Doppler beaming is in fact important is given by the observed luminosity dependence of the β_{app} distribution.

4.4 Luminosity dependence

Figure 3 shows the dependence of the observed β_{app} on observed monochromatic luminosity, calculated assuming isotropic emission. While low β_{app} can be found at any observed luminosity, there seems to be a striking correlation of the largest β_{app} with observed luminosity: the upper envelope rises considerably. However, it will be very desirable to verify this by measuring internal proper motions in a VLBI sample with a considerably wider spread of flux densities, so as to improve the coverage of lower observed luminosities, and also to break the strong redshift – flux density correlation.

Figure 3 suggests that most or all of our flat-spectrum sources at the high observed luminosity end ($\sim 10^{34}$ erg s^{-1} Hz^{-1}) conform roughly to the simple model: they have highly relativistic jets, and are in our sample because their observed luminosity is considerably enhanced by Doppler beaming, and most of them are from a parent population 2 or 3 orders of magnitude down in intrinsic (isotropic) luminosity. But furthermore, the absence of similarly high β_{app}

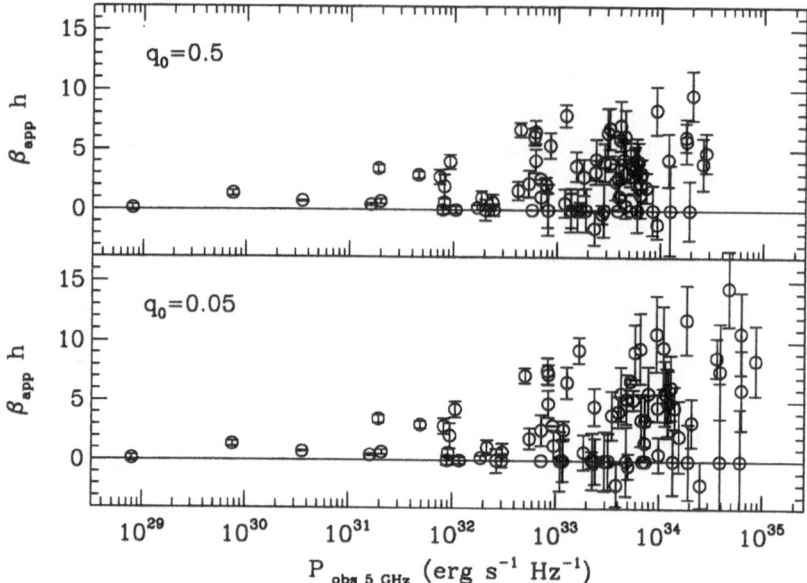

Fig. 3. The observed apparent velocity distribution for 81 objects in the homogeneous PR+CJ flat-spectrum sample, illustrating that the upper envelope rises as a function of the observed 5 GHz monochromatic luminosity. Upper limits are plotted as error bars on $\beta_{\mathrm{app}} = 0$.

at luminosities $\leq 10^{32}$ erg s^{-1} Hz^{-1} suggests that those objects are rather less beamed, which in turn implies that there is no substantial population down another 2 or 3 orders of magnitude in intrinsic luminosity, from which members can get Doppler beamed up. Thus, it would seem that highly relativistic jets may occur only in objects with a restricted range of 5 GHz intrinsic luminosities, perhaps predominantly near 10^{31} erg s^{-1} Hz^{-1}, and that there is a correlation between intrinsic 5 GHz radio luminosity and Lorentz factor, akin to that often postulated for the low frequency radio luminosity (*e.g.* the FR-I – FR-II division). Clearly, further analysis of the effects seen in Figure 3 is needed, and a combination of source counts and the apparent velocity statistics of a larger sample could provide much insight into the properties of radio jet sources.

Acknowledgements. I am grateful to Wolfgang Kundt for inviting me to his pleasant meeting in Bad Honnef. This review is based on an extension of previous work (Vermeulen & Cohen 1994), and a similar review was presented earlier (Vermeulen 1995). The many new measurements of Caltech-Jodrell survey sources (Polatidis et al. 1995, Taylor et al. 1994) will be published separately after further analysis. I am grateful to all my collaborators in that survey for their indispensable contributions. This work has been supported in part by the USA National Science Foundation under grants AST 88–14554, AST 91–17100, and AST 94-20018. Support to attend the meeting in Bad Honnef was provided by the Stiftung Volkswagenwerk.

References

Alberdi, A., Marcaide, J. M., Marscher, A. P., Zhang, Y. F., Elosegui, P., Gomez, J. L., & Shaffer, D. B. 1993, ApJ, 402, 160

Barthel, P. D., Conway, J. E., Myers, S. T., Pearson, T. J., & Readhead, A. C. S. 1995, ApJ, 444, L21

Cohen, M. H., Barthel, P. D., Pearson, T. J., & Zensus, J. A. 1988, ApJ, 329, 1

Daly, R. A. 1996, in *Energy Transport in Radio Galaxies and Quasars*, eds. P. Hardee, A. Bridle, & A. Zensus (Astronomical Socity of the Pacific: San Francisco), in the press

Ekers, R. D., & Liang, H. 1990, in *Parsec-Scale Radio Jets*, eds. Zensus, J. A., & Pearson, T. J. (Cambridge University Press: Cambridge), p. 333

Gabuzda, D. C. 1995, Proc. Natl. Acad. Sci. USA, 92 (05 Dec 1995), 11393

Gabuzda, D. C., Mullan, C. M., Cawthorne, T. V., Wardle, J. F. C., & Roberts, D. H. 1994, ApJ, 435, 140

Ghisellini, G., Padovani, P., Celotti, A., & Maraschi, L. 1993, ApJ, 407, 65

Giovannini, G. 1996, in *Extragalactic Radio Sources*, proc. IAU Symp. 175, ed. C. Fanti (Kluwer: Dordrecht), in the press

Hough, D. H. 1994, in *Compact Extragalactic Radio Sources*, eds. Zensus, J. A., & Kellermann, K. I. (NRAO: Charlottesville), p. 169

Hughes, P. F., Aller, M. F., & Aller, H. D. 1994, ApJ, 396, 469

Lind, K. & Blandford, R. D. 1985, ApJ, 295, 358

Mutel, R. L., Phillips, R. B., Su, B., & Bucciferro, R. R. 1990, ApJ, 352, 81

Mutel, R. L. 1996, in *Extragalactic Radio Sources*, proc. IAU Symp. 175, ed. C. Fanti (Kluwer: Dordrecht), in the press

Orr, M. J. L., & Browne, I. W. A. 1982, MNRAS, 200, 1067

Padovani, P., & Urry, C. M. 1992, ApJ, 387, 449

Pearson, T. J., & Readhead, A. C. S. 1988, ApJ, 328, 114

Pelletier, G. & Roland, J. 1989, A&A, 224, 24

Polatidis, A. G., Wilkinson, P. N., Xu, W., Readhead, A. C. S., Pearson, T. J., Taylor, G. B., & Vermeulen, R. C. 1995 ApJS 98, 1

Readhead, A. C. S. 1994, ApJ, 426, 51

Readhead, A. C. S. 1995, Proc. Natl. Acad. Sci. USA, 92 (05 Dec 1995), 11447

Scheuer, P. A. G., & Readhead, A. C. S. 1979, Nat, 277, 182

Taylor, G. B., Vermeulen, R. C., Pearson, T. J., Readhead, A. C. S., Henstock, D. R., Browne, I. W. A., & Wilkinson, P. N. 1994, ApJS, 95, 345

Taylor, G. B., Vermeulen, R. C., & Pearson, T. J. 1995, Proc. Natl. Acad. Sci. USA, 92 (05 Dec 1995), 11381

Teräsranta, H., & Valtaoja, E. 1994, A&A, 283, 51

Urry, C. M, & Padovani, P. 1995, PASP, 107, 803

Urry, C. M., Padovani, P., & Stickel, M. 1991, ApJ, 382, 501

Vermeulen, R. C. 1995, Proc. Natl. Acad. Sci. USA, 92 (05 Dec 1995), 11385

Vermeulen, R. C. & Cohen, M. H. 1994, ApJ, 430, 467

Vermeulen, R. C., Bernstein, R. A., Hough, D. H., & Readhead, A. C. S. 1993, ApJ, 417, 541

254

von Montigny, C., Bertsch, D. L., Chiang, J., Dingus, B. L., Esposito, J. A., Fichtel, C. E., Fierro, J. M., Hartman, R. C., Hunter, S. D., Kanbach, G., Kniffen, D. A., Lin, Y. C., Mattox, J. R., Mayer-Hasselwander, H. A., Michelson, P. F., Nolan, P. L., Radecke, H. D., Schneid, E., Sreekumar, P., Thompson, D. J., & Willis, T. 1995, ApJ, 440, 525

Wardle, J. F. C., Cawthorne, T. V., Roberts, D. H., & Brown, L. F. 1994, ApJ, 437, 122

Wehrle, A. E., Cohen, M. H., Unwin, S. C., Aller, H. D., Aller, M. F., & Nicolson, G. 1992, ApJ, 391, 589

Zensus, J. A. 1996, in *Extragalactic Radio Sources*, proc. IAU Symp. 175, ed. C. Fanti (Kluwer: Dordrecht), in the press

Zensus, J. A., Cohen, M. H., & Unwin S. C. 1995, ApJ, 443, 35

Zensus, J. A., & Porcas, R. W. 1987, in *Superluminal Radio Sources*, eds. Zensus, J. A., & Pearson, T. J. (Cambridge University Press: Cambridge), p. 126

The Sub-Parsec-Scale Structure and Evolution of the Jet in Centaurus A

S.J. Tingay[1], D.L. Jauncey[2], R.A. Preston[3], J.E. Reynolds[2], A.K. Tzioumis[2],
J.E.J. Lovell[4], M.E. Costa[5], D.W. Murphy[3], D.L. Meier[3], P.M. McCulloch[4],
D.L. Jones[3], S.W. Amy[2], R.W. Clay[6], P.G. Edwards[7], S.P. Ellingsen[4], R.H. Ferris[2],
R.G. Gough[2], P. Harbison[8], P.A. Jones[9], E.A. King[4], A.J. Kemball[10], V. Migenes[2],
G.D. Nicolson[10], M.W. Sinclair[2], T.D. van Ommen[3], R.M. Wark[2], and G.L White[9]

[1] Mount Stromlo and Siding Spring Observatories, Canberra, ACT 2611, Australia
[2] Australia Telescope National Facility, Epping, NSW 2121, Australia
[3] Jet Propulsion Laboratory, Pasadena, CA 91109, USA
[4] University of Tasmania, Hobart, Tasmania 7001, Australia
[5] University of Western Australia, Nedlands, WA 6009, Australia
[6] University of Adelaide, Adelaide, SA 5005, Australia
[7] Institute of Space and Astronautical Science, Sagamihara, Kanagawa 229, Japan
[8] British Aerospace Australia, Canberra, ACT 2601, Australia
[9] University of Western Sydney, Kingswood, NSW 2747, Australia
[10] Hartebeesthoek Radio Astronomy Observatory, Krugersdorp 1740, South Africa

Abstract. We present dual-frequency, co-eval very long baseline interferometry observations of the sub-parsec-scale radio source in Centaurus A which identify the core of the radio source, the origin of the parsec-scale and kiloparsec-scale radio emission. In addition we present multi-epoch observations at a frequency of 8.4 GHz which show the complex evolution of the sub-parsec-scale morphology over a 3.3 year period. Subluminal motion is seen and structural changes observed on time-scales shorter than four months.

1 Introduction

Centaurus A is the closest active extra-galactic radio source to us, at a distance of approximately 3.5 Mpc (Hui et al. 1993). As such it is a very important target for observations of the small-scale (parsec) and large-scale (kpc) structures in extra-galactic jets. In Centaurus A these jets originate at the nucleus of the peculiar elliptical galaxy NGC 5128 and can be observed in detail at radio wavelengths.

We have been studying the sub-parsec-scale radio jet in Centaurus A since 1982 with the Southern Hemisphere VLBI Experiment (SHEVE) array of telescopes (Preston et al. 1989, Jauncey et al. 1994) and have previously reported on observations at 2.3 and 8.4 GHz (Meier et al. 1989) and more recent observations at 8.4 GHz (Meier et al. 1993, Tingay et al. 1994). Most recently Jauncey et al. (1995) have reported on VLBI observations which allow (for the first time) the unambiguous identification of the core of the radio source and show the complex evolution of the jet on the sub-parsec-scale. We will further discuss these most recent results here.

2   Observations and results

The VLBI observations reported on here were made over a 3.3 year period - from early 1991 up until mid 1994, at frequencies of 4.8 and 8.4 GHz. All of the observations were made with the SHEVE array and the Mark II recording system (Clark, 1973). The data were correlated at the Caltech/JPL block II processor in Pasadena, and subsequently reduced and imaged using the AIPS software and the Caltech VLBI package, which includes the DIFMAP software (Shepard et al., 1994).

2.1   Dual-Frequency, Single-Epoch Data

In November 1992 two observations of Centaurus A were made approximately three days apart, one at 4.8 GHz, the other at 8.4 GHz. Figure 1 shows the two images resulting from these observations. The first thing to note is that the brightest feature in the 8.4 GHz image is almost completely absent in the 4.8 GHz image, whereas there is a good coincidence between other features in the two images. On the basis of this result the brightest feature in the 8.4 GHz image can be identified as a highly inverted spectrum component and the other main features seen in both images (C1 and C2) can be identified as steeper spectrum components. These identifications naturally point to the inverted spectrum component as the core of the radio source, and C1 and C2 components within a sub-parsec-scale jet. The highly inverted spectrum of the core may be due to synchrotron self-absorption, although we cannot distinguish between this and absorption due to free-free processes.

2.2   Single-Frequency, Multi-Epoch Data

Having identified the core of the radio source we are now in a good position to examine the evolution of the sub-parsec-scale structure with multi-epoch monitoring observations. Such a monitoring campaign was completed in mid 1994. In figure 2 we present the results, in the form of a series of images. As well as producing the images from the VLBI data we undertook a detailed analysis of the data by fitting models of Gaussian components to the visibility amplitudes and closure phases (e.g. Tzioumis et al. 1989). This analysis gives us a quantitative description of how the various components in the radio source changed their relative positions, dimensions and flux densities over the 3.3 year period. The main results of this analysis are as follows. We found a subluminal motion of approximately 0.15c for the component C2. We found that a similar subluminal speed was possible for the component C1, although C1 appears to undergo significant internal evolution during the series of observations. This internal evolution is difficult to separate from the possible subluminal motion. A 0.15c motion is shown in figure 2, superposed onto the components C1 and C2.

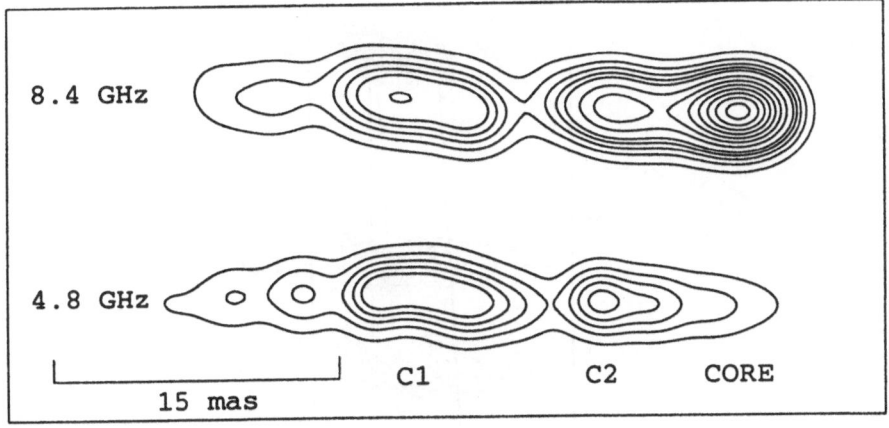

Fig. 1. Images of Centaurus A at 4.8 and 8.4 GHz which were made 3 days apart in November 1992. The images have been convolved with the same restoring beam, a circular Gaussian of 3 mas FWHM. Both images have been rotated 39° anticlockwise. Contour levels are 5, 10, 15, 20, 25, 35, 45, 55, 65, 75, 85, and 95% of the peak brightness of 0.7 Jy/beam

3 Discussion

Centaurus A joins the subset of compact radio sources which exhibit subluminal component motions (see Vermeulen and Cohen, 1994). The components seen in VLBI images are commonly interpreted as patterns on an underlying flow which is the jet. Thus, what we observe in Centaurus A is a subluminal pattern speed. The underlying flow speed of the jet may be much faster, and is very difficult to determine.

One way in which such a component can appear subluminal is by directing the jet (and hence the motion of the component) to be very closely aligned to the observer's line-of-sight. A highly relativistic component can then appear subluminal. This possibility can probably be rejected for Centaurus A, however. On the basis of the kpc-scale morphology of the radio source the chance that the jets are highly aligned with our line-of-sight is small. Now, one way to produce significantly subluminal component motion at angles to the line-of-sight greater than 20° is with intrinsic component speeds less than about 0.5c. At such speeds superluminal component motion can never be produced at any angle to the line of sight. Thus Centaurus A appears to be different from the compact radio sources which do exhibit superluminal component motions. The difference is that the components (patterns speeds) are intrinsically much slower in Centaurus A.

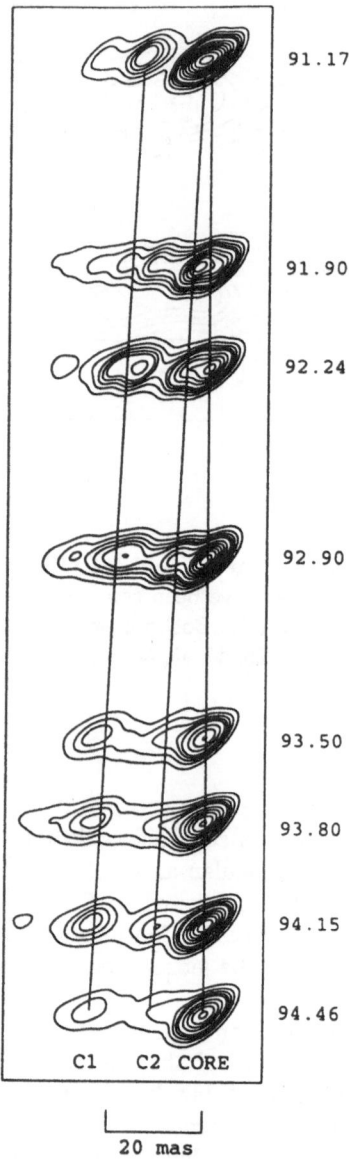

91.17

91.90

92.24

92.90

93.50

93.80

94.15

94.46

C1   C2 CORE

|——————|——————|
20 mas

Fig. 2. The series of images of Centaurus A made at 8.4 GHz with the SHEVE VLBI array over 3.3 years from early 1991. All images have been convolved with an elliptical Gaussian beam of 8.7 × 3.2 mas at a position angle of -83° and plotted on a common flux scale. The series shows change taking place in the sub-parsec-scale structure of Centaurus A. The apparent motions of the components C1 and C2 are indicated. Contour levels are 5, 10, 15, 20, 25, 35, 45, 55, 65, 75, 85, and 95% of the peak brightness of 3.3 Jy/beam.

We are just begining to understand the evolution of the sub-pc-scale radio source in Centaurus A. With continued VLBI monitoring observations we hope to gain some insight into the causes of the evolution we see and consequently be able to make a more complete and detailed comparison between the subluminal sources such as Centaurus A, and superluminal sources such as the BL Lac objects and quasars.

4 Acknowledgements

We express our gratitude to the time assignment committees of the participating observatories, the management of the Deep Space Network, Telstra (Australia) and the European Space Agency. We also thank the staff of the Caltech/JPL correlator. SJT and JEJL acknowledge support via Australian Postgraduate Awards and the Australia Telescope National Facility student program. PGE acknowledges reciept of a Monbusho Fellowship. We also acknowledge support from the Perth Astronomy Research Group and the Australian Research Council. The Australia Telescope is operated as a national facility by the CSIRO. Part of this research was carried out at the Jet Propulsion Laboratory, California Institute of Technology, under contract to the National Aeronautics and Space Administration. SJT also acknowledges generous support from the organisers to attend this workshop.

References

Clark B.G. (1973): Proc.IEEE 61, 1242
Hui X., Ford H.C., Ciardullo R., and Jacoby G.H. (1993): ApJ. 414, 463
Jauncey D.L. et al. (1994): Proc. IAU Symposium No. 158, Very High Angular Resolution Imaging, 131
Jauncey D.L. et al. (1995): Proc.Nat.Acad.Sci. (accepted)
Meier D.L. et al. (1989): AJ 98, 27
Meier D.L. et al. (1993): In 'Sub-arcsecond Radio Astronomy '(Eds R.J. Davis and R.S. Booth), 201 (Cambridge University Press)
Preston R.A. et al. (1989): AJ 98, 1
Shepherd M.C., Pearson T.J., and Taylor G.B. (1994): BAAS 26, 987
Tingay S.J. et al. (1994): Aust.J.Phys. 447, 619
Tzioumis A.K. et al. (1989): AJ 98, 36
Vermeulen, R.C. and Cohen, M.H. (1994): ApJ 430, 467

The Central Engine in the Galactic Nucleus

Wilfred H. Sorrell

University of Missouri-St. Louis

Abstract. The compact central object driving jet-like activity in quasars and active galactic nuclei is generally thought to be an accreting black hole storing a minimum mass of order 10^6 $M_\odot$. If this hypothesis is correct, then almost all nearby galaxies habour a supermassive black hole at their center. The present discussion considers the most recent radio, infrared, hard X-ray, and soft gamma-ray observations of nuclear activity within the central parsec region of our own galaxy. It is concluded that none of these observations demand the presence of a central 10^6 - 10^7 $M_\odot$ black hole.

1 Introduction

Although our understanding of the cosmic jet phenomenon in quasars is still in its infancy, there are two fundamental observations about quasars themselves that carry little or no debate. First, if the redshifts of quasars are produced by the Hubble expansion, then the quasar phenomenon must occur in galactic nuclei where the star concentration is high and where stellar nucleosynthesis of heavy chemical elements has already occurred. The reason is simple. We see solar-like abundances of carbon and oxygen ions in the emission line spectra of quasars. This observation alone tells us that the quasars cannot be a class of primordial objects. Secondly, there were luminous quasars in galactic nuclei billions of years ago, but there exist no such quasars at the center of nearby galaxies here today. According to the Lynden-Bell (1969) hypothesis, almost all the nearby galaxies should have a dead quasar at their center. In this context, a dead quasar is a weakly accreting black hole storing a minimum mass of order 10^6 $M_\odot$ (Chokshi and Turner 1992). The present discussion asks whether the Lynden-Bell hypothesis is correct for the center of our own galactic nucleus.

2 The Galactic Center

Several studies based upon optical and ultraviolet observations have deduced the presence of a supermassive (10^6 - 10^7 $M_\odot$) black hole at the center of nearby spiral and elliptical galaxies (Kormendy 1989 ; Tonry 1989). Unfortunately, the nucleus of our galaxy is unobservable in both the optical and ultraviolet bands because there exists about 30 magnitudes of interstellar dust obscuration along the line of sight. The open windows are in the radio, infrared, hard X-ray, and gamma-ray bands. It was suggested long ago that the nonthermal compact radio source Sgr A* in our galactic nucleus is associated with accretion onto a central massive black hole (Lynden-Bell 1969; Lynden-Bell and Rees 1971). Unlike the

100 pc scale radio sources in quasars and active galactic nuclei, the radio source Sgr A* has dimensions no larger than 1 - 10 AU. A recent 43 GHz VLBI observation shows the Sgr A* source has an elongated component suggesting a jet-like morphology (Krichbaum et al 1993). Black hole accretion models proposed for Sgr A* can account for both the nonthermal radio spectrum and the jet-like morphology (Falcke, Mannheim, and Biermann 1993 ; Duschl and Lesch 1994 ; Narayan, Yi, and Mahadevan 1995). Nevertheless, these models are not unique, and furthermore, the observed properties of the compact nuclear radio source do not require a black hole as an explanation. The observed properties of Sgr A* can be understood equally well on models in which the central object is a nuclear burning disc (Kundt 1995). Such an object stores a mass no larger than 10^2 - 10^3 $M_\odot$.

According to standard theory of dense stellar systems, stars in the core of a galactic nucleus would aggregate around a centrally collapsed mass and thereby form a stellar density cusp. This cusp is generally thought to be a unique signature of a massive black hole and the cusp would show itself as a stellar luminosity spike, which can be detected by using high-resolution measurements of the stellar light distribution. The difficulty here is that measurements of the stellar light distributions are plagued with uncertain corrections for interstellar dust obscuration, and more importantly, intrinsic brightness variations of the nuclear stars observed (cf. Bailey 1980). A better approach would be to measure the number of stars per square arcsecond as a function of distance from the galactic center, and thus avoid the brightness variation problem altogether. Such an approach was taken by Eckart et al. (1993), who studied the central parsec region of our galactc nucleus by using 0.15 arcsecond resolution measurements in near-infrared bands. It should be noted that Eckart et al. (1993) find no evidence for the presence of a central stellar density cusp. Instead, these authors find the stellar surface density follows the behavior of an isothermal (Emden) sphere having a core radius 0.15-0.2 pc. Absence of a stellar density cusp would appear, at first sight, to provide grounds for an argument against a centrally collapsed mass at our galactic center. However, another argument is that stellar collisions destroyed the cusp, and in doing so, created a flat-density profile in core regions. This argument is supported by numerical Fokker-Planck models for dense galactic nuclei (Murphy, Cohn, and Durisen 1989). Recent near-infrared observations also reveal the presence of 20 - 50 hot blue stars in the central parsec region. These stars are young and they show very broad ($600 - 2000$ km·s−1) hydrogen and He I emission line now thought to be produced in dense stellar winds blowing from the hot blue stars themselves (Genzel et al 1994). Ordinary star formation could not have formed the hot blue stars because no high concentration of cold dense gas exists in the central parsec region to make new stars (Jackson et al. 1993). An alternative explanation is that the blue stars formed as a consequence of red dwarf agglomeration (stellar mergers) during frequent stellar collisions in a dense (10^8 $M_\odot$) core (Lee 1989). Frequent stellar collisions might dominate the core dynamics, and formation of binaries should be common.

Several binaries in the central parsec region might consist of a neutron star and a supergiant companion filling its Roche lobe. Roche lobe overflow would lead to accretion onto the neutron star and emission of high-energy radiation in hard X-ray and gamma-ray bands. In this scenario, the accreting neutron stars produce luminosities 10^2 - 10^4 $L_\odot$ in the 100 - 1000 KeV band, consistent with observations. Goldwurm et al. (1994) discussed their results of a deep imaging survey of the galactic center region observed in 35 - 1300 KeV bands. These authors find no evidence for luminous hard X-ray and soft gamma-ray emission from Sgr A*. Observations based on the Sigma-GRANAT telescope data constrain the total 3 - 150 KeV luminosity of Sgr A* to be smaller than $2.5 \cdot 10^{36} erg \cdot s^{-1}$, as Goldwurm et al. (1994) find. This upper limit on the luminosity is much smaller than the Eddington luminosity $1.3 \cdot 10^{44} erg \cdot s^{-1}$ for a 10^4 $L_\odot$ black hole. Hence, if a dead quasar resides at the galactic center, then accretion must proceed either at sub-Eddington rates or with a low radiation efficiency (Abramowicz and Lasota 1995).

It is important to keep in mind that measurements of high-energy radiation have arcminute resolution because of telescope limitations. Such low resolution is insufficient to pindown whether hard X-rays and soft gamma-rays are actually being emitted from Sgr A* itself, or from several compact sources in the near vicinity of Sgr A*. Because of this uncertainty, observations of high-energy radiation (35 - 1300 KeV) are consistent with no hard X-rays and gamma-rays emitted from Sgr A* at all. Hence, no appeal to accretion onto a massive black hole is needed. All high-energy radiation observed from the central parsec region can be understood as emission from accreting neutron stars in close binaries.

The picture developing here is that almost all the violent nuclear activity in the central parsec region can be understood as consequences of stellar collisions, agglomerations, and ordinary stellar evolution. Successive agglomerations of red dwarf stars in a dense nuclear core would continuously build up a small population of heavy blue stars. These blue stars will ultimately explode as supernovae and leave neutron stars in their place. Supernova explosions would expel a sizeable mass of stellar debris from the core and thus account for low gas densities in the central parsec region. Fast winds from recently formed blue stars will dynamically affect the remaining gaseous material by creating fast shocks, ionized gas bubbles, and dense gaseous filaments. Both stellar winds and the tidal forces produced by a dense concentration of stars will act to distort the gases into a variety of geometrical shapes. Hence, the mini-cavity and mini-spiral can be easily understood as gaseous distortions produced by local wind and tidal effects, respectively. As the blue stellar population makes conditions possible to control the gas dynamics by forces in addition to gravitation, it might be misleading to deduce the presence of a centrally collapsed mass from the observed gas motions. The stellar motions are a different matter because they are controlled only by gravitational forces. Recent near-infrared observations of the central parsec region show that the velocity dispersion of both the red giant and blue supergiant populations increases with decreasing distance from the central star cluster IRS 16. Although these observations should be interpreted with considerable caution (Genzel et al. 1994), they appear to provide compelling evidence for the presence of a central black hole storing a mass no larger than $3 \cdot 10^6 \ M_\odot$. This estimate for the maximum mass is comparable to the minimum mass of a dead quasar based upon the Lynden-Bell hypothesis.

Nevertheless, we should also keep in mind that compelling evidence is not the same as the smoking gun. There are at least two difficulties with an interpretation of stellar velocity dispersion data for the central parsec region. First, velocity data for the red giant population are based upon the 2.3 micron CO absorption band observed in the atmosphere of red giant stars (McGinn et al. 1989; Sellgren et al. 1990). The stellar velocity dispersion is estimated from observed CO bandwidths broadened by Doppler motions. The problem is that the strength of the band is observed to dramatically decrease exactly in central nuclear regions where the stellar velocity dispersion appears to increase (Sellgren et al. 1990), and hence, CO bandwidth measurements become increasingly uncertain. Secondly, velocity data for the blue stellar population (the He I stars) are based upon measurements of the broad hydrogen and helium emission lines observed. The question asked here is whether the line broadening is caused mainly by the nuclear gravitational force or by forces (radiation pressure) in addition to gravitation. If He I line broadening is caused mainly by gravitational forces, then the total mass within the central parsec region greatly exceeds the total mass of stellar and gaseous material there. We would then have good dynamical evidence for the presence of a dead quasar at our galactic center (Genzel et al 1994). But, if the He I line broadening is caused mainly by gas motions associated with fast stellar winds driven by radiation pressure forces, then dynamical evidence in fovor of a centrally collapsed mass would be weak. Because of large uncertainties in the data interpretation, none of the present-day observations discussed here require the presence of a dead quasar at our galactic center.

References

Abramowicz, M. A. and Lasota, J.P., Comments on Astrophys.,18,141 (1995).

Bailey, M.E., Mon. Not. R. Astron. Soc., 190, 217 (1980).

Chokshi, A. and Turner, F. L., Mon. Not. R. Astron. Soc., 232, 431 (1992).

Duschl, W.J. aand Lesch, H., Astron. and Astrophys., 286, 431 (1994).

Eckart, A., Genzel, R., Hofmann, R., Sama, B.J., and Tacconi-Garman, L.E., Astrophys, J., 407, L77.

Falcke, H., Mannheim, K., and Biermann, P.L., Astron. and Astrophys., 278, L1 (1993).

Genzel,R., Hollenbach, D., and Townes,C.H., Report on Progress in Physics, 57, 417 (1994).

Goldwurm, A. et al, Nature, 371, 589 (1994).

Jackson, J.M. et al, Astrophys. J., 402, 173 (1993).

Kormendy, J., in Dynamics of Dense Stellar Systems, ed. D. Merritt (Cambridge Univ. Press : Cambridge), p. 31 (1989).

Krichbaum, T.P. et al, Astron. and astrophys., 274, L37 (1993).

Kundt, W. Bad-Honnef Preprint (1995); ApSS, in print.

Lee, H.M., in Dynamics of Dense Stellar Systems, ed. D. Merritt (Cambridge Univ. Press : Cambridge), p.105 (1989).

Lynden-Bell, D., Nature, 223, 690.

Lynden-Bell and Rees, M.J., Mon. Not. R. Astron. Soc., 152, 461 (1971).

McGinn, M.T., Sellgren,K., Becklin, E.E., and Hall, D.N.B., Astrophys. J., 338,824.

Murphy, B.W., Cohn, H.N., and Durisen, R.H., in Dynamics of Dense Stellar Systems,ed. D. Merritt (Cambridge Univ. Press : Cambridge), p.97 (1989).

Narayan, R., Yi., I., and Mahadevan, R., Nature, 374, 623 (1995).

Sellgren, K., McGinn, M.T., Becklin, E.E., and Hall, D.N.B., Astrophys. J., 359, 112.

Tonry, J.L, in Dynamics of Dense Stellar Systems, ed. D. Merritt (Cambridge Univ. Press : Cambridge), p.35 (1989).

Wilfred Sorrell at work

Our Galactic Center

Wolfgang Kundt

Institut für Astrophysik der Universität Bonn

Abstract. My 1990 twin-exhaust model is updated.

At BH, our Galactic Center has been reviewed by Wolfgang Duschl, but unfortunately, his promised writeup for this volume has not reached me. I thus take the opportunity to update my own 1990 model - whose response by the scientific community is (still) outstanding.

The four multifrequency overlay maps in Figure 1 summarize my knowledge on four different length scales: On the largest (halo) scale, the falling high-velocity HI clouds in the upper Galactic hemisphere form a coherent ribbon which can be continuously connected - in both position, line-of-sight velocity, and velocity dispersion - to the 21cm 'jets' of Burton & Liszt (1978) and to the Galactic Center, cf Kundt (1987, 1992). The ribbon approaches us at 60 Km/s in the upper hemisphere, and recedes at 40 Km/s on the other side of the disk; it looks to me like a trail in a Wilson chamber: gas condenses on the edge of the Galactic twin jet, and rains down back into the disk. The twin jet is not necessarily at variance with the (continuum) jet(s) found by Sofue et al (1988) (of unknown line-of-sight velocity), which are probably younger and have rammed a straighter channel of escape (from the Galaxy's potential well) - a phenomenon familiar from the extragalactic radio sources, whose jets tend to straighten with age (Kundt, 1988).

On the somewhat smaller scale of 1 Kpc (third map), I can see (at least) three cylinders (*'chimneys'*) stacked within each other, all centered on the Galactic rotation center. These cylinders appear in the Nobeyama 10 GHz map by Sofue (1989), and can be equally detected on the CO maps by Uchida et al (1992) as well as the longitude-velocity plots by Bally et al (1988), Binney et al (1991), and Sofue (1996) - as both blue- and redshifted emission w.r.t. galactic rotation. I like to interpret them as material having been dragged out by the galactic twin jet some $10^7 yr$ ago as well as having condensed on its boundary, raining back down into the disk. In their middle, they enclose the fourth (pc-scale) cylinder found (at $\leq 160 MHz$) by Yusef-Zadeh et al (1986), and by Kassim et al (1987), and (at 230 GHz) by Chris Salter, cf (Kundt, 1990), which has meanwhile fallen in disgrace because it is absent at other frequencies, but which would morphologically correspond to the VLA jets seen in other (active) galaxies, and which is likewise indicated in the vicinity of Sgr A East.

The last overlay map, on the 10-pc scale, shows Sgr A East and its close environment. Here is where controversial interpretations reach the fortissimo level: what do we see ? Are the different sources - the (non-thermal) point source

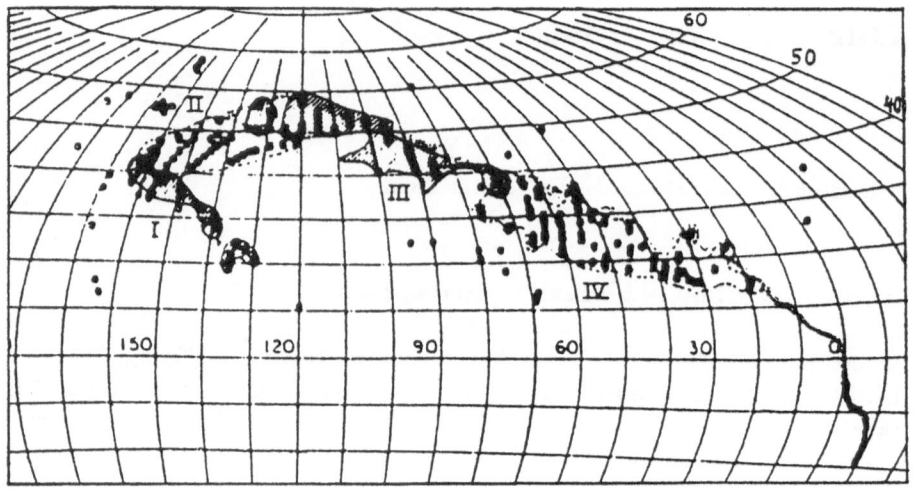

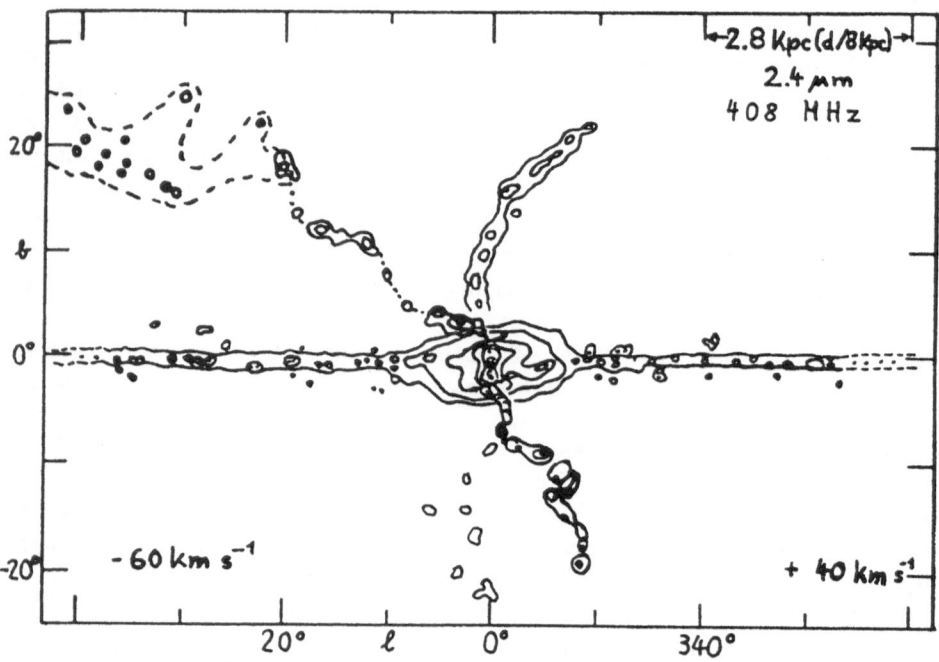

Fig. 1. Multifrequency (radio to IR) view of the Galactic Center on four different angular (length) scales. The first two maps, on scales of 10 and 5 Kpc respectively, show the reported and/or proposed jets from Sgr A . The third and fourth map - on scales of 1 Kpc and 10 pc respectively - highlight their apparent origin: Sgr A* . The distance to the Center has been assumed 8.0 Kpc, acc. to (Reid, 1993). After Kundt (1987, 1990); see text for explanations.

267

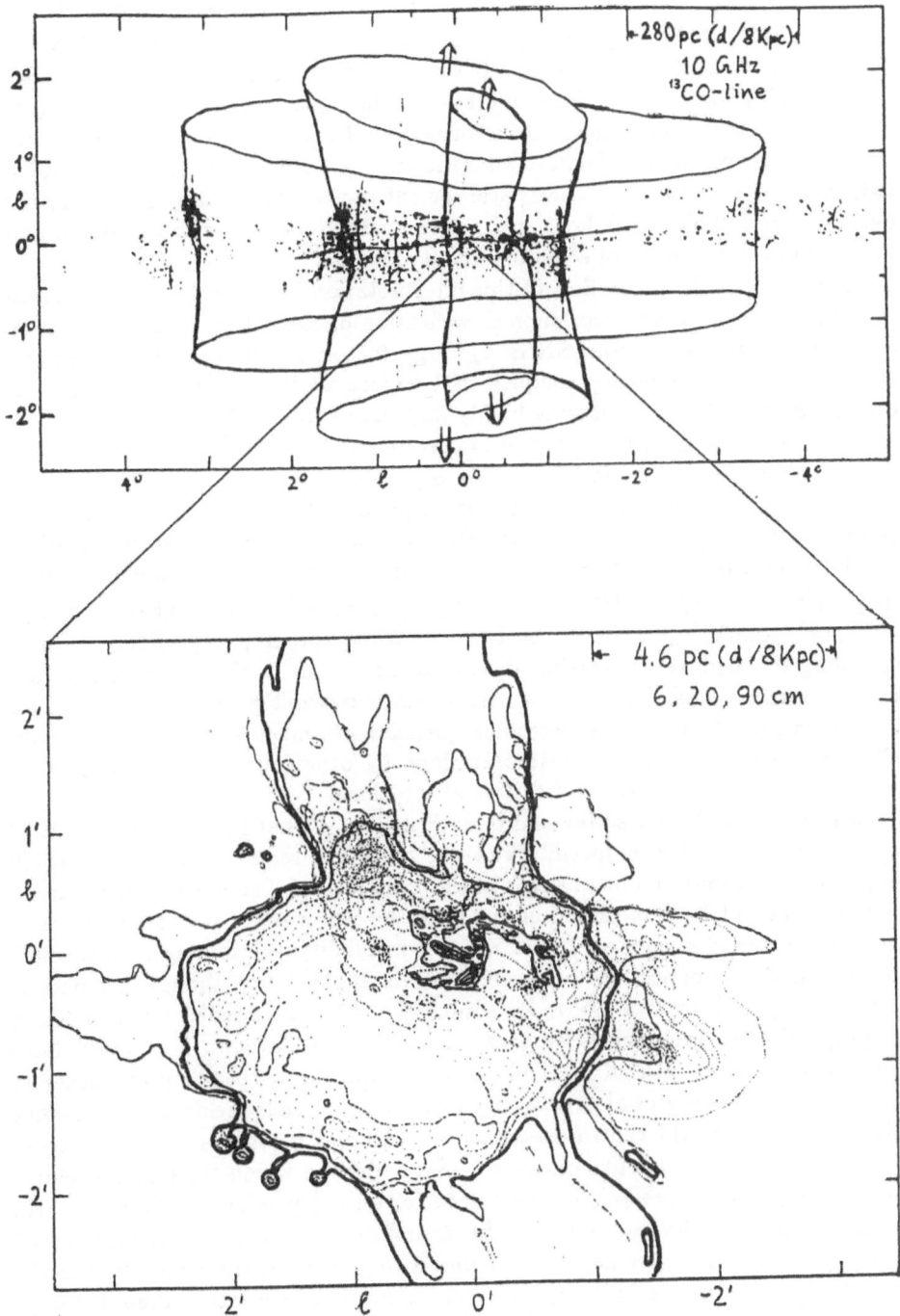

Fig. 1 continued

Sgr A* at the rotation center, the (thermal) 'vortex' Sgr A West, and the SNR-like 'spillover bubble' Sgr A East all stacked within each other: Sgr A* ⊆ Sgr A West ⊆ Sgr A East ⊆ Chimneys ? If they are, they could all have been blown (successively) like a glass blower's craft: by the gigantic pressure of the quasi-weightless pair plasma generated in coronal discharges of Sgr A* . At the same time, we face a straight-forward manifestation of Blandford & Rees's (1974) twin-exhaust scenario: the relativistic jet substance, stored subsonically in the overpressure bubble Sgr A East, escapes to both sides of the disk through a pair of chimneys in the form of a double deLaval nozzle.

The first possible objection to this interpretation has already been discussed in (Kundt, 1990): it is the common classification of Sgr A East as a SN remnant, even though it differs from SNRs by large factors in energy ($10^{1.6}$) and size ($\lesssim 10^{-1}$), and by blowing compact HII-regions at its periphery. Besides, its central location in the Galaxy is highly improbable for a SNR.

A second objection, by Pedlar et al (1989) and Anantharamaiah et al (1991), is more serious: at low radio frequencies, Sgr A East is seen in absorption against Sgr A West. The proposed occultation is good but not perfect, and again in (Kundt, 1990), I have argued that what has been interpreted as 'absorption' can be re-interpreted as 'lack of emission', due to the Razin ('lower') cutoff at frequency $\nu_l = (\gamma \nu_p^3/\nu_B)^{1/2} = 10^{8.7} Hz (\gamma_4 n_4^{3/2}/B_{-2})^{1/2}$, which takes the values $10^{8.1}, 10^{8.7}, and 10^{8.8} Hz$ for Sgr A East, A West, and A*, respectively - corresponding to (slightly) increasing plasma densities n in the 3 sources, on approach of the center. (Sgr A East disappears between 126 and 111 MHz). This interpretation of the 3 lower cutoff frequencies fits much better into the overall scenario than attempted interpretations by other authors (as synchrotron self-absorption, or as thermal free-free absorption).

Much of what I have so far said is a repetition of what I already said in 1990 - of which I am still convinced. But there are a few recent observations which imply modifications of, and changes in the 1990 review. There is (i) the γ-ray result that the 511 KeV annihilation line from the center is not variable and comes from an extended region, not from Sgr A*; variability was mimiced by variable apertures. Part 2 of section 6 thereby loses its applicability, though not part 1. (ii) The apparent size of the radio source Sgr A* is broadened by dispersion, in proportion to wavelength squared; measurements down to $\lambda = 3mm$ have reduced it to a true size of $2r \lesssim 10^{13.2}cm$, perhaps $2r = 10^{12.5}cm$; cf Backer et al (1993), Krichbaum et al (1995). My equation (11) thus yields a much higher plasma density for the central radio source. (iii) The 'snake', analysed by Gray et al (1995), looks to me like one of the many fibres of which the Galactic chimney is composed (figure 1c); I see no need for an independent explanation. (iv) Morris et al (1992) have mapped the linear polarization of the 'arched filaments' in the (thermal) 'bridge region', and found it to be parallel (in projection) to the 'cloud edges'; I still like to interpret them as sheetlike layers near the inner edge of the chimney. Corresponding layers may have been detected by Yusef-Zadeh et al (1993) as blueshifted absorption towards Sgr A East. (v) Maps at 90 cm by Anantharamaiah et al (1991) show the 'sickle' to be on the near side of the 'arc'

(= left edge of chimney), not on the far side (as assumed in 1990). This means (to me) that the central disk is twisted (at separations of 30 pc), looking in projection like a figure 8, and that the (young stars forming the) 'pistol' lie(s) right in the galactic disk, (not way below it). This re-interpretation puts the bright young stars (of the pistol) right where they belong. (vi) New measurements, and indirect determinations of the magnetic field strengths B(r) at various distances from the center have been performed, see Killeen et al (1992). I still defend my fit (2): $B(r) = 3mG(pc/r)^{0.8}$ for $0.3pc \lesssim r \lesssim 10Kpc$. (vii) At larger distances from the center, the gas of the Milky Way must still provide the fuel which, sooner or later, will lead to another Seyfert stage of the nucleus. In 1990, I was not aware of the HI-analysis by Cohen & Davies (1979) which finds a bar in the central Kpc. (viii) Returning to the innermost 1 lyr around Sgr A*, I have noted (in 1990) 3 blown-off windzones by comparing Dan Gezari's map with one by Farhad Yusef-Zadeh and collaborators, in addition to the then known best case of IRS 7, (viz IRS 6, 10, and one more). In the meantime, IRS 7 has been mapped much better by Yusef-Zadeh et al (1992), plus a fifth, somewhat farther blown-off windzone, 0.5 lyr northwest of IRS 7 , again by Yusef-Zadeh (1994). (ix) What else do we know about Sgr A* ? Has it been mapped at IR frequencies (Genzel et al, 1995)? If so, it looks to me like a planetary nebula.

Acknowledgement

I thank Alok Patnaik for a discussion.

References

Anantharamaiah K.R., Pedlar A., Ekers R.D., Goss W.M. (1991): MNRAS **249**, 262.

Backer D.C., Zensus J.A., Kellermann K.I., Reid M., Moran J.M., Lo K.Y. (1993): Sci **262**, 1414.

Bally J., Stark A.A., Wilson R.W., Henkel C. (1988): ApJ **324**, 223.

Binney J., Gerhard O.E., Stark A.A., Bally J., Uchida K.I. (1991): MNRAS **252**, 210.

Blandford R.D., Rees M.J. (1974): MNRAS **169**, 399.

Cohen R.J., Davies R.D. (1979): MNRAS **186**, 453.

Genzel R., Eckart A. Krabbe A. (1995): in *Seventeenth Texas Symposium on Relativistic Astrophysics and Cosmology*, Ann. N.Y. Acad. Sci. **759**, eds. H. Böhringer, G.E. Morfill, J.E. Trümper, New York, p.38.

Gray A.D., Nicholls J., Ekers R.D., Cram L.E. (1995): ApJ **448**, 164.

Killeen N.E.B., Lo K.Y., Crutcher R. (1992): ApJ **385**, 585.

Krichbaum T.P., Schalinski C.J., Witzel A., Standke K., Graham D.A., Zensus J.A. (1995): in *The Nuclei of Galaxies: Lessons from the Galactic Center*, eds. R. Genzel, A.I. Harris, Kluwer, p. xyz.

Kundt W. (1987): Ap&SS **129**, 195.

Kundt W. (1988): in *Hot Spots in Extragalactic Radio Sources*, Lecture Notes in Physics **327**, eds. K. Meisenheimer, H.-J. Röser, Springer, pp. 179, 275.

Kundt W. (1990): Ap&SS **172**, 109.

Kundt W. (1992): Ap&SS **195**, 331.

Morris M., Davidson J.A., Werner M., Dotson J., Figer D.F., Hildebrand R., Novak
  G., Platt S. (1992): ApJ **399**, L63.

Pedlar A., Anantharamaiah K.R., Ekers R.D., Goss W.M., van Gorkom J.H., Schwarz
  U.J., Zhao J.-H. (1989): ApJ **342**, 769.

Reid M.J. (1993): ARA&A **31**, 345.

Sofue Y. (1989): in *The Center of the Galaxy*, IAU **136**, ed. M. Morris, Kluwer, p. 215.

Sofue Y. (1996): PASJ, to appear.

Uchida K.I., Morris M., Bally J., Pound M., Yusef-Zadeh F. (1993): ApJ **410**, L27.

Yusef-Zadeh F. (1994): in *The Nuclei of Normal Galaxies*, eds. R. Genzel, A.I. Harris,
  Kluwer, p. 355.

Yusef-Zadeh F., Melia F. (1992): ApJ **385**, L41.

Yusef-Zadeh F., Lasenby A., Marshall J. (1993): ApJ **410**, L27.

THE MISSING X-RAYS IN SGR A*: EVIDENCE FOR A SUPERMASSIVE BLACK HOLE IN THE GALACTIC CENTER

Heino Falcke [1] [2], Peter L. Biermann [1]

[1]Max-Planck-Institut für Radioastronomie, Auf dem Hügel 69, D-53121
Bonn, Germany
[2]Deptartment of Astronomy, University of Maryland, College Park,
MD 20742-2421, USA, email: hfalcke@astro.umd.edu

Abstract: We present a simple argument that the missing x-ray flux from the Galactic Center source Sgr A* ist not evidence *against* – as claimed by Goldwurm et al. 1994 – but rather indirect evidence *for* the presence of a supermassive black hole. The radio spectrum provides a strict *lower* limit for the size of Sgr A* ($R > 3 \cdot 10^{11}$cm). A more compact source would be completely synchrotron self-absorbed. This size is 10^6 times larger than a stellar-mass black hole, yet the bolometric radio luminosity is comparable to or even larger than the x-ray luminosity where matter accreting onto a *stellar-mass* black hole would inevitably radiate the *bulk* of its luminosity. Hence, either the bulk of the accretion power is radiated in the UV (where the limits are higher), or the accretion has to stop at the radio-scale to avoid producing x-rays brighter than the radio emission. Both would be a natural consequence of a supermasive black hole with $M_\bullet \sim 10^6 M_\odot$.

1. Introduction

SIGMA/GRANAT observations recently showed (Goldwurm et al. 1994) that Sgr A*, the very center of the Galaxy, does not emit significant hard X-ray radiation ($L(35-150\mathrm{keV}) < 3.5 \cdot 10^{35}$ erg/sec) limiting the total X-ray luminosity to $\lesssim 2.5 \cdot 10^{36}$ erg/sec as detected by Art-P at somewhat lower energies. In the same paper this low luminosity was compared with the 10^8 times higher Eddington luminosity of a supermassive black hole (mass $M_\bullet \sim 10^6 M_\odot$) and the hard X-ray spectrum of stellar-mass black holes and concluded that there is "possible evidence against a massive black hole at the Galactic Centre" (title) and that Sgr A* "clearly does not behave like a scaled-down active galactic nucleus".

Since then the situation has become even worse: Koyami (1994) reported ASCA observations of the Galactic Center (GC) where it is found that two sources – a hard and a soft source – exist in the GC separated by 1'. While the soft source coincides with the ROSAT source (Predehl & Trümper 1994) and hence is within 10" of Sgr A*, the 'hard' ASCA source should correspond to the hard GC source detected by Art-P. The hard source therefore might not be Sgr A* but a nearby x-ray binary, thus reducing the x-ray luminosity of Sgr A* even further. This would also relax the need for a high intrinsic absorption in Sgr A* (Predehl & Trümper 1994) as there would be no need to fit the low ROSAT flux to the high Art-P flux anymore. If the 'hard' source is not Sgr A* it can well be behind the Sgr A complex and therefore the soft x-rays are so strongly obscured that the 'hard' source was not detected by ROSAT and hard and soft source were incorrectly identified. The total x-ray luminosity of Sgr A* in the ROSAT band with normal extinction would be not more than $L_x \sim 1 - 2 \cdot 10^{34}$ erg/sec (Predehl 1995, priv. com.) and if there is more at higher energies it proably would be largely contaminated by the nearby hard x-ray binary (Koyama 1995, priv. comm; Maeda et al. 1996). Even though one probably should await further analysis of the x-ray data, it seems quite likely that the total x-ray luminosity is at best a few hundred $L_\odot$ or less.

Here I want to argue that this extremely low x-ray luminosity contradicts the presence of a low mass back hole at the position of Sgr A*, but that together with the radio spectrum it is consistent with the presence of a supermassive black hole.

2. Size limit from the radio spectrum

Sgr A* was known first as a compact flat-spectrum radio source (Balick & Brown 1974) somewhat similar to those in the nuclei of active galaxies. It is now clear that this radio spectrum extends into the submm regime (maximum flux of $F_{\nu_{max}} \sim 3.5$ Jy at $\nu_{max} \sim 10^{12}$ Hz) with an inverted ($\alpha \simeq +1/3$) spectrum at lower frequencies and a steep cut-off towards the IR. This corresponds to a total radio luminosity of a few 100 $L_\odot$ ($L(\text{radio} - \text{submm}) \sim 10^{36}$ erg/sec). An upper limit of $R \leq 2 \cdot 10^{13}$ cm to the size of Sgr A* at λ3mm is given by VLBI observations (Krichbaum et al. 1995). One can, however, easily derive a strict *lower limit* to the size of Sgr A* from its spectrum.

If we approximate the electron distribution in the source by a quasi monoenergetic electron distribution with energy $\gamma_e m_e c^2$, as required at least for the submm regime by the inverted spectrum and the sharp cut-off towards the IR, we have the simple condition that the synchrotron self-absorption frequency ν_{ssa} has to be lower than the characteristic peak frequency ν_{max} (i.e. $\nu_{ssa} < \nu_{max}$), otherwise the source would not be visible in the radio.

For a spherical blob with radius R this translates into a minimum condition for the size for a given peak flux and peak frequency as observed in Sgr A* (see Falcke 1995 for more details).

$$R_{\text{sync}} > 3 \cdot 10^{11} \text{cm} \quad k^{-1/17} \left(\frac{F_{\nu_{\text{max}}}}{3.5 \text{Jy}}\right)^{8/17} \left(\frac{\nu_{\text{max}}}{10^{12} \text{Hz}}\right)^{-16/51} \left(\frac{\nu_{\text{ssa}}}{10^{12} \text{Hz}}\right)^{-35/51}$$

This expression is almost independent of the equipartition parameter k and depends only on observable quantities. This limit is also consistent with earlier independent estimates (Gwinn et al. 1991). If Sgr A* were more compact, then the source would be completely self-absorbed and unable to produce the observed radio flux, but of course can ν_{ssa} be much smaller and hence the size be larger. One should also note that a much smaller, self-absorbed size would lead to substantial synchrotron-self Compton x-ray emission. For comparison it is interesting to note that the gravitational radius of a black hole is $R_g = 1.5 \cdot 10^{11} M_\bullet / 10^6 M_\odot$ cm and therefore Sgr A* is only slightly larger than a supermassive black hole with $10^6 M_\odot$, but $10^6 - 10^3$ times larger than a 1-1000 $M_\odot$ black hole.

3. Why not a low-mass black hole?

The large radio size on its own is not an argument against a low-mass black hole. But if Sgr A* were indeed a low-mass black hole candidate powered by accretion, then even in Bondi-Hoyle accretion some excess angular momentum would lead to the formation of an accretion disk (e.g. Ruffert & Melia 1994), where the bulk of the disk luminosity would be radiated in the x-rays (Shakura & Sunyaev 1973). In the case of Sgr A* *the measured non-thermal radio-submm luminosity would be comparable or even larger than the x-ray accretion disk luminosity* (this is a problem for the Ozernoy 1992 and the Mastichiadis & Ozernoy 1994 model).

Consequently, one would have to argue that for some obscure reasons the gravitational energy in the accretion disk is not dissipated primarily into x-rays but mainly into non-thermal synchrotron radiation, such that the submm regime reflects the true peak in the spectral energy distribution. This, however, implies an important constraint to the size of the emission region. As the bulk of the gravitational energy in the accretion process has to be dissipated within the inner $\sim 10 R_g$ one would predict

$$R_{\text{sync}} \sim 10 R_g = 1.5 \cdot 10^8 (M_\bullet / 10^2 M_\odot) \text{cm}$$

and this small size is obviously in contradiction with the lower limits on the Sgr A* radio-submm size, given above, by a margin of at least 3 orders of magnitude. Hence, for a black hole with $M_\bullet \ll 10^6 M_\odot$ the x-rays are far too low compared to the radio and the radio size is far too large for being the primary energy channel.

The only alternative way to have a small mass black hole, produce the radio emssion at the large scale and avoid producing the x-rays would be to postulate an almost dissipationless transport of energy within a jet. However, this requires an efficieny of $> 99\%$ conversion of gravitational energy into directed jet power which is implausible (see Falcke et al. 1993b; Donea & Biermann 1996)

This problem is more easily resolved if Sgr A* indeed were a supermassive black hole. In this case the accretion disk would radiate mainly in the UV where the limits on the luminosity are much higher (Falcke et al. 1993a, Zylka et al. 1995), with the radio and the X-ray luminosities being secondary emission components representing only a few per cent of L_{disk}. Proposed physical models for the radio emisson are synchrotron emission from a jet (Falcke et al. 1993b), magnetic bremsstrahlung in Bondi-Hole accretion (Melia 1994), or synchrotron radiation from an advection-dominated disk (Narayan et al. 1995). However, the latter two models have to cope now with the lowered limit for the x-ray luminosity of Sgr A* and need to be adjusted (see also Falcke & Heinrich 1994 and Falcke 1996a for a longer discussion of some of the models). Considering the radio emission, Fig. 3 in Falcke (1996, this volume) shows that Sgr A* can well be understood as a scaled down AGN.

References

Balick, B., Brown, R.L. 1974, ApJ 194,265

Falcke H. 1996, in: "Unsolved Problems of the Milky Way", IAU Symp. 169, L. Blitz & P. Teuben (ed.), Kluwer, Dordrecht, p. 163

Donea A., Biermann P.L. 1996, A&A subm.

Falcke H., Heinrich O. 1994, A&A 292, 430

Falcke, H., Biermann, P. L., Duschl, W. J., Mezger, P. G. 1993a, A&A 270, 102

Falcke, H., Mannheim, K., Biermann, P. L. 1993b,A&A 278, L1

Goldwurm, B. et al. 1994, Nat 371, 589

Gwinn, C.R., Danen, R.M., Middleditch, J., Ozernoy, L.M., Tran, T.Kh. 1991, ApJ 381, L43

Krichbaum T.P., Schalinski C.J., Witzel A., Standke K., Graham D.A., and Zensus J.A. 1995, in "The Nuclei of Normal Galaxies: Lessons from the Galactic Center", eds. Genzel R. & Harris A.I., Kluwer, Dordrecht

Koyama, K. 1994, in: New Horizon of X-ray Astronomy - first results from ASCA, Universal Academy Press, p. 181

Maeda Y., Koyama K., Sakano M., Takeshima T., Yamauchi S. 1996, to appear in PASJ

Mastichiadis A., Ozernoy L.M. 1994, ApJ 426, 599

Melia F. 1994, ApJ 426, 577

Narayan R., Yi I., Mahadevan R. 1995, Nat 374, 623

Ozernoy L. 1992, in AIP Conf. Proc. 254, Testing the AGN Paradigm, ed. S.S. Holt et al., New York, p. 40,44

Predehl P., Trümper J. 1994, A&A 290, L29

Ruffert M., Melia F. 1995, A&A 288, L29

Shakura N.I., Sunyaev R.A. 1973, A&A 24, 337

Zylka, R., Mezger, P.G., Ward-Thomson, D., Duschl, W., Lesch, H. 1995, A&A 297, 83

Numerical Simulations of Supersonic Jets: the Cocoon Emission

Silvano Massaglia [1], Gianluigi Bodo [1], Attilio Ferrari [1,2],

Paola Rossi [1]

[1] Osservatorio Astronomico di Torino, Strada Osservatorio 20, I-10025
Pino Torinese, Italy
[2] Dipartimento di Fisica Generale dell'Università, Via Pietro Giuria 1,
I-10125 Torino, Italy

Abstract: We present the results of numerical simulations of the propagation of supersonic jets and of their interaction with the external medium. We compute also the evolution of the distribution function of a population of relativistic electrons moving with the fluid and subject to adiabatic and synchrotron losses in a passively advected magnetic field. The results are discussed in connection with the formation and morphology of lobes in extragalactic radiosources.

1 Introduction

The propagation of a supersonic jet shot into an ambient medium has been studied by many authors since the pioneering work by *Norman* et al. [1] and many different ingredients have been introduced in these studies: they include variability of the injection properties of the jet (*Clarke & Burns* [2]), variation of the physical parameters of the ambient medium along the jet propagation path (*Norman, Burns & Sulkanen* [3]), nonadiabaticity of the flow (*Blondin, Fryxell & Königl* [4]), fully 3-D geometry (*Norman, Stone & Clarke* [5], *Hardee & Clarke* [6], *Hardee, Clarke & Howell* [7]), MHD effects (*Clarke, Norman & Burns* [8]; *Lind* et al. [9]), and relativistic effects (*Martí, Müller & Ibáñez* [10]; *Duncan & Hughes* [11]; *Martí* et al. [12]). In spite of these efforts many aspects of the problem are still not well understood, because of the complexity of the jet-cocoon structure, and in a recent paper (*Massaglia, Bodo and Ferrari* [13], hereinafter Paper I) we have discussed new features of the interaction that develop in the high Mach number regime.

In this paper we extend the results of Paper I, focussing our attention on the radiative effects and the morphologies that one may expect to observe in connection with particular choices of the physical parameters. For doing this we have introduced in the computational scheme a "passive" magnetic field and a distribution of relativistic test particles, subject to synchrotron losses and to adiabatic expansion, for which we follow the temporal evolution.

The outline of the paper is the following. In the next Section 2 we discuss the physical problem; the results of the calculations are discussed in Section 3. Problems and needs of this kind of investigations are reported in Section 4.

2 The Physical Problem

We study the dynamics of a supersonic, cylindrical, axisymmetric jet continuously injected into a medium initially at rest. We solve numerically the full set of adiabatic, inviscid fluid equations for mass, momentum, and energy conservation,

$$\frac{\partial \rho}{\partial t} + \nabla \cdot (\rho v) = 0 \,, \tag{1}$$

$$\frac{\partial v}{\partial t} + (v \cdot \nabla)v = -\nabla p / \rho \,, \tag{2}$$

$$\frac{\partial p}{\partial t} + (v \cdot \nabla)p - \Gamma \frac{p}{\rho} \left[\frac{\partial \rho}{\partial t} + (v \cdot \nabla)\rho \right] = 0 \tag{3}$$

where the fluid variables p, ρ and v are, as customary, the pressure, density, and velocity, respectively; Γ is the ratio of the specific heats.

In order to follow the jet particles in the external environment, we solve an additional advection equation for a scalar field f:

$$\frac{\partial f}{\partial t} + (v \cdot \nabla)f = 0 \,. \tag{4}$$

The initial spatial distribution for this tracer is designed to demarcate the jet alone; thus, we set f initially equal to one inside the jet, and to zero outside; in the following evolution, f is set to one also for the newly injected jet fluid. By this means, we can distinguish between the matter which is initially part of the jet or is afterwards injected into the jet, and that which is part of the external medium.

2.1 Radiation Treatment

As we have stated in the Introduction, we follow the evolution of the distribution function of a population of relativistic electrons, passively advected by the fluid, and subject to adiabatic and synchrotron losses and to acceleration in shocks. Random acceleration processes are neglected. Following *Kardashev* [14], its evolution equation, without shock acceleration, can be written in the form:

$$\frac{DF}{Dt} = \frac{\partial}{\partial E} \left[\left(-\alpha E + \beta E^2 \right) F \right] \,,$$

where F is the distribution function, the coefficient $\alpha = -(\nabla \cdot \boldsymbol{v})/3$ includes the adiabatic expansion, and the term $\beta E^2 = bB^2 E^2$ stands for the synchrotron losses. D/Dt represents the Lagrangian derivative $\partial/\partial t + \boldsymbol{v} \cdot \nabla$. These kinetic equations can be solved, yielding [14]

$$F(E,t) = K E^{-\gamma} \left[1 - E\, e^{-a_2}\, a_1 \right]^{\gamma-2} e^{(\gamma-1)a_2} ,$$

subject to the initial condition $F(E,0) = K_0 E^{-\gamma}$, where

$$a_1 := \int_0^t \beta\, e^{a_2} dt , \tag{5}$$

and

$$a_2 := \int_0^t \alpha\, dt , \tag{6}$$

and the integrals are performed following the trajectory of a fluid element. We can therefore write two evolution equations for a_1 and a_2 of the form

$$\frac{Da_1}{Dt} = \beta\, e^{a_2}$$

and

$$\frac{Da_2}{Dt} = \alpha .$$

These two equations are solved together with the system of the fluid equations. Initially the scalars a_1 and a_2 are set equal to zero. In addition, we consider a systematic shock acceleration, prescribing that at the shock position the particle energy is increased by a given factor w proportional to shock compression. The specific form of the shock acceleration coefficient w bears some arbitrariness due to the poor knowledge of the physical processes involved. We proceed as follows: at first, to locate the position of shocks on the grid at every temporal step, we adopt the criterion of looking for places where the pressure jump between two adjacent grid points exceeds a threshold value s (in the actual computations we chose $s = 3$). The second step is to fix the amount of the acceleration in every shock, where Eq. (6) simply becomes

$$a_2 = w . \tag{6'}$$

To do this we set w equal to the pressure jump in a shock and impose an upper limit at $w_{max} = 5$. Different choices of w_{max} (> 0) modify the details of the results, but the general behaviour remains unchanged.

In this way the energy of each electron varies as

$$E = \frac{E_0 e^{a_2}}{1 + E_0 a_1} . \tag{7}$$

and the resulting spectrum will have a cutoff at an energy

$$E_{\text{cutoff}} = \frac{e^{a_2}}{a_1} . \tag{8}$$

The passive magnetic-field evolution can be followed by noting that in two dimensions the field can be represented by the gradient of a potential A and that the induction equation, written in terms of the potential, assumes the form of an equation for the advection of a scalar field:

$$\frac{\partial A}{\partial t} + (\boldsymbol{v} \cdot \nabla)A = 0 \quad,$$

where A is the magnetic potential. We have chosen the potential A initially linear with the radial coordinate, this is equivalent to setting an initial magnetic field uniform and in the z direction ($\boldsymbol{B} = B_0\ \boldsymbol{z}$). The tracer f, defined by Eq. (4), distinguishes between the jet material, with the associated high energy particles, and the external gas and follows them during all the evolution.

2.2 The Numerical Scheme

The numerical scheme, the grid, the code adopted are discussed in Paper I. In the present calculations we take advantage of the particular setup of Paper I, and we perform the calculation in a reference frame where the jet's head is approximately at rest. Therefore, in the initial configuration, the external medium has a uniform velocity

$$V_{\rm h} = \frac{v_{\rm j}}{1 + \sqrt{\nu}} \ , \tag{9}$$

where $v_{\rm j}$ is the jet velocity in the "laboratory frame" and ν is the ratio of the external to the jet density, and $V_{\rm h}$ is an approximated advance velocity of the jet's head obtained applying momentum conservation in the front region of the jet (see Paper I). This moving frame is adopted in the computations, but afterwards we will discuss the results obtained, translating them back to the reference frame where the external medium is at rest, i.e. to the "laboratory frame".

The system of equations was written in non-dimensional form measuring all lengths in units of the jet radius a, and time in units of the sound crossing time $t_{\rm cr} = a/v_{\rm sound}$ ($v_{\rm sound}$ is the initial isothermal sound speed on the jet's axis: $r = 0$, $t = 0$). Also density and pressure are expressed in units of their values at $r = 0$ and $t = 0$ while the velocities are expressed in terms of $v_{\rm sound}$. With this choice of non-dimensionality, the control parameters are then reduced to the jet (internal) Mach number

$$M = \frac{v_z(r = 0, t = 0)}{\sqrt{\Gamma}\ v_{\rm sound}} \ ,$$

and the density ratio ν (Γ is the ratio of specific heats).

Thus, the jet initially occupies a cylinder of length L, in pressure equilibrium with the external medium, and the initial flow structure has the following form:

$$v_z(r) = \begin{cases} \dfrac{v_z(r = 0)}{\cosh[(r)^m]} - V_{\rm h} & , \quad z \leq L \ , \\ -V_{\rm h} & , \quad z > L \ . \end{cases}$$

where m is a "steepness" parameter for the shear layer separating the jet from the external medium. The density radial dependence has the form:

$$\frac{\rho(r)}{\rho(r=0)} = \nu - \frac{\nu - 1}{\cosh[(\eta r)^n]},$$

with $\eta = 0.75$, $m = 8$ and $n = 2m$; this implies a narrower and smoother radial extension of the "density" jet with respect to the "velocity" jet. The reason for this choice is to obtain a smooth radial profile of the momentum density ρv_z.

For comparing results obtained with different parameters ν and M we have introduced a "normalization" time, t_{norm}, defined as the time employed to cover the unit distance (i.e. the jet radius) moving at the velocity V_h (Eq. 9). Therefore we have $t_{\text{norm}}^{-1} = M/(1 + \sqrt{\nu})$ and $\tau = tM/(1 + \sqrt{\nu})$.

The integration has been performed in cylindrical geometry and the domain of integration ($0 \leq z \leq D$, $0 \leq r \leq R$) is covered by a grid of 750×250 grid points with non-uniform spacing (see Paper I). The axis of the beam is taken coincident with the bottom boundary of the domain ($r = 0$), where symmetric (for p, ρ and v_z) or antisymmetric (for v_r) boundary conditions are assumed. At the top boundary ($r = R$) and right boundary ($z = D$) we choose free outflow conditions, imposing for every variable Q null gradient ($dQ/d(r, z) = 0$).

The numerical scheme adopted is of PPM (Piecewise Parabolic Method) type and is particularly well suited for studying highly supersonic flows with strong shocks (*Woodward & Colella* [15], see also *Bodo* et al. [16], [17]).

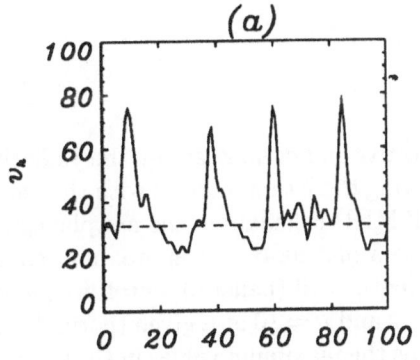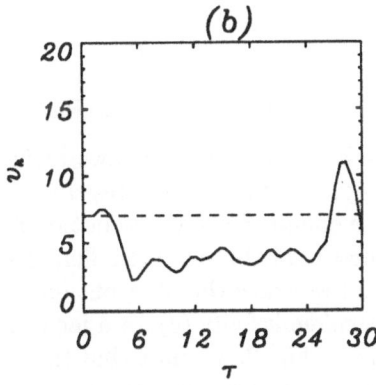

Fig. 1. Velocity of the jet head as a function of time. Panel (a): $M = 100$, $\nu = 10$; panel (b): $M = 100$, $\nu = 300$

3 Results

3.1 Dynamical Evolution

The dynamics of the interaction of the jet head with the ambient medium and its dependence on M and ν has been extensively discussed in Paper I, where we have explored the plane (ν, M) with $\nu = 3, 10, 30, 100, 300$ and $M = 3, 10, 30, 100, 300, 1000$. We summarize here the main features of this interaction:

a. Jets with high M (≥ 30) and low ν (≤ 30) have higher head velocities. In this case the backflow compression of the jet behind the head yields the formation of strong biconical shocks that transmit the thrust to the head, increasing the ram pressure on the front region; afterwards the compression reflects at the jet axis and a recurrent process leads to recurrent impulsive accelerations of the head (Fig. 1a).

b. Jets with low M ($M < 30$), or high M ($M > 30$) and $\nu > 30$, have lower head velocities. In this case the backflow compression is much weaker and the resulting thrust is not sufficient to accelerate the head (Fig. 1b).

The critical parameter is the inclination of biconical shocks: shocks that have a small inclination angle on the axis are successful in accelerating the jet head.

The cocoon morphology reflects the head's evolution. The two classes of jets listed above lead to two different cocoon morphologies: class a) jets form "spearhead" cocoons (Fig. 2) while class b) jets lead to "fat" cocoons (Fig. 3).

3.2 Radiation Distribution

The adopted scheme allows to find qualitative indications about the radiative properties of the structures that arise following the jet interaction with the ambient medium. Since we do not perform a full MHD calculation, the morphological structure of the magnetic field evolves as in a plasma-$\beta \gg 1$ approximation. In Fig. 2 we show the distribution of the magnetic field (panel a), cutoff frequency (b) and emissivity (c) for a jet with $M = 100$ and $\nu = 10$ at a given (normalized) time. In Fig. 2a we note that the field attains the maximum values in the shocked region that surrounds the overpressured cocoon, which expands supersonically, and in the biconical shocks along the jet; conversely in the region internal to the cocoon the magnetic field is tangled by the turbulent velocity field, leading to a highly inhomogeneous structure with filaments surrounded by regions almost devoid of field. This behaviour reflects on the cutoff frequency distribution (Fig. 2b), that is also maximal on the cocoon contour, on the bow shock at the head, and in the initial part of the jet that is continuously replenished by fresh particles not yet affected by radiative losses, as can be seen in Fig. 2c where the emissivity reaches high values around the cocoon. We note that the region that contours the cocoon is mostly made up of ambient fluid particles shocked by the expanding cocoon. According to our prescription, this is a site of relativistic particle acceleration and compressed magnetic field (Fig. 2a).

In Fig. 3a-c) we show the corresponding images for $M = 100$ and $\nu = 300$. We can note the different morphological shape of the cocoon, but the radiation distribution follows a pattern similar to the previous case.

4. Summary and Conclusions

We have performed numerical simulations of the interaction of a supersonic jet with the ambient medium. The introduction of a passive magnetic field and of a relativistic particle population, of which we follow the temporal evolution, allowed us to compute the synchrotron radiation properties. An extensive coverage of the parameter space, especially towards very high Mach numbers, show two typical coocoon morphologies, depending on Mach number and density ratio.

We finally recall that the particular setup chosen for the initial configuration allows us i) to describe and classify the cocoon structures, and ii) to follow the jet-environment interaction over long evolution time scales. Both these possibilities are hampered when injecting the jet from the left boundary, as usually done in the literature (see e.g. Martí et al. (1995) and references therein).

References

1.   M. L. Norman, L. Smarr, K. H. A. Winkler, M. D. Smith: A&A **113** 285 (1982)
2.   D. A. Clarke, J. O. Burns: ApJ **369** 308 (1991)
3.   M. L. Norman, J. O. Burns, M. E. Sulkanen: Nature **335** 146 (1988)
4.   J. M. Blondin, B. A. Fryxell, A. Königl: ApJ **360** 370 (1990)
5.   M. L. Norman, J. M. Stone, D. A. Clarke: AIAA, Aerospace Sciences Meeting, 29th, Reno, NV, Jan. 7-10, p. 12 (1991)
6.   P. E. Hardee, D. A. Clarke: Apj **400** 9 (1992)
7.   P. E. Hardee, D. A. Clarke, D. A. Howell: Apj **441** 644 (1995)
8.   D. A. Clarke, M. L. Norman, J. O. Burns: ApJ **311** L63 1986
9.   K. R. Lind, D. G. Payne, D. L. Meyer, R. D. Blandford: ApJ **344** 89 (1989)
10.   J. Mª Martí, E. Müller, J. Mª Ibáñez: A&A **281** L9 (1994)
11.   G. C. Duncan, P. A. Hughes: ApJ **436** L119 (1994)
12.   J. Mª Martí, E. Müller, J. A. Font, J. Mª Ibáñez: ApJ **448** L105 (1995)
13.   S. Massaglia, G. Bodo, A. Ferrari: A&A (Paper I) in press (1995)
14.   N.S. Kardashev: Soviet Astron. **6** 317 (1962)
15.   P. R Woodward, P. Colella: J. Comp. Phys. **54** 174 (1984)
16.   G. Bodo, S. Massaglia, A. Ferrari, E. Trussoni: 1994, A&A **283** 655 (1994)
17.   G. Bodo, S. Massaglia, P. Rossi, R. Rosner, A. Malagoli, A. Ferrari: A&A in press (1995)

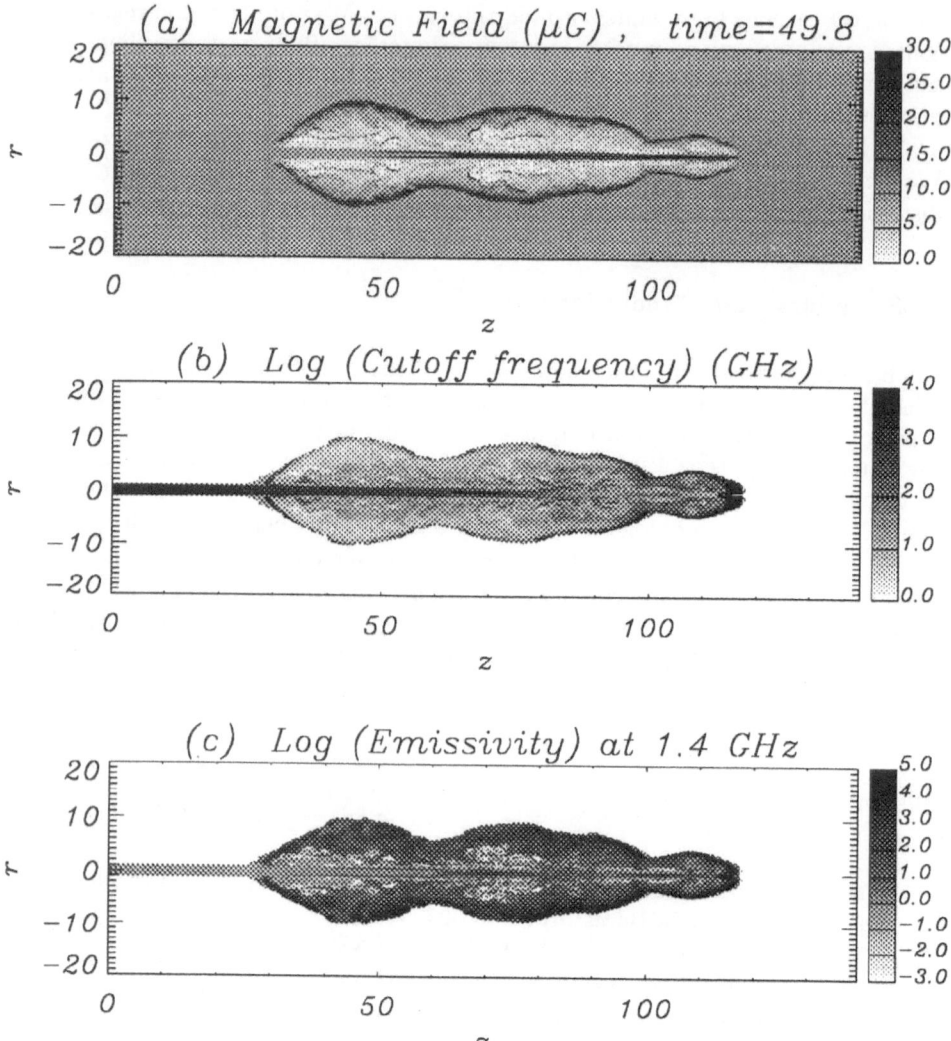

Fig. 2. Magnetic field intensity distribution (in units of the initial value) at a given time, for a jet with $M = 100$ and $\nu = 10$ (panel a); behaviour of the cutoff frequency obtained (in GHz) assuming the following initial values: jet radius $a = 1$ kpc, magnetic field $B_0 = 10^{-5}$ G, spectral index of the particle distribution $\gamma = 2.2$, sound speed $c_s = 10^8$ cm s^{-1} (panel b); emissivity distribution at 1.4 GHz (in arbitrary units) (panel c).

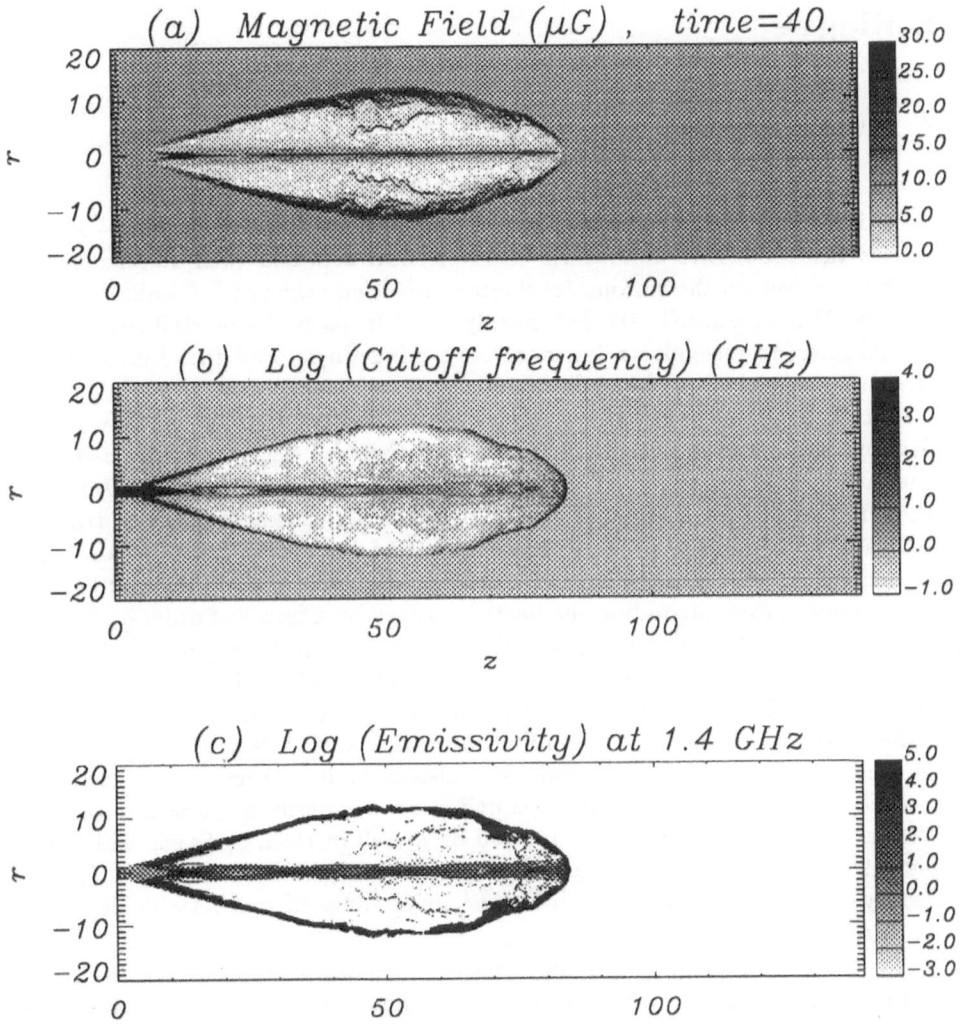

Fig. 3. The same as in Fig. 1 for $M = 100$ and $\nu = 300$

Epilogue

W. Kundt

This volume of the 'Lecture Notes in Physics' is - loosely speaking - an update of the 1986 Erice Course on 'Astrophysical Jets and their Engines'. How much consensus on the various Jet Sources has been achieved ? A ballott improvised by Martin Gaskell revealed: hardly any ! It needs the sensibilized view of an 'oldtimer' to recognize a lot more open-mindedness, and friendliness in the present volume than in the earlier ones.

Let me begin with the hardest item of all: in–situ acceleration, i.e. the mechanism of upgrading the Lorentz factors of essentially all electrons (and positrons) from a few to a few times 10^6 or more, in low-density regions of the Universe, far from the Central Engine. (Note that cosmic-ray physicists want to transfer - via analogous multistep acceleration - most of the shock energy to the ions, instead of the electrons). The mechanism is thought to work via many small elastic boosts, each upgrading the particle energy by a factor of order $\delta^2 \lesssim 1 + \beta$ (where $\delta := 1/\gamma(1 - \beta cos\Theta)$ is the Doppler factor). A factor of 10^6 therefore requires at least $13.8/ln(1+\beta) \gtrsim 20/\beta$ successful reflections by a heavy shock of velocity βc , whereby both inelasticities and losses have been ignored. The snag is the expectation that this upgrading in particle energy will not only work in the - calculable - test-particle limit, but also at high conversion efficiencies, in excess of 10%, say, when inelasticities and losses can easily degrade the transfer. Worries of this kind have been expressed by myself in 1984, by Sarris and Krimigis (1985, in application to the Earth's bowshock), and by Sam Falle (1990), also by Pesses (1979) who highlights the importance of magnetic fields in the shock region.

Even when magnetic fields are ignored, it is my understanding that few-particle interactions are time-reversible (i.e. show Poincaré cycles) - as are the fundamental equations of physics - but that complexities grow rapidly with the number of interacting particles, and that macroscopic physics deals with their coarse-grained behaviour (in phase space) which obeys the second law of thermodynamics. At any particular time, only exponentially few particles can drain large energies from a multiply interacting system, (not a power-law 'tail'). More quantitatively, this behaviour can be read off the energy integral ΔW extended over the history of a generic particle, as formulated in my lecture on jet formation. For this reason, I have occasionally used the term 'Münchhausen effect' - after Hieronimus von Münchhausen, dem Lügenbaron (1720-1796) - for what I consider a violation of the Second Law.

Should in–situ acceleration be a forbidden effect, a number of proposed jet-formation mechanisms in this book would likewise be eliminated, such as Jonathan Ferreira's disk-ejection structures (like the ones by Königl and Kartje, 1994), and the 'astrophysical plasma gun' of John Contopoulos (unless he would load it with pair plasma) - both of which depend, moreover, on magnetic-field configurations of non-fluctuating sign, and involve postulated return currents - further Vasilii Gvaramadze's cumulative phenomena (all of which appear highly contrived anyway), the 'bullets' of the SS 433 system in the standard model reviewed by René Vermeulen, and even the simulated jets of Silvano Massaglia which tap their radiation from a shock-boosted test-particle population. Would anybody's model qualify? Indeed, relativistic particle acceleration on the Sun is thought to happen via localized magnetic reconnections, and Harald Lesch considers this process in the corona of the central galactic disk.

Has the workshop arrived at a universal jet-formation mechanism? There has been a consensus - as far as I could notice - that the jets from very young stars, (young) binary neutron stars, and probably (young) white dwarfs show phenomena which look so similar to the jets from AGN that a universal formation mechanism may well be at work. Paradoxically, a 'universal engine' is proposed by Heino Falcke which involves a black hole; his mechanism cannot, a fortiori, be universal, or else does not really require - or even function with - a black-hole central engine. In contrast, at Erice in 1986 there was the suggestion that the central engine may be - in all four cases - the combination of starlike nuclear burning with a rapidly rotating magnetosphere. This suggestion has been followed up, at BH and in this volume, under the name of a 'Burning Disk' by Peter Scheuer, Wilfred Sorrell, and by myself - with perhaps insignificantly different conclusions - for the central engines of AGN which, unlike YSOs, may simply consist of a continuation of ordinary galactic disks all the way in to their centers: as the rotation center is approached, densities, temperatures, revolution rates, and magnetic-field strengths grow such that both nuclear burning, and coronal activities are enhanced beyond their normal (quasi-stationary) levels. Nuclear detonations eject the ashes from the central disk - observed in the form of the BLR with perhaps an enormous ($\lesssim 10^2$-fold) iron overabundance (D.Turnshek and J.Wampler, 1988; figure 1 of B.Wills and M.Brotherton, above) - and coronal discharges create the (relativistic) radiation cavity which feeds the jets - perhaps via the 'twin-exhaust mechanism' proposed by Blandford and Rees in 1974 - whenever the pair plasma manages to escape fast enough through quasi-vacuum channels before being burnt in situ. The necessary inverse-Compton losses of the escaping pair plasma are observed as the huge hard-γ-ray luminosities ($\lesssim$ TeV) reported by Stefan Wagner.

The scheme just described has by no means been the consensus of this workshop: Stefan Wagner discusses - and criticises - the hadronic interpretation of the hard γ-rays proposed by Mannheim and Biermann, which requires a subsequent in-situ upgrading of the electron energies. Beverley Wills, in her search for an explanation of the dichotomy between radio-loud and radio-quiet (-silent) QSOs, prefers a dependence on the tilting angle between the inner and outer disk (to a differing confinement by the circumnuclear gas). Various authors use model-dependent language in describing the observations, like 'disk (accretion) power' for what is believed to be emitted by the central engine, or 'net mass inflow' in the BLR for delayed blue line wings (or inverted P-Cygni profiles), or beam 'precession' for meandering. Garrelt Mellema discusses a large number of models for the jet production inside planetary nebulae - all of which he points out to have problems - but is shy of pursuing his own numerical result (Icke et al, 1992) which indicates that the jets require a much lighter medium than the rest of the PN in order to be focused like, e.g., in Fg 1 (his figure 8). And Jochen Eislöffel talks of 'turbulent' motions, and of 'entrainment' when reviewing outflows from young stars, instead of multicomponent motions, and of 'episodic mass ejections' instead of (perhaps) stationary relativistic beams with inverse-Compton losses, blowing a Strömgren-type preceding bowshock.

Large leptonic bulk Lorentz factors all the way from the hot core of an AGN are considered by Klaus Meisenheimer, with a sceptical but open-minded attitude. His reservations are considered in my own contribution on jet formation. René Vermeulen calls them the "un-appealing" solution of his superluminal-velocity-distribution problem. They seem at variance with the occasionally reported subluminal motion of jets on the VLBI scale, discussed by Steven Tingay, which may, however, have fallen trap of the stroboscopic effect: in order to resolve superluminal motion on the lyr scale, one has to observe at weekly intervals. A corresponding remark applies to the stellar jets resolved by the HST.

An important detail of the extragalactic jet phenomenon is discussed by Martin Gaskell: do AGN involve binary central engines? In 1986 I found the idea quite convincing, e.g. in view of the high rate of binarity in star formation (from massive accretion disks). But the binaries he requires are all uniformly close (0.1 lyr), and kinematically noisy. Could all these double-peaked broad emission lines be the result of preferred self-scattering at near-systemic velocities? As the central disks are expected to be (extremely) optically thick, $\tau \gtrsim 10^{11}$, ejection from them is viewed in the shape of an approaching hemisphere; figure 1 in (Kundt,1988) is misleading! Active disks seen face-on are thus expected to show exclusively blue-shifted emission whereas those seen edge-on would show both blue- and red-shifted emission, with their line centers removed by self-scattering. Such a re-interpretation of Martin's taxonomy may be worth pursuing.

Another important detail presented at BH are Bob Fosbury's pencil beams from AGN, with opening angles of a few degrees, illuminating halo condensations like search lights. What mechanism creates such narrow beams? Are they the expected inverse-Compton radiation from the - largely invisible - jet? They remind me of the preceding bowshocks reported by Jochen Eislöffel for several stellar sources.

This epilogue reflects my own views - assembled over a twenty-year baseline - and may (or may not) be very misleading. Clear, however, is the fact that both in this volume and in parallel ones, the jet phenomenon has not found a consistent, unanimous explanation.

References

Falle S.A.E.G., 1990: in *Neutron Stars and their Birth Events*, NATO ASI C **300**, W. Kundt ed., p. 303.

Icke V., Mellema G., Balick B., Eulderink F., Frank A., 1992: Nat **355**, 524.
Königl A., Kartje J.F., 1994: ApJ **434**, 446.
Kundt W., 1984: JAp&A **5**, 277.
Kundt W., 1988: Ap&SS **149**, 175.
Mannheim K., Biermann P.L., 1992: A&A **253**, L21.
Pesses M.E., 1979: Proc. 16th ICRC at Kyoto II, O.G. 9-1-8, p.33.
Sarris E.T., Krimigis S.M., 1985: ApJ **298**, 676.
Turnshek D., 1988: in *QSO Absorption Lines: Probing the Universe*, J.C. Blades, D.A. Turnshek, C.A. Norman eds., Cambridge Univ. Press, p. 17.

List of Participants

Antonucci	Robert	Physics Dept., Univ. of California Santa Barbara, CA 93106 antonucci@sbphy.physics.ucsb.edu	USA
Bogovalov	Sergei	Moscow Engineering Physics Inst. Kashiroskoje shosse 31, Moscow 115409 bogoval@photon.mephi.ru	Russia
Camenzind	Max	Landessternwarte, Königstuhl, D-69117 Heidelberg mcamenzi@hp2.lsw.uni-heidelberg.de	Germany
Contopoulos	John	Astron. Dept., The Univ. of Chicago 5640 South Ellis Ave., Chicago, IL 60637 conto@rosserv.gsfc.nasa.gov	USA
Dimitrov	Bogdan	Sofia (1000) 12 'stara planina' str manoff%bgearn.bitnet@vm.gmd.de	Bulgaria
Duschl	Wolfgang	Inst. f. Theor. Phys. der Universität Tiergartenstr. 15, D-69121 wjd@platon.ita.uni-heidelberg.de	Germany
Eislöffel	Jochen	Thüringer Landessternw. Tautenburg Karl-Schwarzschild-Obs., D-07778 Tautenbg. eisloeff@tls-tautenburg.de	Germany
Falcke	Heino	Dept. of Astron., Univ. of Maryland College Park, MD 20742-2421 p617hfa@mpifr-bonn.mpg.de	USA
Ferreira	Jonathan	Landessternwarte Königstuhl, D-69117 Heidelberg jferreir@hp7.lsw.uni-heidelberg.de	Germany
Fischer	Daniel	MPIfR Bonn Auf dem Hügel 69, D-53121 Bonn dfischer@solar.stanford.edu	Germany
Fosbury	Robert	Space Telescope, ESO Karl-Schwarzschild-Str. 2, D-85748 Garching rfosbury@eso.org	Germany
Gaskell	Martin	Dept. Phys. and Astron., Univ. of Nebraska Lincoln, NE 68588-0111 gaskell@unlinfo.unl.edu	USA
Gvaramadze	Vasilii	Abastumani Observatory, Rep. of Georgia Krasin Str.19, Ap 81, Moscow 123056 dzogin%esoc1.bitnet@vm.gmd.de	Russia
Ikhsanov	Nazar	Central Astron. Obs. of Russian Acad. Sci. Pulkovo 65, St. Petersburg 196140 nri@pulkovo.spb.su	Russia

Krichbaum	Thomas	MPIfR Bonn Auf dem Hügel 69, D-53121 Bonn p459kri@mpifr.bonn.mpg.de	Germany
Kundt	Wolfgang	Inst. f. Astrophysik der Universität Auf dem Hügel 71, D-53121 Bonn wkundt@astro.uni-bonn.de	Germany
Leahy	Patrick	NRAL Jodrell Bank, Univ. of Manchester Macclesfield, Cheshire SK119DL jpl@jb.man.ac.uk	England
Lesch	Harald	Inst. f. Astron. und Astroph. der Univ. Scheinerstr. 1, D-81679 München lesch@hal1.usm.uni-muenchen.de	Germany
Martí	José	Departamento de Astron. y Astrof., Univ. Valencia, 46100 Burjassot (Valencia) martij@evalvx.ific.uv.es	Spain
Massaglia	Silvano	Osservatorio Astronomico di Torino Strada Osserv. 20, I-10025 Pino Torinese massaglia@ph.unito.it	Italy
Meisenheimer	Klaus	MPIfA Heidelberg Königstuhl, D-69117 Heidelberg meise@mpia-hd.mpg.de	Germany
Mellema	Garrelt	Stockholm Observatory S-13336 Saltsjöbaden garrelt@astro.su.se	Sweden
Mirabel	Felix	SApCE - Saclay F-91191 Gif-sur-Yvette Cedex mirabel@araine.saclay.cea.fr	France
Scheuer	Peter	Mullard Radio Astron. Obs.,Cavendish Lab Madingley Road, Cambridge CB3OHE pags@mrao.cam.ac.uk	England
Schlickeiser	Reinhard	MPIfR Bonn Auf dem Hügel 69, D-53121 Bonn p337@sch@fs1.mpifr-bonn.mpg.de	Germany
Sorrell	Wilfred	Dept. Phys. and Astron., Univ. of Missouri 8001 Nat. Bridge Rd., St. Louis, MO-63121 swsorre@slvaxa.umsl.edu	USA
Tingay	Steven	Mt. Stromlo and Siding Spring Observts. Canberra, ACT 2611 tingay@merlin.anu.edu.au	Australia
Vermeulen	René	Dept. of Astron., Caltech Pasadena, CA 91125 rcv@astro.caltech.edu	USA
Wagner	Stefan	Landessternwarte Heidelberg Königstuhl, D-69117 Heidelberg swagner@mail.lsw.uni-heidelberg.de	Germany
Wills	Beverley	Astron Dept., Univ. of Texas Austin, TX 78712 bev@panic.as.utexas.edu	USA

SUBJECT INDEX

Springer-Verlag
and the Environment

We at Springer-Verlag firmly believe that an international science publisher has a special obligation to the environment, and our corporate policies consistently reflect this conviction.

We also expect our business partners – paper mills, printers, packaging manufacturers, etc. – to commit themselves to using environmentally friendly materials and production processes.

The paper in this book is made from low- or no-chlorine pulp and is acid free, in conformance with international standards for paper permanency.

Lecture Notes in Physics

For information about Vols. 1–434
please contact your bookseller or Springer-Verlag

New Series m: Monographs